CHEMICAL HYDROMETALLURGY

Theory and Principles

CHEMICAL HYDROMETALLURGY
Theory and Principles

A R BURKIN
Imperial College, UK

Imperial College Press

Published by

Imperial College Press
57 Shelton Street
Covent Garden
London WC2H 9HE

Distributed by

World Scientific Publishing Co. Pte. Ltd.
5 Toh Tuck Link, Singapore 596224
USA office: 27 Warren Street, Suite 401-402, Hackensack, NJ 07601
UK office: 57 Shelton Street, Covent Garden, London WC2H 9HE

Library of Congress Cataloging-in-Publication Data
Burkin, A. R. (Alfred Richard)
   Chemical hydrometallurgy : theory and principles / A.R. Burkin.
   p. cm.
   Includes bibliographical references and index.
   ISBN-13 978-1-86094-184-9
   ISBN-10 1-86094-184-2
   1. Hydrometalurgy. I. Title.
   TN688 .B78 2001
   669'.028'3--dc21

                         2002276235

British Library Cataloguing-in-Publication Data
A catalogue record for this book is available from the British Library.

First published 2001
Reprinted 2009

Preface

Hydrometallurgy is concerned with methods of producing metals, and some compounds, by reactions which take place in water or organic solvents. Until about 1950 commercial hydrometallurgical practice was confined to the recovery of relatively few metals, but since then the situation has changed dramatically. The factors which have contributed most towards this are the growth of the uranium industry, the acceptance of solvent extraction as a process suitable for industrial use on a large scale, the development of techniques of leaching and reduction at temperatures up to 240°C at moderate pressures, and the demand for numerous less-common metals and other elements.

During the same period, since 1950, the application of chemical thermodynamics to pyrometallurgical methods of metal extraction, particularly in relation to the accumulation of data on slag–metal equilibria, caused fundamental changes in that field of technology also. As a consequence of these changes in all areas of extractive metallurgy it became necessary to give graduates a much deeper knowledge of the fundamental, scientific basis of their speciality than had been necessary previously.

After graduating in chemistry, spending three years in industry on war work and six years as a university lecturer in inorganic chemistry, I joined the Mining Department of Imperial College in 1952 to teach the chemical aspects of mineral processing. As hydrometallurgical processes became more important it became clear that the technology would be more appropriately taught in the Metallurgy Department where the teaching of extractive metallurgy was based firmly on physical chemistry. I therefore transferred to Professor F.D. Richardson's section of that department in 1960.

In order to support my lecture courses I required a textbook and wrote "The Chemistry of Hydrometallurgical Processes," which was published in 1966. This book is the successor to that and is concerned with the application of modern experimental methods and chemical science to the development of hydrometallurgical processes. Numerous papers from the literature are discussed and in some cases the experimental procedures are described. Since it is felt that students should not believe that everything published in the literature is correct some critical discussion is included.

A.R. Burkin

Contents

Chapter 1

Ore Minerals

1.1. Types of Ore Mineral

In a hydrometallurgical process leaching of a metal from an ore is carried out at a relatively low temperature, below about 200°C. The rate of reaction may be slow and will depend on the nature of the phase containing the metal. For example under acidic oxidizing conditions copper is dissolved very much more rapidly from chalcocite, Cu_2S, than it is from chalcopyrite, $CuFeS_2$. Therefore in a hydrometallurgical process, the nature of the minerals present is important whereas when a copper sulphide concentrate is fed into a smelter this is not the case.

The type of process used to recover a metal from an ore depends on the chemical nature of the mineral containing it. Many minerals are treated in hydrometallurgical processes but there are few classes of such minerals. They are shown, with some examples, in Table 1.1.

1.2. Types of Crystal Lattice

1.2.1. *Introduction*

Textbooks on physical chemistry give accounts of crystal chemistry and specialist texts are available such as that of Wells.[1] These should be consulted for information concerning the arrangements of atoms or ions in the different

1

Table 1.1. Some ore minerals of different types.

Silicates	Sulphides	Oxides	Others
Bertrandite (Be)	Cinnabar (Hg)	Anatase (Ti)	Bastnasite (rare earths)
Beryl (Be)	Copper minerals	Baddeleyite (Zr)	Fluorapatite (P)
Lateritic ores (Ni, Co)	Galena (Pb)	Bauxite (Al)	Gold metal
Lepidolite (Li)	Molybdenite (Mo)	Cassiterite (Sn)	Gold tellurides
Magnesite (Mg)	Nickel minerals	Chromite (Cr)	Monazite (rare earths)
Olivine (Mg)	Silver minerals	Columbite-tantalite (Nb, Ta)	Phosphate rock (P)
Petalite (Li)	Sphalerite (Zn)	Ilmenite (Ti)	Platinum group metals and alloys
Spodumene (Li)		Oxidized products of sulphides (Cu, Pb)	
Thorite (Th)		Pyrochlore (Nb)	Silver halides
Zircon (Zr)		Pyrolusite (Mn)	Silver metal
		Rutile (Ti)	Xenotine (rare earths)
		Scheelite (W)	
		Thorianite (Th)	
		Uraninite (U)	
		Wolframite (W)	

types of crystal structure. This chapter deals with some general matters of importance in understanding the behaviour of minerals of different kinds during chemical processing.

Almost all minerals are crystalline. The fundamental fact about the crystalline state is that the atoms are arranged in an ordered three dimensional pattern known as the "space lattice". This is formed by repetition of a unit cell which is a group of atoms characteristic of a particular kind of crystal which, when extended in three dimensions leads to the macroscopic crystal. A group of atoms smaller than the unit cell does not contain all the information necessary to produce the structure of the macroscopic crystal.

The shape of the unit cell and hence the constants needed to define it depends on the symmetry of the crystal. The maximum number of constants required is six, three lengths of sides of the unit cell, written a, b and c to correspond to the crystal axes with those symbols, and the angles between the axes, taken in pairs. These angles are written as a, b and γ, between bc, ca and ab respectively. The choice of axes in the triclinic system is arbitrary but in the other systems is determined by the symmetry elements, which are of no concern here.

Unit cells are figures of simple geometrical shape, bounded by plane faces. Usually they contain only a few atoms and so are small with sides only a few Ångstroms long. However sometimes there is a degree of order which is repeated only at relatively great distances; this leads to a superlattice structure which may have a hundred or more atoms in the unit cell.

The atoms in a unit cell are packed together in a manner to minimise the energy of the group. This means that there will be preferred distances of closest approach between them. Since in many minerals the atoms are best regarded as being ions, these distances are described as ionic radii. In the case of an element which can have several valencies, each kind of ion has a different ionic radius. The outer regions of ions are volumes of electronic charge and heavier elements tend to be larger than lighter; however for elements of similar atomic number, anions are larger than cations.

In some solids, an oxide of a light metal such as MgO for example, the size of O^{2-} is so much greater than that of Mg^{2+} that the interionic distances in the lattice are determined by the distance of closest approach of the oxide ions, the magnesium ions being small enough to go into the spaces between the spherical O^{2-} ions. This is known as close packing and in general such lattices are made up of layers of spherical anions, arranged so that each occupies as little space as possible, packed one on top of another. There are two ways in which this packing can occur, one leading to the hexagonal close packed and the other to the cubic close packed structure. The ratio of the radii of the anions and cations determines whether or not close packing is possible.

It is often convenient to consider a mineral lattice as an array of anions with cations occupying holes between them. Some holes will be between four anions, usually arranged tetrahedrally around the centre point of the hole, the so-called tetrahedral holes. Others will be between six anions arranged octahedrally around the centre point, the octahedral holes. In general the arrangement of anions around a cation will be the most symetrical in three dimensions, that is 3, 4, 6 or 8 ions will be arranged at the apices of a triangle, tetrahedron, octahedron or cube around the central ion.

In oxides such as MgO the bond between cation and anion can be regarded as ionic in character although physical properties indicate not entirely so. The lattice structure is determined by electrostatic interactions between oppositely charged spheres and MgO has the rock salt (NaCl) structure in which each ion is surrounded octahedrally by six of the other kind. In the cases of many other solids the bonds between cations and anions have a degree of covalent character and the arrangement of one kind of atom around another is determined by the orientations of the covalent bonds in space. This is the case in some metal sulphides, many of which are semiconductors. The properties of the solids are then best considered in terms of electron energy levels rather than individual bonds between atoms.

1.2.2. *Silicate Structures*

The earth's crust is made up almost entirely of silicates and silica, the rock forming minerals and the breakdown products of them which result from the weathering of rocks. The basis of the structure of silicates is the SiO_4^{4-} group in which four oxygen atoms are arranged tetrahedrally around the silicon atoms, the Si–O distance being about 1.6 Å and the distance between centres of oxygen atoms about 2.6 Å. There are some minerals which are orthosilicates, with lattices built up from simple SiO_4^{4-} ions and cations. Most silicates however are based on polymeric silicate groups, the simplest of which is $Si_2O_7^{6-}$, and which proceed to rings, chains, doubled chains (bands) and sheets, some of which are illustrated in Fig. 1.1. In addition there are the three dimensional frameworks, in the various forms of silica. The tetrahedral group AlO_4 has very nearly the same size as the SiO_4 group and so aluminium can replace silicon in the silicate anion structure, giving rise to the aluminosilicates.

Silicate and aluminosilicate lattices can be looked upon as being tetrahedra of oxide ions containing silicon or aluminium ions in the tetrahedral holes, packed together so as to form polyhedra which contain holes of different coordination numbers and sizes. These latter contain cations having appropriate ionic radii, in sufficient number to neutralise the ionic charge of the anionic silicate structure. Clearly as aluminium replaces some silicon more cationic charge is necessary. There are a number of ions of similar size which can occupy octahedral holes in silicate structures. Many metals were present in the magma from which silicate minerals crystallised so that isomorphous substitution is common in silicate minerals. This is the reason why many of them do not have the "ideal" composition of a chemical compound. Aluminium cannot only replace silicon in tetrahedral holes but also occupy octahedral holes and it is this dual role which makes the compositions of some aluminosilicates appear to be exceedingly confusing.

If aluminium replaces some silicon in a three dimensional framework silica structure, the composition remaining $(Si, Al)O_2$, the solid becomes negatively

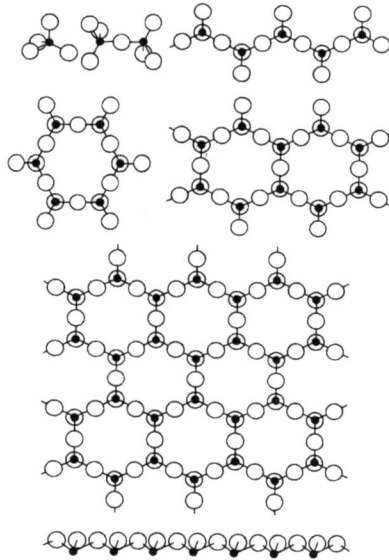

Fig. 1.1.   Silicate structures, from top left. (1) Orthosilicate $(SiO_4)^{4-}$ ion. (2) Pyrosilicate $(Si_2O_7)^{6-}$ ion. (3) Chain structure $(SiO_3)^{2n-}$. (4) Ring $(Si_6O_{18})^{12-}$ group structure, present in beryl. (5) Doubled chain (band) structure based on $(Si_4O_{11})^{6-}$ groups. (6) Sheet structure silicon–oxygen network with cross section of the sheet below.

charged and positive ions must be present in holes in the structure. Zeolites and felspars are examples of such materials. If there are suitable diffusion paths in a zeolite for metal ions to pass into and out of the structure when it is fully wetted by water then ion exchange can occur. In other cases a metal present in a silicate can only be extracted by breaking up the silicate structure. This occurs naturally during weathering under certain conditions, resulting in silicate species as well as metals passing into solution in any water present if chemical conditions are suitable. If reprecipitation occurs due to the water passing to a region where the chemical conditions, for example pH or redox potential, change a silicate of indeterminate composition and containing several metals may form. This may subsequently crystallise, usually to give a poorly crystalline solid which can have a wide range of composition because of substitution of one metal in the "ideal" composition of the mineral by others present which have suitable ionic radii. This occurs during the laterisation process which forms the "oxide ores" of nickel.

Olivine, an important mineral in the basic, or mafic, rock peridotite is the first silicate to crystallise from magna of appropriate composition and tends to preferentially incorporate Ni^{2+} ions into its lattice if nickel is present. The regions in which this occurred to a significant extent became nickel provinces. Olivine and related minerals have the general formula M_2SiO_4 where M is a divalent metal having a suitable ionic radius to allow it to be accommodated in holes between six oxygen atoms. The "ideal" composition of olivine itself may be written $9Mg_2SiO_4.Fe_2SiO_4$, a representation which can lead to confusion. The crystal lattice of olivine contains discrete SiO_4^{4-} tetrahedra packed together with Mg^{2+} ions between them in octahedral holes betwen oxygen atoms. Approximately one in ten of the Mg^{2+} ions is replaced by Fe^{2+}. Olivine is not a "double salt" or mixture of two silicates. Zircon, $ZrSiO_4$, is also an orthosilicate and because the ionic radii of Zr and Hf are very similar, due to the lanthanide contraction, the mineral always contains hafnium in its lattice also.

1.2.3.   *Oxides*

Two types of oxide are important in hydrometallurgy, the binary oxides which contain only one metal and the complex oxides which contain two or more. The majority of structures of both types are essentially ionic and the metal atoms have large coordination numbers, often 6 or 8. The most important structures of binary oxides of interest here are given in Table 1.2. The coordination numbers of the rutile structure, taken as an example, indicate that each metal atom has a group of 6 oxygen atoms around it and each oxygen has 3 metal atoms.

Table 1.2.

Formula type	Name of structure	Coordination numbers		Examples
		M	O	
MO_3	ReO_3	6	2	WO_3
MO_2	Rutile	6	3	TiO_2, MnO_2, SnO_2
	Fluorite	8	4	ZrO_2, ThO_2, UO_2
MO	Sodium chloride	6	6	MgO, CaO, NiO
	Wurtzite	4	4	ZnO
M_2O_3	Corundum	6	4	Al_2O_3, Fe_2O_3, Cr_2O_3
M_2O	Cuprite	2	4	Cu_2O

In addition to those in the table there are the silicate structures which were discussed in Sec. 1.2.2; layer structures of MoO_3, As_2O_3, PbO; chain structures of HgO, SeO_2, Sb_2O_3 and molecular structures of RuO_4 and OsO_4. Due to its molecular structure, OsO_4 volatilises at 130°C and is separated from other platinum group metals by this means. RuO_4 melts at 25.5°C and decomposes at 108°C. These figures may be compared with a melting point and boiling point for SiO_2 (quartz) of 1610°C and 2230°C; for α-Al_2O_3 (corundum) of 2015°C and 2980°C and a melting point for Cu_2O (tenorite) of 1326°C, values characteristic of solids with highly ionic bond character.

In the examples of binary oxides shown in Table 1.2, all metal ions in a compound have the same valency and have the same environment. There are a few examples of such compounds where some of the metal atoms are in a different environment from the others but they do not concern us here. If a binary oxide contains the metal in two oxidation states there are two kinds of metal ion, Pb^{II} and Pb^{IV} in Pb_3O_4 for example, and these may be in different environments, having 3- and 6-coordination numbers respectively in this case. Complex oxides can also have such a structure, which is regular like that of a simple binary oxide. Compounds with a single metal in two valency states should be classified as complex oxides, however.

In complex oxides more than one metal is present and in some complex oxide structures the environments of the different metal ions are so different that the size difference or charge difference between the ions necessary for the structure to be stable is not possible for a single metal in two valence states. Thus there are a number of structures for complex oxides which are not available for binary oxides. One of the most important of these is the spinel structure. The mineral spinel, $MgAl_2O_4$ crystallises in a cubic system and has 8 formula units in the unit cell.

The spinel structure, AB_2Z_4 is taken up by a large number of compounds, Z being a divalent anion, A a divalent metal and B a trivalent metal. The anion Z may be O, S, Se or Te. In complex oxides with the spinel structure the following metals have been found

A — Mg, Cr, Mn, Fe, Co, Ni, Cu, Zn, Cd, Sn

B — Al, Ga, In, Ti, V, Cr, Mn, Fe, Co, Rh

although only certain combinations of the metals are known. The basis of the spinel structure is a set of layers of O^{2-} ions each of radius 1.32 Å. These layers

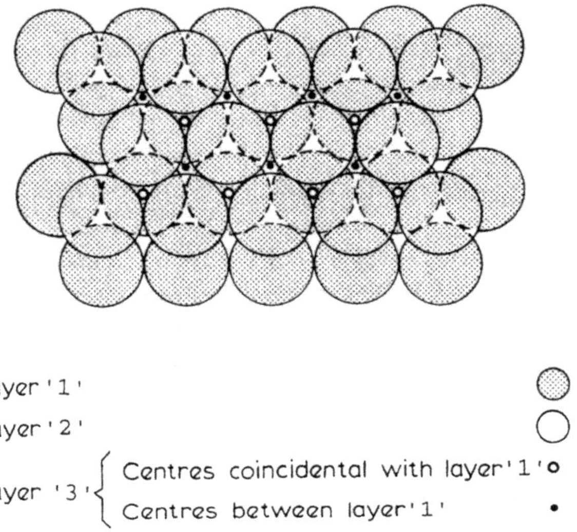

Layer ' 1 '

Layer ' 2 '

Layer ' 3 ' { Centres coincidental with layer' 1 ' o

Centres between layer' 1 '   •

Fig. 1.2.   Stacking sequence of layers of close packed oxide ions. Layer 1 is in the plane of the paper. Layer 2 is stacked on layer 1. If the centres of oxide ions in layer 3 lie above the centres of oxide ions in layer 1 as shown by small open circles the stacking sequence is 121, the oxygens are cubic close packed giving the spinel structure. If the centres of the oxide ions in layer 3 lie above the voids where three oxide ions in layer 1 meet, shown by small solid circles, layer 3 is spacially different from layers 1 and 2, the stacking sequence is 123 and a hexagonal close packed oxide lattice is formed.

are stacked so as to be almost close packed in a face centred cubic arrangement; that is oxygens of layers 1, 3, 5 etc. are in lines with respect to one another while those of the intermediate layers are also in lines, equidistant from the former oxygen atoms, see Fig. 1.2. Two kinds of hole exist between the oxygen atoms in the spinel structure, octahedral of which there are 32 in the unit cell, and tetrahedral, of which there are 64. In spinel itself all of the Mg^{II} ions are in tetrahedral (or A) sites and all of the Al^{III} are in octahedral (or B) sites. Electrical charge balance has to be maintained and there are far more holes than metal ions in the structure. Only 8 of the tetrahedral and 16 of the octahedral sites are filled. This is known as the "normal" spinel structure. The positions occupied by the metal ions are regularly arranged and they form additional face centred cubic sublattices interlaced with the oxygen lattice.

   In ferrites Fe^{III} is substituted for Al^{III} in the lattice while Mg^{II} may be replaced by a divalent cation having an ionic radius in the range 0.6 to 1.0 Å.

The "normal" spinel structure, all M^{II} ions in A sites and all Fe^{III} in B sites, is adopted by some ferrites such as $ZnFe_2O_4$ and $CdFe_2O_4$, which are both paramagnetic. Other ferrites have the "inverse spinel" structure in which all the M^{II} ions are in B sites so that 8 of the Fe^{III} ions are in A sites and 8 in B sites. Examples of compounds with this inverse spinel structure are Fe^{II}, Co^{II} and Ni^{II} ferrites, which are all strongly magnetic. The ions in the octahedral sites are probably statistically distributed between the total number of such sites available although some long range ordering of the cations in $Fe^{2+}Fe_2^{3+}O_4$ at temperatures below 120 K is indicated by the marked decrease of electrical conductivity which occurs below that temperature.

The normal and inverse spinel structures are best regarded as extremes, intermediate structures being known. In $MnFe_2O_4$, 80% of the Mn^{II} ions are in A sites and 20% in B sites, Fe^{III} making up the balance in the A and B sites. In $MgFe_2O_4$ only 10% of the magnesium is in A sites. The distribution of divalent metals in ferrites is not determined by their ionic radii since the largest ions, Cd and Mn, are found predominantly in the smaller tetrahedral sites. Adoption of the normal or inverse structure seems to be due to any pronounced preference for a particular environment, for example for zinc to be tetrahedrally coordinated and nickel to be octahedrally coordinated.

One of the most important metals for a hydrometallurgist is iron, because of its ubiquity in ores and the need to dispose of it. When ferric iron precipitates from an aqueous solution at atmospheric temperature due to a rise in pH it does so as in the form of amorphous $Fe(OH)_3$. When this dries to a solid cake it slowly crystallises as a-FeOOH, goethite which can be dehydrated to give a-Fe_2O_3, hematite. The oxides and oxide hydroxides of iron (III) and an outline of their structures are listed in Table 1.3.[2] Although Fe_3O_4 and γ-Fe_2O_3 have spinel structures and can readily be transformed into one another there appears to be little similarity between the structures of the others. It has been shown however[3] that many of the transformations between the substances are topotactic; that is they occur in such a way that the structure of the product has a definite relationship with that of the original solid. Sequences of such transformations are shown in Fig. 1.3.

The oxides Fe_3O_4 and γ-Fe_2O_3 are closely related, having the spinel structure, the cubic unit cell containing 32 cubic close packed oxygen atoms having an edge length approximately 8.5 Å. FeO has the sodium chloride structure in which both cations and anions have octahedral coordination so that it has 32 Fe^{II} ions in octahedral sites by anology with the spinels; Fe_3O_4 has 8 Fe^{III}

Table 1.3. Summary of the crystal structures of iron oxides and oxide hyroxides.

Composition	Mineral name	System	Cell constants	Notes
α-Fe_2O_3	hematite	Trigonal (hexagonal unit cell)	$\alpha = 5.035$ Å $c = 13.72$	h.c.p. oxygen Fe^{3+} in octahedral positions
γ-Fe_2O_3	maghemite	Cubic (spinel)	$\alpha = 8.33$–8.38	Spinel, various degrees of ordering
Fe_3O_4	magnetite	Cubic (fcc-inverse spinel)	$\alpha = 8.397$–8.394	Inverse spinel
α-FeOOH	goethite	Orthorhombic	$a = 10.0$ $b = 3.03$ $c = 4.64$	Based on h.p.c. oxygen
β-FeOOH	akaganeite		$a = 10.48$ $c = 3.06$	α-MnO_2 structure
γ-FeOOH	lepidocrocite	Orthorhombic	$a = 3.06$ $b = 12.4$ $c = 3.87$	Based on distorted c.c.p. oxygen
δ-FeOOH			$a = 2.941$ $c = 4.58$	h.c.p. oxygen disordered CdI_2 structure

Key: c.c.p. — cubic close packed
    h.c.p. — hexagonal close packed
    f.c.c. — face centred cubic

ions in tetrahedral sites and 8 Fe^{II} + 8 Fe^{III} ions in octahedral sites; γ-Fe_2O_3 has $21_{1/3}$ Fe^{III} ions distributed statistically in the 24 types of site utilised in Fe_3O_4. Careful oxidation of Fe_3O_4 produces γ-Fe_2O_3 which is converted back into Fe_3O_4 by heating to 250°C in vacuo. The stepwise oxidation of FeO to γ-Fe_2O_3 shown in Fig. 1.3 can be described as follows: oxygen is added to the lattice of FeO as new layers of close packed O atoms while some iron is oxidised to Fe^{III}; cations diffuse into this addition to the oxide lattice leading to a decrease in the concentration of Fe in a unit volume of solid.

Hematite and the hydroxy-oxides, other than γ- and β-FeOOH have structures which can also be described in terms of stacked layers of O or OH ions which are almost close packed. The iron atoms fit into the interstitial holes without substantial distortion of the oxygen lattice. The topotactic transformations take place because the cubic close packed and hexagonal close packed structures of the oxide lattice are very closely related and solids can readily

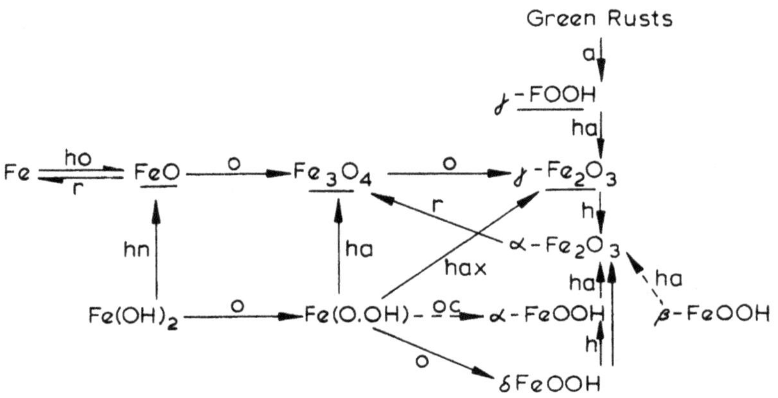

Green Rusts

γ-FOOH

Fe $\underset{r}{\overset{ho}{\rightleftharpoons}}$ FeO $\xrightarrow{\;o\;}$ Fe$_3$O$_4$ $\xrightarrow{\;o\;}$ γ-Fe$_2$O$_3$

α-Fe$_2$O$_3$

hn   ha   hax

Fe(OH)$_2$ $\xrightarrow{\;o\;}$ Fe(O.OH)- $\overset{oc}{\longrightarrow}$ α-FeOOH β-FeOOH

δFeOOH

Fig. 1.3.   Transformations between oxides and hydroxides of iron.[3] Solid arrows indicate topotactic transformations; broken arrows nontopotactic transformations. Compounds underlined have oxygen atoms in cubic close packed configuration; the others in hexagonal close packing except β-FeOOH which has a more comple structure.

Key   h = heating   n = in nitrogen or in vacuum
       a = atmospheric oxidation   o = on oxidation
       c = in alkali   r = on reduction
                                   x = in excess

switch from one packing to the other. Figure 1.2 shows a close packed layer of oxide ions in the plane of the paper, layer 1, stippled circles. A second layer of close packed oxide ions stacked on the first is shown by open circles, layer 2. If a third layer is to be stacked on the second it is possible to place the ions either directly above the ions in the first layer or above the voids where three oxygens in that layer meet. In the diagram the centres of the ions in the third layer would lie in the positions marked by small open and solid circles respectively. If the third layer is stacked so that the oxygens lie above those of the first layer the stacking sequence is 12 because layer 3 is identical with layer 1 and the oxygens are cubic close packed. If the third layer oxygens lie above the holes in layer 1 then the third layer is spacially different from layer 1, the stacking sequence then is 123, and a hexagonal close packed oxide lattice is formed. Changing from one to the other requires only that planes of oxide ions should slip so as to change the stacking sequence.

Certain transformations have been omitted by the authors[3] from Fig. 1.3 because the conditions necessary for causing them to occur are not known.

An example quoted is the transformation of Fe_3O_4 to a-FeOOH. It has been deduced that this occurs in nature because pseudomorphs of a-FeOOH after Fe_3O_4 have been found. Many examples of pseudomorphous replacement are known, some of them being relevant to ore minerals. For example maghemite is described[4] as being formed under certain conditions of oxidation from magnetite or from limonite gels (containing goethite) by a process analogous to its synthesis in the laboratory through moderate heating of lepidocrocite. Maghemite is also formed by decomposition of pyrite (FeS_2) by heating it in slightly alkaline aqueous systems under oxidising conditions. It has been suggested that this is due to replacement of sulphur in the pyrite lattice by oxygen, a different type of topotactic transformation. This is dealt with in Sec. 5.3.2.

1.2.4.   *Sulphides*

The sulphides of the alkali and alkali earth metals are ionic in character and have structures which resemble those of oxides. Sulphides of other metals have bonds with substantial covalent character and many of them have physical properites which resemble those of alloys rather than salts. They are semiconductors, have a metallic lustre and high reflectivity and, particularly in the case of the transition metal sulphides, many do not have compositions corresponding to the normal valencies of the metals. Examples are NiAs and Co_3S_4. Covellite, CuS, rather than being Cu^{II} sulphide has a structure in which part of the sulphur has paired to form S_2^{2-} ions and the solid is a diamagnetic metallic conductor, copper being present as Cu^+.

Description of sulphides in terms of purely ionic models, the ions being considered as charged spheres of a particular radius, can help to predict types of structure and cation substitution likely to occur, as in the case of oxides. However, it provides no information about the electronic structures of the solids or their properties which depend on the behaviour of electrons in them. Although the valence bond approach still provides useful information a major advance in understanding transition metal sulphides took place when crystal field theory was applied to them in the 1960's. Subsequently the molecular orbital and the band theories have been applied. The latter is applicable to multiatomic assemblies in principle but it is difficult to use it quantitatively with crystals having more than a few atoms per unit cell. These aspects of the chemistry of mineral sulphides are of peripheral importance in hydrometallurgy at present, although the position may change as greater theoretical insight is gained. A specialist text is available.[5]

The phase diagrams of many binary metal sulphides are exceedingly complex. In the copper-sulphur system the following exist: (Sec. 5.3.3): $Cu_{1.993-2.000}S$, chalcocite; $Cu_{1.934-1.965}S$, djurleite; $Cu_{1.79-1765}S$, digenite; $Cu_{1.75}S$, anilite; $Cu_{1+x}S$ where x is 0.05–0.10, blaubleidender covellite; $Cu_{1.000}S$, covellite. In addition three more minerals have been discovered recently: spionkopite, $Cu_{1.4}S$; yarrowite, $Cu_{1.125}S$ and geerite, $Cu_{1.60}S$. All of these solids change to other forms or to phases of other compositions at particular elevated temperatures. The low temperature orthorhombic form of chalcocite, a-Cu_2S has a unit cell containing 96 formula units with $a = 11.881$ Å, $b = 27.323$ Å and $c = 13.491$ Å. The sulphur atoms are in hexagonal close packing but the exact distribution of the copper atoms in the holes is not known. In some sulphides the metal atoms are in an ordered array, in others they are distributed statistically.

Because the metal–sulphur bonds in many sulphides have significant covalent character, involving electron sharing, a particular metal will be able to form only a limited number of bonds and they will form in specific directions in space relative to one another. The most common arrangements are four bonds directed tetrahedrally and six bonds directed octahedrally. The metal atoms are present in tetrahedral or octahedral holes in the sulphur framework respectively.

More than one metal may be present in a sulphide phase. A number of copper–iron sulphides occur as minerals; the most common is chalcopyrite, $CuFeS_2$, the others being bornite, Cu_5FeS_4 and cubanite, $CuFe_2S_3$. Chalcopyrite has a relatively simple lattice structure based on the zinc blende, ZnS and ultimately the diamond lattice. Carbon forms four tetrahedrally arranged covalent bonds; in the diamond crystal each atom is linked to four equidistant neighbours and the linking extends throughout the whole crystal. If alternate carbon atoms are replaced by Zn and S the zinc blende structure results, each zinc atom being surrounded tetrahedrally by four sulphur atoms and each sulphur by four zinc atoms. If the zinc is now replaced by alternate copper and iron atoms the chalcopyrite structure is produced. If this replacement is done using a single unit cell of ZnS the crystal of chalcopyrite is not produced by repeating it in three dimensions. The unit produced from one unit cell of ZnS must be repeated in one direction to form the unit cell of chalcopyrite. Thus the structure of chalcopyrite is based on a superstructure of ZnS, zinc blende. Quaternary compounds with the tetrahedral structure also occur; for example the structure of stannite, Cu_2FeSnS_4, is produced by replacing half of the iron atoms in two unit cells of chalcopyrite by tin.

Metals requiring 6-coordination cannot occur in zinc blende or other tetrahedral structures but are commonly found to assume the nickel arsenide (NiAs) structure. In this, each atom has six nearest neighbours of the other kind but whereas an arsenic atom is surrounded by six nickel atoms at the apices of a trigonal prism the nearest neighbours of a nickel atom are six arsenic atoms arranged octahedrally around it. In addition each nickel atom has two other nickel atoms sufficiently close to be bonded to it. In NiAs there are six arsenic atoms with Ni–As interatomic distances 2.43 Å and two nickel atoms with Ni–Ni distance 2.52 Å. The nickel arsenide structure has the property of being able to take up in solid solution a considerable excess of the transition metal, resembling typical alloys in this respect.

Pyrrhotite, iron (II) sulphide, has the NiAs structure but does not have the stoichiometric composition FeS, being iron deficient in an intact sulphur lattice. These sulphides are considered in Sec. 5.3.2. The structures of pyrite and marcasite (both FeS_2) are different. They contain discrete S_2^{2-} groups in which the sulphur atoms are joined by a covalent bond, the S–S distance being 2.10 Å. The pyrite structure is based on the sodium chloride lattice (face centred cubic) the Fe atoms and the centres of the S_2 groups taking the places of Na and Cl.

The nickel arsenide structure is adopted by many solids of composition MX in which M is a transition metal, particularly a Group VIII metal and X is an element in the late B subgroups, Sn, As, Sb, Bi, S, Se, Te. The pyrite structure is characteristic of solids of composition MX_2 of the same elements. In substances such as arsenopyrite, FeAsS, the lattice has lower symmetry but is still based on the pyrite structure. The marcasite structure is a less symmetrical arrangement of the same structural units, Fe and S_2, based on the sodium chloride lattice but having lower symmetry. Some special cases occur, however, for example MoS_2 has a layered structure so that like graphite it can be used as a lubricant.

Chapter 2

Equilibrium in Aqueous Solutions

2.1. Introduction

The molecules and ions which together make up an aqueous solution interact with one another in various ways. The extent to which they do so and the changes in behaviour which result are of the greatest importance in solution chemistry. In general the ions and molecules of a solute are solvated by water and interaction occurs between ions of opposite sign or between ions and neutral molecules. The two extreme methods of describing the interaction, or association, are:

(i) The modern forms of equations which have evolved from the Debye–Hückel theory.[6] The original Debye–Hückel equation applies only to very dilute solutions of electrolytes and was based on a model in which ions were regarded simply as charged particles having no size. The excess Gibbs energy of the system, G^{ex} is calculated by considering electrostatic interactions between the ions and their close neighbours.

(ii) To regard the interactions as chemical equilibria and describe the extent to which they occur by means of experimentally determined equilibrium constants. The excess Gibbs energy of the system controls the values of these constants. In this method the species are treated as chemical entities and the products of association are often referred to as complex species.

Chemical equilibria are described in terms of the thermodynamic activities of the species taking part in the reactions. These activities depend on but are not usually equal to the concentrations of the species because of deviations from ideal behaviour in the system. These deviations are the origin of the excess Gibbs energy and are all included in a single coefficient, the activity coefficient. Each unit of concentration has a different activity coefficient for a particular solution of a solute. The most obvious chemically satisfactory unit of quantity is the mole fraction x_i, the ratio of the number of moles of component i to the total number of moles of all components present

$$x_i = n_i / \sum_j n_j. \tag{2.1}$$

It is almost never used for aqueous solutions or for describing the thermodynamics of such solutions.

In chemical practice the concentration of a solute is most conveniently expressed in molarity units M, c_i, g moles of i per litre of solution. Its advantage is that the water content of a sample of the solution does not have to be determined. Its disadvantage is that it depends on temperature and to a minor extent on pressure. When the composition scale must be independent of those factors, for example, when considering the thermodynamics of a solution, the molality scale is used. This gives m_i, g moles of i per kg of solvent. In the case of water there are $1000/18.0152 = 55.51$ g moles of water in 1 kg of it, so the mole fraction of a sole solute i is $x_i = m_i/(m_i + 55.51)$. The activity of i, $\{i\}$ is given by its concentration $[i]$ in any unit multiplied by the appropriate activity coefficient.

$$a_i = m_i \gamma_i = x_i f_i = c_i \gamma_i. \tag{2.2}$$

If the component i were, for example, sodium chloride its molal activity coefficient in a particular solution would be written γ_{NaCl}. When dealing with an equilibrium reaction in which Cl^- plays a role but Na^+ does not, for example

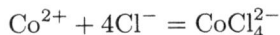

$$Co^{2+} + 4Cl^- = CoCl_4^{2-}$$

we might be interested in relating solvent extraction behaviour of cobalt to the activities of the species taking part in the reaction. This would involve writing single ion activity coefficients such as γ_{Cl^-}. These have no thermodynamic validity because the definition of a component is that it is an independently variable constituent of a solution. Thus in the $NaCl$–H_2O system there are two components and Na^+ and Cl^- are merely constitutents of the solution. Modern methods of calculating activity coefficients (Sec. 2.2.2) do lead to single ion activity coefficients, however, as does the original Debye–Hückel theory in Eq. (2.22). For a salt, the individual single ion activities can be combined so as to give the activity coefficient for the salt, $\gamma_\pm$, Eqs. (2.23) and (2.24).

In the case of dilute solutions of electrolytes in water it is often assumed that the activity of the water is unity. This is not correct in thermodynamic terms and can lead to difficulties in calculations. In general when a solution is in equilibrium with its vapour the chemical potential of each component of the solution must be equal to the potential of that component in the vapour.

$$\mu_i^{\text{solution}} = \mu_i^{\text{vapour}} = \mu_i^\circ + RT \ln p_i \tag{2.3}$$

where p_i is the fugacity of i over the solution. This can usually be taken as being the partial vapour pressure of i. If the solution obeys Raoult's law,

$$p_i = p_i^\circ x_i \tag{2.4}$$

where p_i° is the vapour pressure of the pure component i at the temperature of the solution. Thus (2.3) becomes

$$\mu_i^{\text{solution}} = \mu_i^\circ + RT \ln p_i^\circ + RT \ln x_i \,. \tag{2.5}$$

Then

$$d\mu_i = RT \, d \ln x_i \,. \tag{2.6}$$

All very dilute solutions behave ideally, apart from electrolytes in water which do so only at very high dilutions, as discussed below. In more concentrated solutions the deviations from ideality are allowed for by multiplying the mole fraction by the activity coefficient f_i. For the solvent, component 1, the activity coefficient is defined by

$$\mu_l^{\text{solution}} = \mu_l^\circ + RT \ln p_l^\circ + RT \ln f_l x_l \tag{2.7}$$

and

$$d\mu_l = RT \, d \ln f_l x_l \,. \tag{2.8}$$

In the case of aqueous solutions of electrolytes this definition of the activity coefficient is not convenient for the solute, which is not volatile and so has no vapour pressure in equilibrium with the solution, and does not obey Raoult's law. Instead Henry's law can be used. For the solute, component 2 of the solution, this can be written as

$$P_2 = k_2 x_2 \quad \text{or} \quad p_2 = k_2' m_2$$

leading to

$$\mu_2^{\text{solution}} = \mu_2^\circ + RT \ln k_2 + RT \ln f_2' x_2 \tag{2.9}$$

$$d\mu_2 = RT \, d \ln f_2' x_2 \tag{2.10}$$

and

$$\mu_2^{\text{solution}} = \mu_2^\circ + RT \ln k_2' + RT \ln \gamma_2 m_2 \tag{2.11}$$

$$d\mu_2 = RT \, d \ln \gamma_2 m_2 \, . \tag{2.12}$$

This is the theoretical basis for the use of activity coefficients to obtain activities of substances, which are required for thermodynamic calculations, from their concentrations.

An alternative way of dealing with deviations from ideality which is frequently employed for the solvent is to use Bjerrum's "osmotic coefficient" ϕ in the equation

$$\mu_1^{\text{solution}} = \mu_1^\circ + RT \ln p_1^\circ + \phi RT \ln x_1 \, . \tag{2.13}$$

Thus, from (2.7)

$$\phi \ln x_1 = \ln f_1 x_1 \, . \tag{2.14}$$

Comparing (2.3) and (2.7)

$$p_1 = p_1^\circ f_1 x_1 \, . \tag{2.15}$$

The activity coefficient is the quantity by which the vapour pressure of the solvent over an ideal solution must be multiplied to give its vapour pressure over the real solution of the same concentration. Thus it is used in general physical chemistry in formulae relating the elevation of boiling point, depression of freezing point and osmotic pressure of liquids to the solute concentration.

The molal osmotic coefficient ϕ is defined by

$$\ln a_s = -\frac{vmW_s}{1000} \phi$$

where a_s is the activity of the solvent and W is the molecular weight, m is the molality of the solute and in the case that this is an electrolyte, v is the number of ions produced when one formula unit of it dissociates in solution. In the case of water

$$\ln a_w = -\frac{18.0152\, vm}{1000}\, \phi.$$

As an example, at 25°C, a 2 m solution of potassium chloride has a water activity of 0.9364; $v = 2$ and $m = 2$ so that

$$\ln 0.9364 = -0.06571 = -0.07206\phi$$

$$\phi = 0.912.$$

If the solution contains more than one solute, the equation for $\ln a_w$ can be used if the term vm is replaced by a summation over all the solutes present.

The activity of the solvent can be determined by measuring its partial vapour pressure over a solution and comparing it with that over an ideal solution of the same concentration. This has in the past been done experimentally using the isopiestic method in which solutions of the salt being studied are left to come to equilibrium with solutions of reference substances for which accurate values of the osmotic coefficients are known over a wide range of concentrations. Reference substances used are $NaCl$, KCl, $CaCl_2$, H_2SO_4 and for nonelectrolytes, sucrose. Known quantities of these substances are placed in containers, in solutions of appropriate concentrations and are left at constant temperature to come to equilibrium with solutions of the salts under investigation. This is achieved more quickly if the chambers containing the pairs of solutions are evacuated. At equilibrium the vapour pressure of the two solutions is the same and the containers are weighed to obtain the concentrations of the solutions. The osmotic or activity coefficient of the water can be used to calculate the activity of the solute by a variety of analytical and graphical methods. Equation (2.11) gives the chemical potential of a solute in a solution, k' being the proportionality constant when the partial pressure of a volatile solute is related to the molality of one which is nonvolatile. When the solute is an electrolyte (2.11) can be written, omitting k',

$$\mu_2 = \mu_2^\circ vRT \ln \gamma_2 m_2.$$

Changes in the chemical potential of the solute as its concentration is changed are related to those of the solvent through the Gibbs–Duhem equation.

Integration of this leads to

$$\int_{m_1}^{m_2} \frac{\phi - 1}{m} \, dm + \phi m_2 - \phi m_1 = \ln \gamma m_2 - \ln \gamma m_1 \, .$$

If a polynomial expression of water activity as a function of molality of the electrolyte is available the integral can be evaluated numerically. An example of such a polynomial is given in Sec. 2.2.4. The isopiestic method is suitable for solutions of concentration 0.1 M to saturation. It is discussed extensively in Refs. 7 and 8.

A more recently developed method is to use an electrostatically charged micron-size (about 20 μm) droplet or particle levitated in an electrodynamic balance and to weigh it by using the force due to a dc field which exactly balances the gravitational force on the droplet or particle. This is allowed to come to equilibrium with a stream of air flowing past it, the relative humidity being calculated from the absolute humidity of the air, measured using a dew point hygrometer, and the temperature of the particle, measured by a thermistor and appropriately corrected. The advantages of this method are: (i) Because the size of the sample of solution is so small it reaches equilibrium with its surroundings very quickly. (ii) Measurements of the water activity as a function of concentration are limited in practice only by the time it takes to establish and measure the relative humidity in the chamber. (iii) An entire experiment, with measurements over a wide range of humidities can be carried out in a day. (iv) Because the levitated droplets are not in contact with any surface their concentration can be increased by using progressively less humid air to give very highly supersaturated solutions, due to the absence of foreign nuclei. The water activity in such solutions can be measured and in addition the factors influencing spontaneous nucleation can be investigated.[9–11] The disadvantage of the technique is that it is not as accurate as the isopiestic method.

2.2.   Ionic Activities

2.2.1.   *Debye–Hückel Theory*

The activity coefficients of three electrolytes of different valency types are shown in Fig. 2.1 as a function of concentration and it can be seen that the curves show an infinite negative gradient as zero concentration is approached. Nonelectrolytes such as glycine and sucrose may have either rising or falling

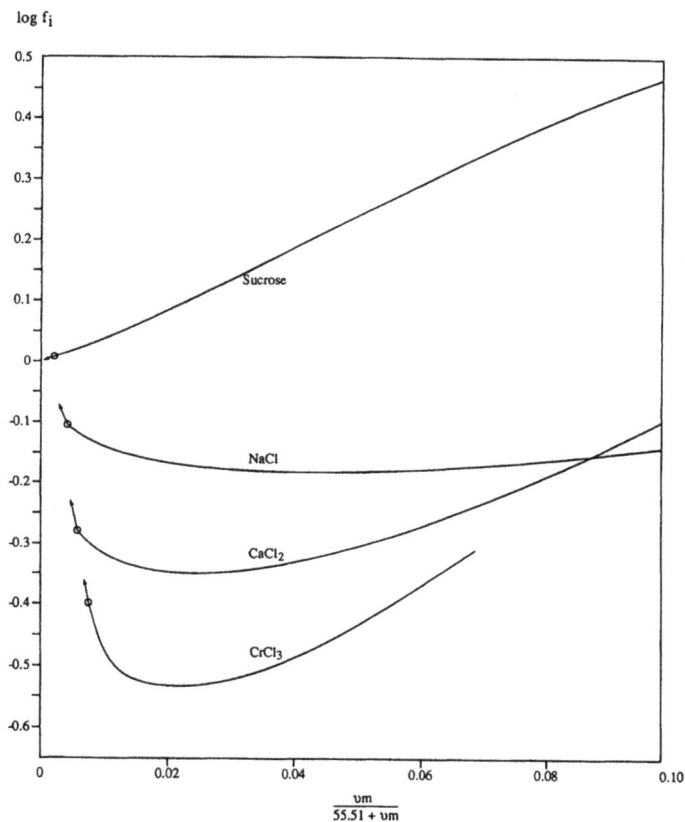

Fig. 2.1. Mole fraction activity coefficients as a function of concentration.

activity coefficients as zero concentration is approached but the values are usually close to unity and lie approximately on straight lines. The deviation of solutions of nonelectrolytes from ideality is due to short range forces such as van der Waals forces etc. whereas the nature of the shapes of the curves for the electrolytes is a consequence of long range forces. Debye and Hückel[6] in 1923 and 1924 suggested that these forces are due to interionic attractions and developed an equation relating the activity coefficient of a salt and a function of its concentration. The equation is valid solely for very dilute solutions and only a general summary of the theory is given here. Full accounts are available.[7,8] It had been suggested previously by several workers that interionic attraction was the cause of the long range forces but no satisfactory mathematical treatment had been put forward.

The basis of the Debye–Hückel theory is that the electrolyte dissociates in solution to give ions which may be regarded as point charges. The chemical nature of the ions is immaterial but the amount of charge per unit volume of solution controls the extent of interaction between charges. Salts of different valency types produce different numbers of ions in solution, a 1:1 electrolyte such as NaCl produces 2; a 2:1 or a 1:2 electrolyte such as $CaCl_2$ or Na_2SO_4 produces 3; a 3:1 or 1:3 electrolyte such as $LaCl_3$ or Na_3PO_4 produces 4 and so on. The "ionic strength" of a solution, I, is defined as

$$I = \frac{1}{2} \sum_i m_i z_i^2 \tag{2.16}$$

where m_i and z_i are the molal concentration and valency of the ion i.

Every ion is considered to be surrounded by an ionic atmosphere of opposite sign. An individual positive ion in a dilute solution of a binary electrolyte will be surrounded by both negatively and positively charged ions, but because of electrostatic attraction and repulsion, rather, there will, be more negative ions close to the central positive ion than other positive ions, when taking the average figure each time. This imbalance of charge is the ionic atmosphere, which exerts an attractive force on the central ion.

Since the whole solution contains an equal quantity of positive and negative charge, the total charge of the ionic atmosphere surrounding the central ion being considered is equal in magnitude and opposite in sign to the charge on the central ion. It is found that the effect of the ionic atmosphere on the ion is equivalent to that of a single point charge of the same magnitude as that on the ion, placed at a distance $1/\kappa$ from it, where

$$\kappa = \left(\frac{4\pi e^2}{\varepsilon_r kT} \sum_i n_i z_i^2 \right)^{1/2} \tag{2.17}$$

e is the charge on an electron, ε_r the dielectric constant (relative permittivity) of the liquid, k the Boltzmann constant, T the absolute temperature, n_i the number per ml and z_i the valency of ions of type i. If the concentration of these ions is m_i g ions per litre, then

$$n_i = m_i N / 1000$$

where N is Avogadro's number. Taking ε_r as 78.6, the value for water at 25°C, and T as 298

$$1/\kappa = 4.31 \times 10^{-8}/\left(\sum_i m_i z_i^2\right)^{1/2} \text{cm}. \tag{2.18}$$

This distance is known as the thickness of the ionic atmosphere and depends on the ionic strength of the solution, Eq. (2.16)

$$1/\kappa = 4.31 \times 10^{-8}/\sqrt{(2I)} \text{cm}. \tag{2.19}$$

The ionic atmosphere causes a potential ψ_i to exist at the point of space occupied by the central ion which it surrounds. If the charge on this ion is $z_i e$ then

$$\psi_i = \frac{z_i e \kappa}{\varepsilon_r}.$$

The central ion possesses a certain amount of energy due to the work expended in charging it to this potential, equal to half the product of its charge and the potential. For 1 g ion of the kind of ion considered as acting as centres of ionic atmospheres, N individuals are charged, so the energy is

$$E_i = -\frac{N z_i^2 e^2 \kappa}{2\varepsilon_r}. \tag{2.20}$$

This excess energy possessed by ions, due to their being surrounded by ionic atmospheres, is assumed to be the cause of the deviation of solutions of electrolytes from ideal behaviour.

The chemical potential of an ion i in an ideal solution is

$$\mu_i = \mu_i^\circ + RT \ln x_i$$

where R is the gas constant. For a nonideal solution

$$\mu_i = \mu_i^\circ + RT \ln a_i = \mu_i^\circ + RT \ln x_i + RT \ln f_i.$$

The difference between the two equations, $RT \ln f_i$, is the difference in the free energy accompanying the addition or removal of 1 g ion of the given species from a large volume of an ideal and a real solution. It is equated with E_i. Thus

$$\ln f_i = -N z_i^2 e^2 \kappa / 2RT\varepsilon_r. \tag{2.21}$$

Substituting for κ, using molalities of ions, and transforming to $\log_{10}$ gives

$$- \log \gamma_i = \left[\frac{N^2 e^3 \sqrt{(\pi/1000)}}{2.303 R^{3/2}} \right] \frac{z_i^2 \sqrt{(\sum_i m_i z_i^2)}}{(\varepsilon_r T)^{3/2}}.$$

The term in large [square] brackets consists of universal constants and may be replaced by A'. Using the definition of ionic strength (2.16)

$$- \log \gamma_i = A''(\varepsilon_r T)^{-3/2} z_i^2 \sqrt{I} \qquad (2.22)$$

where

$$A'' = A'/\sqrt{2}.$$

At 25°C in water, with $\varepsilon_r = 78.6$, $A = A''(\varepsilon_r)^{-3/2} = 0.509$.

The Debye–Hückel theory and Eqs. (2.22) and (2.25), the latter derived below, represent what is frequently called the Debye–Hückel limiting law, applicable to dilute solutions. It states that the extent of the departure of an ion from ideal behaviour in a given solvent is governed by the density of charge in the solution as expressed by the ionic strength. It is independent of the chemical nature of the ions. The theory leads directly to a single ion activity coefficient in (2.22).

The individual activities of ions and so the individual ion-activity coefficients cannot be measured and, as stated above, have no thermodynamic meaning. They may, however, be related to the measurable mean activity of the ions. If one formula unit, or "molecule", of a binary electrolyte dissociates into a total of v ions, of which v_+ are cations and v_- anions, then the mean ion-activity coefficient $\gamma_\pm$ is related to the individual ionic-activity coefficients by:

$$\gamma_\pm = \sqrt[v]{(\gamma_+^{v_+} \gamma_-^{v_-})} \qquad (2.23)$$

or

$$\log \gamma_\pm = (v + \log \gamma_+ + v_- \log \gamma_-)/(v_+ + v_-).$$

If the valencies of the ions are z_+ and z_-

$$\log \gamma_\pm = (z_- \log \gamma_+ + z_+ \log \gamma_-)/(z_+ + z_-). \qquad (2.24)$$

Thus

$$-\log \gamma_{\pm} = A z_+ z_- \sqrt{I}.$$ (2.25)

2.2.2. Extension to More Concentrated Solutions

The theoretical treatment of electrolytic solutions outlined above, based on the concept of the ionic atmosphere and leading to the limiting law of Debye and Hückel, provides a statement of the law of the force of attraction between ions in a solution. Harned and Owen[7] emphasise this fact by stating that "the application of statistical mechanics to the theory of electrolytes has clarified the profound basis of the theory and has served to prove the exact validity of the limiting law". They also state that by use of the concepts of electrostatics, hydrodynamics and statistical mechanics an exact theory has been developed which will describe the properties of solutions of electrolytes where the forces of attraction between ions predominate. They clearly and comprehensively demonstrate that this is the case. In doing so they also reveal that the exact validity of the limiting law begins to break down for 1:1 strong electrolytes in solutions more concentrated than about 10^{-3} M, and for other classes of electrolytes in solutions of similar ionic strength. In certain cases, for example zinc sulphate and some other 2:2 electrolytes, the electrical conductivity of solutions of them decreases much more rapidly with increasing concentration than is required by the Onsager conductance equation, even at concentrations below 10^{-4} M. This behaviour is described as being characteristic of weak electrolytes and possibly due to incomplete dissociation into ions.[7]

In most processes and procedures in which solutions of electrolytes are used much higher concentrations than 10^{-3} M are required, and if thermodynamic principles are to be applied knowledge of activity coefficients is necessary. A number of attempts to extend the limiting law to deal with more concentrated solutions were made. Shortly after Debye and Hückel published their theory N. Bjerrum put forward a modification developed from consideration of the factors which determine the extent of ion association or, more particularly, the formation of ion pairs due to interionic attraction. The most simple model was assumed. The ions were supposed to be rigid unpolarizable spheres in a medium of fixed dielectric constant and the formation of chemical bonds by electron sharing was excluded. He suggested that if the sum of the radii of the two ions of the electrolyte, å, exceeds the minimum value given by $r_{min} = |z_+ z_-| e^2 / 2\varepsilon_r kT$ then the limiting law is applicable, but when it is less than this the ions should be regarded as being associated in pairs. In this paired state they contribute nothing to the electrical energy of the "central" ion due to its ionic atmosphere. The extent of ion pair formation increases very rapidly with decreasing r and the value of an "equilibrium constant" can be calculated. The final result is the Extended Debye–Hückel equation, the sign of the charges z_+ and z_- being neglected as indicated above by short vertical lines in the equation for r_{min},

$$- \log \gamma_{\pm} = A z_+ z_- \sqrt{I}/(1 + B \mathring{a} \sqrt{I})$$ (2.26)

where B is another constant. If the solution contains more than one electrolyte it is thermodynamically inconsistent to use this equation for calculating activity coefficients,[12] so that it is of restricted use. In addition, the closest distance of approach of the ions is not known and although $\mathring{a}$ is sometimes referred to as the ion-size parameter, it is not the sum of two ionic radii as determined by any other means.

In order to eliminate such pseudo-fundamental parameters, equations of the type

$$- \log \gamma_{\pm} = \frac{A z^2 \sqrt{I}}{1 + \sqrt{I}} - \beta I$$ (2.27)

have been used where $z_+ = z_-$, and β is an empirical constant for the particular system, chosen by fitting the best line to the available data. An equation with no adjustable parameters

$$- \log \gamma_{\pm} = A z^2 \sqrt{I}/(1 + \sqrt{I})$$ (2.28)

was suggested and although it is not particularly accurate it can be used for solutions containing several electrolytes. An extension of it, the Davies equation[13] which is (2.27) with $\beta = 0.2$

$$- \log \gamma_{\pm} = A z^2 \left(\frac{\sqrt{I}}{1 + \sqrt{I}} - 0.2 I \right)$$ (2.29)

gives good agreement with measured values of mean ion-activity coefficients in dilute solutions, the average deviation being about 2% in 0.1 M solutions, less in more dilute solutions. This was an empirical equation based on an examination of published data. Provided sufficient experimental results are available for $\gamma_{\pm}$ of a salt to permit adjustable parameters such as $\mathring{a}$ and β to be calculated accurately, such equations can be used to calculate (or perhaps interpolate) $\gamma_\pm$ for concentrations of simple 1:1 electrolytes of low valency up to about 1 M. In general however none of the equations can be regarded as satisfactory.

Suggested alternatives were correlation equations in which an additional parameter array $\sum c_i I^i$ replaced βI in (2.27) but a large number of coefficients were often needed and it was difficult to extend the treatment to mixed electrolyte solutions. In 1936, it had been suggested[14] that the value of β should be made dependent on ionic strength and it was shown that this did improve the performance of equations similar to (2.27).

2.2.3. *Extension to Mixed Electrolyte Solutions*

Because of the gradual realisation that a knowledge of the thermodynamic properties of aqueous electrolyte solutions of potential importance was becoming essential, a considerable amount of work was carried out on this problem during the 1960's and 1970's, particularly in the USA. In addition to process chemistry this was related to the chemistry of seawater, geothermal waters and other systems at high temperatures. During the 1970's several expressions were published which attempted to represent the existing data on activity coefficients of electrolytes and osmotic coefficients of water in the solutions and all earlier equations such as those mentioned above must be regarded as having been superseded. The recently proposed equations and many other recent developments in practically useful thermodynamics have been reviewed.[15–17]

Activity or osmotic coefficients of most pure electrolyte solutions of interest have been measured at temperatures around 25°C. Fitting equations which represent this information are convenient for interpolation for these solutions. On the other hand there are experimental data for only a limited number of cases of mixed electrolyte solutions, whereas there is an enormous range of solution compositions of actual or potential interest. Clearly it would be of great value to be able to predict activity or osmotic coefficients in multicomponent electrolyte solutions from data obtained from solutions containing single electrolytes. This is now possible using the procedure developed by Pitzer and his coworkers, or one of the other methods mentioned below.

Pitzer's Method

Pitzer's procedure presently is the most widely used. It was described in a series of papers.[18–27] In Part 1,[18] it was stated that the objective was "to develop equations which reproduce the measured properties substantially within experimental accuracy, which are compact and convenient in that only a very few parameters need to be tabulated for each substance and the mathematical calculations are simple, which have appropriate form for mixed electrolyes as well as for solutions of a single solute, and whose parameters have physical meaning as far as possible". It should be noted that when used to calculate osmotic or activity coefficients it is necessary only to substitute numbers for symbols in the relevant equations and carry out simple arithmetic calculations. For solutions containing several electrolytes these become numerous and a computer is used. Some fully worked examples are given in Sec. 2.2.4.

Pitzer's equations are based on the Debye–Hückel limiting law which is taken as representing the effects of long range interactions between ions (Sec. 2.2.1 and Fig. 2.1). Other terms are added to represent effects of short range interactions without having to assume that discrete chemical species are formed, such as uncharged ion pairs in solutions of 2:2 electrolytes. Relatively strong interactions occur between cations and anions of higher valencies and, as is to be expected, there are cases where the behaviour of the electrolytes can be treated either by a method such as that of Pitzer or by use of equilibrium constants based on the proposed existence of discrete, complex species (Sec. 2.3).

Pitzer[18] proposed that the general form of the equation for the total excess Gibbs energy for a solution containing n_w kg of solvent and $n_i, n_j, \ldots$ moles of solute species $i, j, \ldots$ should be taken to be

$$\frac{G^{ex}}{RT} = n_w f(I) + \frac{1}{n_w} \sum_{ij} \lambda_{ij}(I) n_i n_j + \frac{1}{n_w^2} \sum_{i,j,k} \mu_{ijk} n_i n_j n_k \qquad (2.30)$$

where G^{ex} is the difference between the real Gibbs energy of the solution and its ideal Gibbs energy.

In this equation, $f(I)$ is a function of ionic strength, solvent properties and temperature and expresses the effect of the long range electrostatic forces. Thus it is related to the Debye–Hückel theory. $\lambda_{ij}(I)$ represents the effect of short range forces between species i and j and is a function of ionic strength. Thus it is related to the term βI in (2.27). A term for triple ion interaction is included but it was assumed that μ_{ijk} would not be dependent on ionic strength.

Equations for the osmotic and activity coefficients were then written using the appropriate derivatives of G^{ex}.

$$\varphi - 1 = -\frac{\partial G^{ex}/\partial n_w}{RT \sum_i m_i}$$

$$= \frac{(If' - f) + \sum_{i,j}(\lambda_{ij} + I\lambda'_{ij})m_i m_j + 2\sum_{i,j,k}\mu_{ijk}m_i m_j m_k}{\sum_i m_i} \qquad (2.31)$$

$$\ln \gamma_i = \frac{1}{RT}\frac{\partial G^{ex}}{\partial n_i}$$

$$= \frac{z_i^2}{2}f' + 2\sum_j \lambda_{ij}m_j + \frac{z_i^2}{2}\sum_{j,k}\lambda'_{jk}m_j m_k + 3\sum_{j,k}\mu_{ijk}m_j m_k \qquad (2.32)$$

where $f' = \mathrm{d}f/\mathrm{d}I$; $\lambda'_{ij} = \mathrm{d}\lambda_{ij}/\mathrm{d}I$ and $m_i = n_i/n_w$ etc.

Further development of the equations, taking into account the valencies of ions and numbers of ions produced on dissolution of electrolytes of different types (Sec. 2.2.1) indicated that each type would have to be examined separately, and mixed electrolytes considered as a different case.

This examination consisted of taking the available data on the osmotic coefficients of a particular class of single electrolytes and evaluating by the method of least squares the values of coefficients of equations derived from (2.31) and (2.32) which gave the best fit. The values of osmotic coefficients used were usually those given by Robinson and Stokes[8] after critical review.

The results obtained for many single electrolytes which show no association were presented in Part 2.[19] These salts have one or both ions univalent. The equations used for G^{ex}, ϕ and γ are

$$(G^{ex}/n_w RT) = f^{Gx} + m^2(2\upsilon_M \upsilon_X)B_{MX}^{Gx} + m^3[2(\upsilon_M \upsilon_X)^{3/2}]C_{MX}^{Gx} \qquad (2.33)$$

$$\varphi - 1 = |z_M z_X|f^\varphi + m\left(\frac{2\upsilon_M \upsilon_X}{\upsilon}\right)B_{MX}^\varphi + m^2\frac{2(\upsilon_M \upsilon_X)^{3/2}}{\upsilon}C_{MX}^\varphi \qquad (2.34)$$

$$\ln \gamma = |z_M z_X|f^\gamma + m\left(\frac{2\upsilon_M \upsilon_X}{\upsilon}\right)B_{MX}^\gamma + m^2\frac{2(\upsilon_M \upsilon_X)^{3/2}}{\upsilon}C_{MX}^\gamma \qquad (2.35)$$

$$f^{Gx} = -A_\varphi(4I/b)\ln(1 + bI^{1/2}) \qquad (2.36)$$

$$f^\varphi = -A_\varphi\frac{1^{1/2}}{1 + bI^{1/2}} \qquad (2.37)$$

$$f^\gamma = -A_\varphi \left[\frac{I^{1/2}}{1 + bI^{1/2}} + \frac{2}{b} \ln(1 + bI^{1/2}) \right] \tag{2.38}$$

$$B_{MX}^{Gx} = \beta_{MX}^{(0)} + \frac{2\beta_{MX}^{(1)}}{\alpha^2 I} [1 - e^{-\alpha I^{1/2}}(1 + \alpha I^{1/2})] \tag{2.39}$$

$$B_{MX}^\varphi = \beta_{MX}^{(0)} + \beta_{MX}^{(1)} e^{-\alpha I^{1/2}} \tag{2.40}$$

$$B_{MX}^\gamma = 2\beta_{MX}^{(0)} + \frac{2\beta_{MX}^{(1)}}{\alpha^2 I} [1 - e^{-\alpha I^{1/2}}(1 + \alpha I^{1/2} - (1/2)\alpha^2 I)] \tag{2.41}$$

$$C_{MX}^{Gx} = (1/2)C_{MX}^\varphi \tag{2.42}$$

$$C_{MX}^\gamma = (3/2)C_{MX}^\varphi \tag{2.43}$$

v_M and v_X are the numbers of M and X ions in the formula of the salt
z_M and z_X are the valencies of M and X
$v = v_M + v_X$
n_w is the number of kg of solvent
m is the conventional molality
b and α are numbers selected to fit equations to the data for ϕ or γ.

f^{Gx} is the term in the equation for $(G^{ex}/n_w RT)$ which takes account of long range electrostatic forces. As given in (2.36) it is determined by the Debye–Hückel coefficient A, derived from (2.22), but it was shown empirically in Part 1 that better results were obtained if the coefficient for the osmotic function was used. This is given by Ref. 19

$$A_\phi = \frac{1}{3} \left(\frac{2\pi N d}{1000} \right)^{1/2} \left(\frac{e^2}{\varepsilon_r kT} \right)^{3/2} = \frac{1}{3} A_\gamma .$$

It should be noted that A_γ here does not include 2.303 in the denominator, as does the value based on Eq. (2.22), 0.509 at 25°C. This is because A_ϕ is used in Eqs. (2.33) and (2.34) which are not logarithmic. The density of water, d, is also included so that the value of A_ϕ takes account of the change in volume with temperature change. This affects the density of charge in the solution. A_ϕ has the value 0.391 at 25°C.[28] f^{Gx} also includes the value of the ionic strength of the solution and a constant b. In order to maintain simple equations for mixed electrolytes the value for b must remain the same for all solutes and in Part 1 the value 1.2 was selected. The equivalent terms taking account of long range forces in the equations for $\phi - 1$ (2.34) and $\ln \gamma$ (2.35), f^ϕ and f^γ also involve A_ϕ, I and b. Note that the superscripts to f indicate which equation f refers to.

The coefficient λ_{ij} in (2.30) taking account of the effect of short range forces between species i and j is represented in Eq. (2.33) for $(G^{ex}/n_w RT)$ by B_{MX}^{Gx}, the superscript Gx again showing that the value of B for the salt MX refers to the equation for $(G^{ex}/n_w RT)$. The form of the equations used for deriving values of B can be seen most clearly from (2.40). B replaces the second virial coefficient in (2.27) and its value for MX is defined by two parameters $\beta_{MX}^{(0)}$ and $\beta_{MX}^{(1)}$ which are obtained by curve fitting to the osmotic coefficient data for MX. The values $\beta_{MX}^{(0)}$ and $\beta_{MX}^{(1)}$ are characteristic for MX and can be used to calculate $(\phi - 1)$ and $\ln \gamma$, using (2.40) and (2.41) to obtain B_{MX}^ϕ and B_{MX}^γ. B is a function of ionic strength and also involves a number, α. The value $\alpha = 2$ was found to be satisfactory for all salts considered in Ref. 19.

The third virial coefficient is defined by C_{MX}^ϕ and values to be used in (2.33) and (2.35) can be calculated from (2.42) and (2.43). This third virial coefficient is usually small and in some cases negligible.

The numerical factors such as $(2v_M v_X)$, multiplying the second and third virial coefficients in Eqs. (2.33) to (2.35) are needed in order to retain the simple meaning that the virial coefficients represent the short range interaction of pairs and triplets of ions.

The strong electrolytes considered in Ref. 19 are said to total 227 in all and comprise inorganic compounds of 1:1, 2:1, 3:1, 4:1 and 5:1 types, salts of carboxylic acids (1:1 type), tetraalkylammonium halides, sulphonic acids and sulphonates (1:1 and 2:1 types) and some other 1:1 type organic salts. The tabulated data give values of $\beta^{(0)}$, $\beta^{(1)}$ and C^ϕ, for each electrolyte, the maximum molality for which agreement is attained up to 0.01 in ϕ or for which data were available, and the standard deviation of fit. The source of the data is also given.

When 2:2 electrolytes were considered,[20] it was found necessary to include an additional term in (2.40) to give

$$B^\phi = \beta^{(0)} + \beta^{(1)} \exp(-\alpha_1 I^{1/2}) + \beta^{(2)} \exp(-\alpha_2 I^{1/2}). \qquad (2.44)$$

Constants α_1 and α_2 are introduced, having the values 1.4 and 12.0 respectively. The parameter $\beta_{MX}^{(2)}$ allows data for electrolytes of high valency types to be fitted without specifically considering discrete, complex, species to be present.

Calculation of ϕ for Salts in Mixed Electrolyte Solutions

Equations (2.31) and (2.32), for solutions of single electrolytes, indicate three terms to be important; the first dealing with long range electrostatic interactions as a function of ionic strength, the second takes account of short range interactions between the anion and cation and the third with those between three ions. When two electrolytes are present together it would be expected that equations for $(\phi-1)$ and $\ln\gamma$ would contain the same first term, since the Debye–Hückel equation takes no account of the chemical nature of the ions, only their charges and concentrations. It would also be expected that as a first approximation the interactions between the ions present in the mixed electrolyte solution would be the same as those in solutions containing only one salt in each, resulting in a summation of effects in the mixed solution. However, if Eqs. (2.31) and (2.32) are to be used to calculate $(\phi-1)$ or $\ln\gamma_i$ for a mixed electrolyte solution it is necessary to compare the results obtained with experimentally determined data and to add other parameters to improve the fit of the equations where necessary.

This was done by Pitzer and Kim.[21] They wrote for the osmotic coefficient of a mixed electrolyte solution

$$\phi - 1 = \left(\sum_i m_i\right)^{-1} \left\{ 2If^\phi + 2\sum_c \sum_a m_c m_a \left[B_{ca}^\phi + \frac{(\sum mz)}{(z_c z_a)^{1/2}} C_{ca}^\phi \right] \right.$$

$$+ \sum_c \sum_{c'} m_c m_{c'} \left[\theta_{cc'} + I\theta'_{cc'} + \sum_a m_a \psi_{cc'a} \right]$$

$$\left. + \sum_a \sum_{a'} m_a m_{a'} \left[\theta_{aa'} + I\theta'_{aa'} + \sum_c m_c \psi_{caa'} \right] \right\} \qquad (2.45)$$

where f^ϕ was defined in (2.37); B_{MX}^ϕ in (2.40) is written B_{ca}^ϕ; c and c' are indices covering all cations while a and a' cover all anions; μ is the term for triple ion interaction in Eq. (2.30) and λ represents the effect of short range forces.

$$C_{MX}^\phi = 3 \left[\left| \frac{z_X}{z_M} \right|^{1/2} \mu_{MMX} + \left| \frac{z_M}{z_X} \right|^{1/2} \mu_{MXX} \right] \qquad (2.46)$$

$$\theta_{MN} = \lambda_{MN} - \left(\frac{z_N}{2z_M} \right) \lambda_{MM} - \left(\frac{z_M}{2z_N} \right) \lambda_{NN} \qquad (2.47)$$

$$\psi_{MNX} = 6\mu_{MNX} - \left(\frac{3z_N}{z_M}\right)\mu_{MMX} - \left(\frac{3z_M}{z_N}\right)\mu_{NNX}. \qquad (2.48)$$

In (2.45), $(\sum mz) = \sum_c m_c z_c = \sum_a m_a z_a$.

The first term in (2.45) in the main brackets, $2If^\phi$ is the Debye–Hückel term, the second term comprises a double sum over molalities and the second and third virial coefficients for pure electrolytes, obtained as discussed above. The final two terms include the differences between the second and the third virial coefficients for unlike ions of the same sign, M and N, from the appropriate averages for like ions and were expected to be small. Thus the principal effects of mixing electrolytes arise from differences in the parameters $\beta^{(0)}$, $\beta^{(1)}$ and C^ϕ for the pure electrolytes. The parameters θ and ψ were expected to have only a small effect, if any.

Calculation of γ for Salts Present in Mixed Electrolyte Solutions

For the activity coefficient of the salt MX in a mixed electrolyte solution Pitzer and Kim wrote

$$\ln \gamma_{MX} = |z_M z_X| f^\gamma$$

$$+ \left(\frac{2v_M}{v}\right) \sum_a m_a \left[B_{Ma} + \left(\sum mz\right)C_{Ma} + \left(\frac{v_X}{v_M}\right)\theta_{Xa}\right]$$

$$+ \left(\frac{2v_X}{v}\right) \sum_c m_c \left[B_{cX} + \left(\sum mz\right)C_{cX} + \left(\frac{v_M}{v_X}\right)\theta_{Mc}\right]$$

$$+ \sum_c \sum_a m_c m_a \{|z_M z_X| B'_{ca}$$

$$+ v^{-1}[2v_M z_M C_{ca} + v_M \psi_{Mca} + v_X \psi_{caX}]\}$$

$$+ \frac{1}{2} \sum_c \sum_{c'} m_c m_{c'} \left[\left(\frac{v_X}{v}\right)\psi_{cc'X} + |z_M z_X|\theta'_{cc'}\right]$$

$$+ \frac{1}{2} \sum_a \sum_{a'} m_a m_{a'} \left[\left(\frac{v_M}{v}\right)\psi_{Maa'} + |z_M z_X|\theta'_{aa'}\right]. \qquad (2.49)$$

f^γ is given by Eq. (2.38)

$$B_{MX} = \beta^{(0)}_{MX} + \frac{2\beta^{(1)}_{MX}}{\alpha^2 I}[1 - (1 + \alpha I^{1/2})e^{-\alpha I^{1/2}}] \qquad (2.50)$$

$$B'_{MX} = \frac{2\beta^{(1)}_{MX}}{\alpha^2 I^2}\left[-1 + \left(1 + \alpha I^{1/2} + \frac{1}{2}\alpha^2 I\right)e^{-\alpha I^{1/2}}\right] \tag{2.51}$$

$$C_{MX} = \frac{C^\phi_{MX}}{2|z_M z_X|^{1/2}} . \tag{2.52}$$

The terms in Eqs. (2.45) and (2.49) which involve θ or ψ have an insignificant effect on the calculated values of $\phi - 1$ and $\ln\gamma_\pm$ except in the cases of salts in which short range interaction forces are particularly strong. These usually involve ions having valencies of 3 or higher, but 2 in a few cases. Thus these terms can frequently be omitted, at least in initial calculations.

γ for Single Ions

Pitzer's approach to the calculation of activity coefficients is based on the Debye–Hückel Eq. (2.22) which gives the value for the single ion i. If the ionic-activity coefficients for M and X are not combined but are calculated separately the following equations are obtained.

$$\ln\gamma_M = z_M^2 f^\gamma + \sum_a m_a\left[2B_{Ma} + \left(2\sum_c m_c z_c\right)C_{Ma}\right]$$

$$+ \sum_c\sum_a m_c m_a(z_M^2 B'_{ca} + |z_M|C_{ca})$$

$$+ \sum_c m_c\left(2\theta_{Mc} + \sum_a m_a\psi_{Mca}\right) + \frac{1}{2}\sum_a\sum_{a'} m_a m_{a'}\psi_{Maa'} \tag{2.53}$$

$$\ln\gamma_X = z_X^2 f^\gamma + \sum_c m_c\left[2B_{cX} + \left(2\sum_a m_a z_a\right)C_{cX}\right]$$

$$+ \sum_c\sum_a m_c m_a(z_X^2 B'_{ca} + |z_X|C_{ca})$$

$$+ \sum_a m_a\left(2\theta_{Xa} + \sum_c m_c\psi_{Xac}\right) + \frac{1}{2}\sum_c\sum_{c'} m_c m_{c'}\psi_{Xcc'} . \tag{2.54}$$

The terms f^γ, B, B' and C are those given in Eqs. (2.38) and (2.50)–(2.52). The activity coefficient of the salt MX is obtained from the values for the single ions by use of Eq. (2.23)

$$\gamma_\pm \sqrt[v]{(\gamma_+^{v_+} \gamma_-^{v_-})}. \tag{2.23}$$

Pitzer incorporated the appropriate values of the stoichiometric coefficients, $(2v_M v_X/v)$ for the parameters B_{MX}^ϕ and B_{MX}^γ, $[2(v_M v_X)^{3/2}/v]$ for C_{MX}^ϕ and C_{MX}^γ, in the equations for ϕ and $\ln\gamma$, (2.34) and (2.35). The values of $\beta^{(0)}$, $\beta^{(1)}$ and C^ϕ tabulated by Pitzer include these coefficients and they must be removed before the parameters are used in the equations for single ion activity coefficients. Thus for $SrCl_2$ the values listed are: $(4/3)\beta^{(0)}$, 0.3810; $(4/3)\beta^{(1)}$, 2.223; $(2^{5/2}/3)C^\phi$, -0.00246. Values of C^γ are derived from C^ϕ. The values of the coefficients to be used in the equations for single ion activity coefficients are: $\beta^{(0)}$, 0.2858; $\beta^{(1)}$, 1.667; C^ϕ, -0.001305.

Studies of Systems

Pitzer and Kim calculated the values of ϕ or $\ln\gamma_\pm$ for 52 binary electrolyte mixtures having a common ion using appropriate values of the parameters for the pure electrolytes and obtained the difference between them and the experimental values of ϕ or $\ln\gamma$. From these differences values of θ and ψ were derived and are reported. Thus θ and ψ are used as additional curve-fitting parameters for mixed electrolyte solutions. Similar calculations were also carried out for 11 binary mixtures without a common ion. It was confirmed that all values of θ and ψ are relatively small.

Most of the calculations were repeated[22] with two additional terms E_θ and $E_{\theta'}$ included in the fitting equation. In some systems the improvement of fit was appreciable and in the cases of HCl–SrCl$_2$; HCl–BaCl$_2$ and HCl–MnCl$_2$ significant, while in the case of HCl–AlCl$_3$ inclusion of the additional terms was essential. In these systems the activity coefficient of HCl had been measured and was used in the calculations. The extensive data for the osmotic coefficients of rare earth chlorides, nitrates and perchlorates have also been fitted to the Eq. (26) and good agreement was obtained.

An extensive set of data was published in 1968–1969 on systems of geo-chemical interest[29,30] involving the ions Na^+, Mg^{2+}, Cl^- and SO_4^{2-} and this was fitted to high ionic strengths using Pitzer's equation,[31] as was additional data for osmotic coefficients at $25°C$ for binary mixtures formed from NaCl, Na$_2$SO$_4$, CuCl$_2$ and CuSO$_4$. Other systems of geochemical interest and work on modelling seawater are referred to in Ref. 17.

The properties of solutions of single substances of particular practical importance such as sulphuric[24] and phosphoric[23] acids, sodium chloride[25,32] and sodium sulphate[33,34] have also been expressed in terms of Pitzer's equations. The complete analysis[32] of the published data for the $NaCl$–H_2O system extended to 300°C and 100 MPa pressure and required 28 parameters including those for the aqueous standard state. Below 100°C a different set of 20 parameters was required. The temperature dependence of the activity coefficient is dependent on the enthalpy of the substances involved in the deviation from ideal behaviour and when using Pitzer's equations it is convenient to have the temperature coefficients for the parameters used. The detailed equations for calculating these have been derived[25] and parameters have been published for a number of electrolytes[27] which permit calculation for temperatures "not too different from 25°C". This limitation is due to the fact that heat capacity data for most solutes is only available for temperatures at or near 25°C and this data is required for calculating the parameters. The thermodynamics of the effect of temperature on G^{ex} whether expressed in terms of osmotic or activity coefficients or in terms of equilibrium constants is dealt with in Sec. 2.3.2. The thermodynamic behaviour of solutions at high temperatures ($> 150°C$) is considered in Sec. 3.6.

Pitzer developed his equations in terms of long range and short range forces between ions and on the basis that these forces do not lead to the formation of discrete chemical species in the solution. That means that there is no interaction between the electrons of the ions leading to chemical bonding between them. In consequence the Pitzer equation in its normal form cannot be used to describe the behaviour of systems involving nonionic substances. Neither should it be used when there is evidence that associated species do in fact exist, based on physical measurements of spectra, electrical conductance, ultrasonic properties etc. See Sec. 2.4.

It has, however, been used to deal with vapour-liquid equilibria in the systems NH_3–CO_2–H_2O, NH_3–SO_2–H_2O and NH_3–H_2S–H_2O, which are weak electrolytes.[35,36] Edwards *et al.* extended the Pitzer equation to give correlation to higher concentrations than it can deal with in its usual form, up to 10 to 20 mol kg^{-1} and between 0 and 170°C. For these weak electrolytes the concentrations correspond to ionic strengths about 6 M.

The methods used to modify Pitzer's equation in both papers have been criticised by Chen and his coworkers.[37–39] The procedure preferred by them is to add terms to the basic virial form of the Pitzer equation to account for molecule–ion and molecule–molecule interactions. A set of new parameters is then defined, being functions of the virial coefficients. Thus the Pitzer equation is extended, not modified, to account for the presence of molecular solutes. The interpretation of the terms and parameters of the original Pitzer equation is unchanged so that it is considered by Chen *et al.* to suffer from all of the limitations of a virial expansion type equation. That is the equation parameters denoting the short range interactions between species are "arbitrary, highly temperature dependent and are characteristic of the solvent". The authors themselves use the "Local Composition Model".

Another equation which has been used to correlate the mean activity coefficients of strong electrolytes in aqueous solutions up to ionic strength about 6 was proposed by Bromley.[40]

$$\log \gamma_\pm = \frac{A|z_+z_-|I^{1/2}}{1 + I^{1/2}} + \frac{(0.06 + 0.6B_m)|z_+z_-|I}{[1 + (1.5I)(|z_+z_-|^{-1})]^2} + B_m I . \tag{2.55}$$

The value of the parameter B_m is a function of temperature. Values of parameters are given by Bromley and in Ref. 15.

2.2.4. Application of Pitzer's Equations; Some Worked Examples

Calculation of ϕ and a_w for NaCl Solutions

Pitzer's equations will first be used to calculate the activity of water in solutions of sodium chloride by means of Eq. (2.34) for $\phi - 1$, f^ϕ being replaced by (2.37) and B^ϕ_{MX} by (2.40). This gives

$$\phi - 1 = -z_{Na^+}z_{Cl^-}A_\phi\frac{I^{1/2}}{1 + bI^{1/2}} + m\frac{2(v_{Na^+}v_{Cl^-})}{v}$$

$$\times (\beta^{(0)}_{NaCl} + \beta^{(1)}_{NaCl}e^{-\alpha I^{1/2}}) + m^2\frac{2(v_{Na^+}v_{Cl^-})^{3/2}}{v} C^\phi_{NaCl} .$$

A_ϕ is the Debye–Hückel parameter for the osmotic coefficient, 0.3910 at 25°C.[25]

$b = 1.2$, this value having been selected for use with all substances in aqueous solution at 25°C.

$\alpha = 2$, this value having been found satisfactory for all substances listed in Ref. 19. It may be adjusted for each substance if desired.

For NaCl as a neutral electrolyte $v_{Na^+} = v_{Cl^-} = 1$ and $v = v_{Na^+} + v_{Cl^-} = 2$. The values of the parameters for NaCl,[19] $\beta^{(0)} = 0.0765$; $\beta^{(1)} = 0.2664$; $C^\phi = 0.00127$.

1 m *solution of NaCl*. $I = 1$

$$\phi - 1 = -\left(0.391 \times \frac{1}{2.2}\right) + 1 \times (0.0765 + 0.2664e^{-2}) + 0.00127$$

$$= -0.1777 + (0.0765 + 0.0361) + 0.00127$$

$$= -0.0638.$$

2 m *solution of NaCl*. $I = 2$

$$\phi - 1 = -\left(0.391 \times \frac{\sqrt{2}}{1 + 1.2\sqrt{2}}\right) + 2(0.0765 + 0.02664e^{-2\sqrt{2}})$$

$$+ (4 \times 0.00127) = -0.0154.$$

Similar calculations for 4, 6, 10 and 13.6 molal solutions give $\phi - 1 = 0.1152$, 0.3057, 0.6389 and 1.0118 respectively. The activity of water in each solution can be calculated from the equation (Sec. 2.1).

$$\ln a_w = -\frac{18.0125 \, vm}{1000} \phi$$

giving the results shown in Table 2.1. Values of a_w tabulated by Robinson and Stokes (Ref. 8, Appendix 8.3) after critical evaluation of published data are given for comparison. This book was a major source of the activity coefficient data used by Pitzer and Mayorga in calculating the parameters $\beta^{(0)}$, $\beta^{(1)}$ and C^ϕ.[19] They found no data for solutions more concentrated than 6 m so that their parameters should not, strictly speaking, be used for more concentrated solutions. It should be noted that the equation relating $\ln a_w$ and ϕ can be used when the solution contains more than one solute. The product vm in this definition of ϕ is replaced by a summation over all the solutes present; that is, vm is the total molality of ions of solute electrolytes present.

Water activities in solutions of a number of single and of mixed electrolytes up to high concentration have been measured more recently using the electrodynamic balance technique (Sec. 2.1).[9,10] The water activity of solutions of the single electrolytes studied as a function of molality were expressed as a

Table 2.1. Osmotic coefficient and water activity data for sodium chloride solutions.

Concentration M	1	2	4	6	10	13.6
$\phi - 1$	−0.0638	−0.0154	0.1152	0.3057	0.6389	1.0118
ϕ	0.9362	0.9846	1.1152	1.3057	1.6389	2.0118
$\ln a_w$	−0.03373	−0.07095	−0.1607	−0.2823	−0.5905	−0.9858
$\log a_w$	−0.01465	−0.03081	−0.0698	−0.1226	−0.2565	−0.4281
a_w Pitzer parameters	0.9668	0.9315	0.8515	0.754	0.554	0.373
a_w Ref. 8	0.96686	0.9316	0.8515	0.7598	–	–
a_w Ref. 9	0.966	0.930	0.854	0.755	0.550	0.438
a_w reestimated parameters[9]	0.967	0.930	0.845	0.752	0.566	0.424

polynomial of the form

$$a_w = a_0 + a_1 m + a_2 m^2 + a_3 m^3 + \cdots .$$

The coefficients for this for the single electrolytes studied are given in Table 2.2 and other information in Table 2.3. Published data were used in fitting the polynomial, together with the new values which extended the range of concentrations for which a_w values were known. Parameters for ammonium sulphate solutions up to 36.2 m are not included here. Values of $\gamma_\pm$ for the salts listed were calculated by numerical integration of the Gibbs–Duhem equation, using the polynomial (Sec. 2.1).

It can be seen from Table 2.1 that although the data for sodium chloride used to calculate Pitzer's parameters extended only to 6 m solutions the value of a_w calculated for a 10 m solution, 0.554, is quite close to that calculated from the polynomial, 0.550. For the highest concentration at which measurements were made, 13.6 m, however there is a serious discrepancy between the value of a_w found, 0.438, and that calculated using Pitzer's parameters, 0.373. Similar comparisions for most of the salts listed in Table 2.2 showed[9] that while satisfactory results were obtained in some cases, Pitzer's equation cannot be used with confidence at concentrations much higher than those for which its parameters were estimated. This must be so because there is no method known which permits accurate prediction of the behaviour of concentrated electrolye solutions from their behaviour at low concentrations. With their new data for high concentrations and the polynomial equation, solute activities for such solutions were calculated and used together with data from the literature to estimate new parameters for these substances for Pitzer's equations. These are

Table 2.2. Coefficients of water activity polynomial fit $a_w(m) = a_0 + a_1 m + a_2 m^2 + a_3 m^3 + \cdots$.

Salt	a_0	a_1	a_2	a_3	a_4	a_5	a_6
NaCl	1.0084	−4.939(−2)	8.888(−3)	−2.157(−3)	1.617(−4)	−1.990(−6)	−1.142(−7)
NaBr	0.9996	−3.116(−2)	−2.112(−3)	−9.347(−5)	2.000(−5)	−5.472(−7)	0
KCl	0.9975	−2.173(−2)	−1.053(−2)	4.253(−3)	−7.780(−4)	6.203(−5)	−1.764(−6)
KBr	1.0008	−3.531(−2)	2.490(−3)	−6.729(−4)	5.318(−5)	−8.040(−7)	−2.866(−8)
NH$_4$Cl	0.9968	−2.611(−2)	−1.599(−3)	1.355(−4)	−2.317(−6)	−1.113(−8)	0
Na$_2$SO$_4$	1.0052	−6.484(−2)	3.519(−2)	−1.319(−2)	1.925(−3)	−1.224(−4)	2.870(−6)
(NH$_4$)$_2$SO$_4$	0.9968	−2.969(−2)	1.753(−5)	−3.253(−4)	3.571(−5)	−9.787(−7)	0
CaCl$_2$	0.9947	−6.062(−3)	−4.122(−2)	6.091(−3)	−3.433(−4)	7.009(−6)	0
MnCl$_2$	0.9989	−3.639(−2)	−2.049(−2)	4.286(−3)	−4.137(−4)	1.960(−5)	−3.417(−7)
MnSO$_4$	0.9817	−3.294(−2)	−2.773(−2)	3.483(−3)	−1.773(−4)	3.229(−6)	0
FeCl$_3$	1.0930	−1.782(−1)	3.041(−2)	−6.872(−3)	8.728(−4)	−5.089(−5)	1.096(−6)

As an example for 13.6 M NaCl $a_w = 1.0084 - (0.04939 \times 13.6) + 0.008888(13.6)^2 + \cdots$.

Table 2.3. Water activity polynomial equation.

Salt	Max. molality for which fit is valid	σ_{a_q}	Range of literature data used in fit	Source of literature data used in fit
NaCl	13.6	0.0053	0.5–6.0	8
NaBr	20.1	0.0055	0.1–8.75	8
KCl	12.7	0.0041	0.1–4.8	8
KBr	14.6	0.0025	0.1–5.5	8
NH$_4$Cl	23.2	0.0043	0.1–6.0	8
Na$_2$SO$_4$	12.8	0.0026	0.1–4.0	8
(NH$_4$)$_2$SO$_4$	17.9	0.0027	0.1–5.5	8
CaCl$_2$	14.1	0.0066	0.005–9.0	
MnCl$_2$	12.0	0.0032	0.1–7.699	8
MnSO$_4$	17.0	0.0115	0.1–4.966	8
FeCl$_3$	15.0	0.0085	1.0–15.0	

σ_{a_q} is the standard deviation between the predictions of the polynomial and the experimental data.

Table 2.4. Reestimated parameters for Pitzer's equation.[9]

Salt	$\beta_{MX}^{(0)}$	$\beta_{MX}^{(1)}$	C_{MX}^{ϕ}	Molality range used for parameter estimation
NaCl	0.10820	0.03127	−0.002469	0.1–13.6
NaBr	0.13523	0.02917	−0.003478	0.1–20.1
KCl	0.06577	0.09351	−0.002160	0.1–12.7
KBr	0.06292	0.16046	−0.001981	0.1–14.6
NH$_4$Cl	0.04568	0.20431	−0.001731	0.1–23.2
Na$_2$SO$_4$	0.08610	0.13037	−0.003104	0.1–12.8
(NH$_4$)$_2$SO$_4$	0.04763	0.44459	−0.001311	0.1–17.9
CaCl$_2$	0.41328	0.53043	−0.014250	0.1–14.1
MnCl$_2$	0.25811	2.31108	−0.010540	0.1–12.0
MnSO$_4$	0.33089	3.14630	−0.014731	0.1–17.0
FeCl$_3$	0.31583	10.4224	−0.010078	0.1–15.0

given in Table 2.4. The values for sodium chloride have been used instead of those of Pitzer and Mayorga to calculate a_w for solutions of sodium chloride of the molalities used previously. The results are included in Table 2.1 and it is clear that the fit is greatly improved for the highest concentration.

Calculation of γ_{NaCl}

Calculation of γ_{NaCl} involves the use of Eq. (2.35) with f^γ being replaced by (2.38) and B^γ by (2.41), rearranged to increase clarity, to give with $b = 1.2$

and $\alpha = 2$ as before and writing

$$C^\gamma_{\text{NaCl}} = (3/2)C^\phi_{\text{NaCl}} \tag{2.43}$$

$$\ln \gamma_{\text{NaCl}} = -|z_{\text{Na}^+} z_{\text{Cl}^-}| A_\phi \left(\frac{I^{1/2}}{1 + 1.2I^{1/2}} + \frac{2}{1.2} \ln(1 + 1.2I^{1/2}) \right)$$

$$+ m \frac{2\upsilon_{\text{Na}^+}\upsilon_{\text{Cl}^-}}{\upsilon} \left(2\beta^{(0)}_{\text{NaCl}} + \frac{2\beta^{(1)}_{\text{NaCl}}}{4I} \right) [1 - (1 + 2I^{1/2} - 2I)e^{-2I^{1/2}}]$$

$$+ \frac{3m^2}{2} \frac{(2(\upsilon_{\text{Na}^+}\upsilon_{\text{Cl}^-}))^{3/2}}{\upsilon} C^\phi_{\text{NaCl}}$$

$$\upsilon_{\text{Na}^+} = \upsilon_{\text{Cl}^-} = 1 \quad \upsilon = 2$$

$$\beta^{(0)}_{\text{NaCl}} = 0.0765 \, ; \quad \beta^{(1)}_{\text{NaCl}} = 0.2664 \, ; \quad C^\phi = 0.00127$$

4 m *solution of NaCl.* $I = 4$

$$\ln \gamma_{\text{NaCl}} = -0.391 \left(\frac{2}{1 + 2.4} + \frac{2}{1.2} \ln 3.4 \right)$$

$$+ 4 \left((2 \times 0.0765) + \frac{2 \times 0.2664}{16} [1 - (1 + 4 - 8)e^{-4}] \right)$$

$$+ 0.00127 \left(\frac{3 \times 16}{2} \right) = -1.02749 + 0.75252 + 0.03048 = -0.24449$$

$$\log \gamma_{\text{NaCl}} = -0.244 \times 0.434294 = -0.10618 = \bar{1}.89382$$

$$\gamma_{\text{NaCl}} = 0.7831 \, .$$

The experimental value given by Robinson and Stokes[8] is $\log \gamma_{\text{NaCl}} = -0.1061$.

Calculation of γ_{MnCl_2}

As an example of calculations of $\gamma_\pm$ for a 2:1 electrolyte γ_{MnCl_2} in a 1 m solution will be used.

$$\upsilon_M = 1; \; \upsilon_X = 2; \; \upsilon = 3; \quad \frac{2\upsilon_M\upsilon_X}{\upsilon} = \frac{4}{3}; \quad \frac{2(\upsilon_M\upsilon_X)^{3/2}}{\upsilon} = 2.66667 \, .$$

Pitzer and Mayorga[19] tabulate parameters for $MnCl_2$ which include some of these factors. They give:

$$4/3\beta^{(0)} = 0.4363; \quad 4/3\beta^{(1)} = 2.067; \quad \frac{2^{5/2}}{3} C^{\phi} = -0.03865$$

$$\ln\gamma = -2 \times 0.391 \left[\frac{\sqrt{3}}{1 + 1.2\sqrt{3}} + 1.6667 \ln(1 + 1.2\sqrt{3}) \right]$$

$$+ 1 \times \left(2 \times 0.4363 + \frac{2 \times 2.067}{4 \times 3} [1 - (1 + 2\sqrt{3} - 6)e^{-2\sqrt{3}}] \right)$$

$$+ 1^2 \times \frac{3}{2} \times -0.03865$$

$$\ln\gamma = -1.90552 + 1.23366 - 0.05798 = -0.72984$$

$$\log\gamma = -0.31697 = \bar{1}.42246$$

$$\gamma_{MnCl_2} = 0.4820 .$$

The value given by Robinson and Stokes[8] for a 1 m solution is 0.481.

The parameters $\beta^{(0)}$, $\beta^{(1)}$ and C^{ϕ} were calculated using activity coefficient data up to 2.5 m. They will be used to calculate γ_{MnCl_2} in a 4 m solution.

$$\ln\gamma = -2 \times 0.391 \left[\frac{\sqrt{12}}{1 + 1.2\sqrt{12}} + \frac{2}{1.2} \ln(1 + 1.2\sqrt{12}) \right]$$

$$+ 4 \left(2 \times 0.4363 + \frac{2 \times 2.067}{4 \times 12} [1 - (1 + \{2\sqrt{12}\} - 24)e^{-2\sqrt{12}}] \right)$$

$$+ 4^2 \times \frac{3}{2} \times -0.03865$$

$\ln\gamma = -2.66321 + 3.84032 - 0.9276 = 0.2495$
$\log\gamma = 0.10836$
$\gamma_{MnCl_2} = 1.283$
Robinson and Stokes give the value 1.240.

Mixed Electrolyte Solutions. Calculation of Osmotic Coefficients

For a solution containing two or more electrolytes, Eq. (2.45) is used to calculate the osmotic coefficient. When the appropriate substitutions are made this becomes

$$(\phi - 1) \sum_i m_i = 2 I A_\phi \left[\frac{I^{1/2}}{1 + bI^{1/2}} \right] + 2 \sum_c \sum_a m_c m_a$$

$$\times \left[(\beta_{ca}^{(0)} + \beta_{ca}^{(1)} e^{-aI^{1/2}}) + \frac{\sum mz}{(z_c z_a)^{1/2}} C_{ca}^\phi \right]$$

if only the first two terms are considered. The same values of $b = 1.2$ and $\alpha = 2$ are used as in the calculations for single electrolyte solutions, although the value of α may be changed for any substance desired when alternative sets of parameters are calculated. The parameters $\beta_{ca}^{(0)}$, $\beta_{ca}^{(1)}$ and C_{ca}^ϕ are those for a solution of a single electrolyte (ca) where the cation c has charge z_c and the anion a has charge z_a. This shortened equation and the equivalent equation for $\ln \gamma_{ca}$ are frequently adequate for calculating values of ϕ and $\gamma_\pm$ for mixed electrolyte solutions. The additional terms in (2.45) include parameters θ and ψ which are derived from data obtained from measurements on solutions containing a common ion. Values for a number of mixtures of two electrolytes have been tabulated[21] and the estimated errors if they are taken as zero are given. Clearly if values of θ and ψ are available they can be used where maximum precision is required. However a major advantage of Pitzer's equations is that activities in mixed electrolyte solutions can be calculated using the parameters for the individual salts. Qualifications as to the accuracy of the results for solutions containing ions of higher valency were given in the discussion of the equations, Sec. 2.2.2.

Mixed electrolytes with a common ion. For a solution containing 2 m NaCl + 2 m KCl $I = 4$.

Using Eq. (2.45) and neglecting terms containing Θ or ψ

$$8(\phi - 1) = -2 I A_\phi \left[\frac{I^{1/2}}{1 + 1.2 I^{1/2}} \right] + 2[m_{Na} m_{Cl} [\beta_{NaCl}^{(0)} + \beta_{NaCl}^{(1)} e^{-2I^{1/2}}]$$

$$+ m_{Na} m_{Cl} [4 C_{NaCl}^\phi] + m_K m_{Cl} [\beta_{KCl}^{(0)} + \beta_{KCl}^{(1)} e^{-2I^{1/2}}]$$

$$+ m_K m_{Cl} [4 C_{KCl}^\phi]].$$

The numerical values of the terms are

$$-2 I A_\phi \left[\frac{I^{1/2}}{1 + 1.2 I^{1/2}} \right] = -2 \times 4 \times 0.391 \left[\frac{2}{1 + 2.4} \right] = -1.840$$

$$m_{Na}m_{Cl}[\beta_{NaCl}^{(0)} + \beta_{NaCl}^{(1)}e^{-2I^{1/2}}] = 2 \times 4[0.0765 + 0.2664e^{-4}] = 0.6510$$

$$m_{Na}m_{Cl}[4C_{NaCl}^{\phi}] = 2 \times 4 \times 4 \times 0.00127 = 0.0406$$

$$m_{K}m_{Cl}[\beta_{KCl}^{(0)} + \beta_{KCl}^{(1)}e^{-2I^{1/2}}] = 2 \times 4[0.04835 + 0.2122e^{-4}] = 0.4178$$

$$m_{K}m_{Cl}[4C_{KCl}^{\phi}] = 2 \times 4[4 \times -0.00084] = -0.0269 \,.$$

Thus

$$8(\phi - 1) = -1.840 + 2[0.6510 + 0.0406 + 0.4178 - 0.0269]$$

$$\phi - 1 = 0.0992$$

$$\phi = 1.0992$$

$$\ln a_w - \frac{18.0152 \times 8}{1000}\phi = -0.1584$$

$$\log a_w = -0.0688 = 1.9312$$

$$a_w = 0.854 \,.$$

This value agrees within limits of measurement with the results presented graphically for this solution composition.[10]

Mixed electrolytes with no common ion, KBr–NaCl.   For the mixed electrolyte solution 2 m KBr + 3.2284 m NaCl all of the possible cation–anion interactions must be used in the summation of terms in the equation for $\phi - 1$. Pitzer's parameters are[19]

	$\beta^{(0)}$	$\beta^{(1)}$	C^{ϕ}
NaCl	0.0765;	0.2664;	0.00127
NaBr	0.0973	0.2791	0.00116
KCl	0.04835	0.2122	-0.00084
KBr	0.0569	0.2212	-0.00180 .

For this solution $I = 5.2284$ and $(\sum mz) = \sum_c m_c z_c = \sum m_a z_a = 5.2284$ $\sum_i \times m_i = 10.4568$

$$-2IA_{\phi}\left[\frac{I^{1/2}}{1 + 1.2I^{1/2}}\right] = -2 \times 5.2284 \times 0.391 \left[\frac{\sqrt{5.2284}}{1 + 1.2\sqrt{5.2284}}\right] = -2.4971$$

$$m_{\text{Na}}m_{\text{Cl}}[\beta_{\text{NaCl}}^{(0)} + \beta_{\text{NaCl}}^{(1)}e^{-2I^{1/2}}]$$

$$= (3.2284)^2[0.0765 + 0.2664e^{-2\sqrt{5.2284}}] = 0.8260$$

$$m_{\text{Na}}m_{\text{Cl}}\frac{(\sum mz)}{(z_{\text{Na}}z_{\text{Cl}})^{1/2}}\,C_{\text{NaCl}}^{\phi}$$

$$= 3.2284 \times 2 \times 5.8284 \times 0.00127 = 0.478$$

$$m_{\text{Na}}m_{\text{Br}}[\beta_{\text{NaBr}}^{(0)} + \beta_{\text{NaBr}}^{(1)}e^{-2I^{1/2}}]$$

$$= (3.2284 \times 2)[0.0973 + (0.2791 \times 0.0103)] = 0.6468$$

$$m_{\text{Na}}m_{\text{Br}}\left(\sum mz\right)C_{\text{NaBr}}^{\phi}$$

$$= 3.2284 \times 2 \times 5.2284 \times 0.00116 = 0.0392$$

$$m_{\text{K}}m_{\text{Cl}}[\beta_{\text{KCl}}^{(0)} + \beta_{\text{KCl}}^{(1)}e^{-\alpha I^{1/2}}]$$

$$= (2 \times 3.2284)[0.04835 + (0.2122 \times 0.0103)] = 0.3263$$

$$m_{\text{K}}m_{\text{Cl}}\left(\sum mz\right)C_{\text{KCl}}^{\phi}$$

$$= 2 \times 3.2284 \times 5.2284 \times -0.00084 = -0.0284$$

$$m_{\text{K}}m_{\text{Br}}[\beta_{\text{KBr}}^{(0)} + \beta_{\text{KBr}}^{(1)}e^{-\alpha I^{1/2}}]$$

$$= -(2 \times 2)[0.0569 + (0.2212 \times 0.0103)] = 0.2367$$

$$m_{\text{K}}m_{\text{Br}}\left(\sum mz\right)C_{\text{KBr}}^{\phi}$$

$$= 2 \times 2 \times 5.2284 \times -0.00180 = -0.0376$$

$$(\phi - 1)\sum_i m_i = -2.4971 + 2(0.8260 + 0.0478 + 0.6468 + 0.0392$$

$$+ 0.3263 - 0.0284 + 0.2367 - 0.0376)$$

$$(\phi - 1)\sum_i m_i = -2.4971 + 4.1136 = 1.6165$$

$\sum_i m_i = 10.4568$ so that

$$(\phi - 1) = 0.1546$$

$$\ln a_w = -\frac{18.0152 \times 10.4568}{1000}\,\phi = -0.2175$$

$$\log a_w = -0.0944 = \bar{1}.4056$$

$$a_w = 0.805\,.$$

This value agrees within limits of measurement with the results presented graphically for this solution composition (NaCl to KBr ratio) and ionic strength.[10]

Mixed Electrolyte Solutions. Calculations of Activity Coefficients

Like Eq. (2.45), for the calculation of osmotic coefficients, Eqs. (2.49), (2.53) and (2.54) for the calculation of activity coefficients contain terms including parameters θ and ψ which are not used in the specimen calculations presented here.

It is much easier to measure water activities by use of the isopiestic or a related method than it is to measure the activity coefficient of a solute. Thus Pitzer found relatively little data to use with his equations for $\gamma_\pm$. However, activity coefficients for HCl have been measured in a number of solutions containing another metal chloride over a wide range of concentrations and temperature and have been presented in Table (14-2-1A) of Ref. 7.

γ_{HCl} *for a solution containing* **0.01 m HCl + 0.025 m SrCl$_2$.** For this solution $I = 0.085$; $\sum mz = 0.12$; $\alpha = 2$ and $b = 1.2$ as usual. Since only one anion is present only two salts are involved in the calculation, their Pitzer parameters being:

$$\text{HCl}: \beta^{(0)} = 0.1775\,; \qquad \beta^{(1)} = 0.2945\,; \qquad C^\phi = 0.00080$$

$$\text{SrCl}_2: \frac{4}{3}\beta^{(0)} = 0.3810\,; \qquad \frac{4}{3}\beta^{(1)} = 2.223\,; \qquad \left(\frac{2^{5/2}}{3}\right)C^\phi = -0.00266\,.$$

Using Eq. (2.49) gives the following Eq. (2.56) in which the terms are numbered.

$\ln \gamma_{\text{HCl}} =$
(I)

$$-A_\phi \left[\frac{I^{1/2}}{1 + 1.2I^{1/2}} + \left(\frac{2}{1.2} \right) \ln(1 + 1.2I^{1/2}) \right]$$

(II)

$$+ \frac{2}{2} \left[m_{\text{Cl}} \left[\beta_{\text{HCl}}^{(0)} + \left(\frac{2\beta_{\text{HCl}}^{(1)}}{4I} \right) \{1 - (1 + 2I^{1/2})e^{-2I^{1/2}}\} + 0.12C_{\text{HCl}} \right] \right]$$

(IIIa)

$$+ \frac{2}{2} \left[m_{\text{H}} \left[\beta_{\text{HCl}}^{(0)} + \left(\frac{2\beta_{\text{HCl}}^{(1)}}{4I} \right) \{1 - (1 + 2I^{1/2})e^{-2I^{1/2}}\} + 0.12C_{\text{HCl}} \right] \right]$$

(IIIb)

$$+ m_{\text{Sr}} \left[\beta_{\text{SrCl}_2}^{(0)} + \left(\frac{2\beta_{\text{SrCl}_2}^{(1)}}{4I} \right) \{1 - (1 + 2I^{1/2})e^{-2I^{1/2}}\} + 0.12C_{\text{SrCl}_2} \right]$$

(IVa)

$$+ m_{\text{H}}m_{\text{Cl}} \left[z_{\text{H}}z_{\text{Cl}} \left(\frac{2\beta_{\text{HCl}}^{(1)}}{4I^2} \right) \left\{ -1 + \left(1 + \alpha I^{1/2} + \frac{1}{2}\alpha^2 I \right) e^{-\alpha I^{1/2}} \right\} \right.$$

$$\left. + \frac{1}{v}(2v_{\text{H}}z_{\text{H}}C_{\text{HCl}}) \right]$$

(IVb)

$$+ m_{\text{Sr}}m_{\text{Cl}} \left[z_{\text{H}}z_{\text{Cl}} \left(\frac{2\beta_{\text{SrCl}_2}^{(1)}}{4I^2} \right) \left\{ -1 + \left(1 + \alpha I^{1/2} + \frac{1}{2}\alpha^2 I \right) e^{-\alpha I^{1/2}} \right\} \right.$$

$$\left. + \frac{1}{v}(2v_{\text{H}}z_{\text{H}}C_{\text{SrCl}_2}) \right]. \tag{2.56}$$

Substituting the numerical values into the terms gives
(I)

$$-0.391(0.2160 + 0.5000) = -0.2800$$

(II)

$$0.06\left[0.1775+\left(\frac{0.5890}{0.34}\right)0.1164\right]+(9.6\times10^{-5})=0.0228$$

(IIIa)

$$0.01\left(0.1775+\left(\frac{2\times0.2945}{0.34}\right)0.1164\right]+(9.6\times10^{-5})=0.0039$$

(IIIb)

$$0.025\left[0.3810+\left(\frac{2\times2.223}{0.34}\right)0.1164\right]+0.12(-0.00246)=0.0473$$

(IVa)

$$0.01\times0.06\left[\left(\frac{2\times0.02945}{0.0289}\right)(-0.0214)\right]+\frac{1}{2}(2\times0.00080)=0.00077$$

(IVb)

$$0.025\times0.06\left[\left(\frac{2\times2.223}{0.0289}\right)(-0.0214)\right]+\frac{1}{3}(2\times-0.00246)=-0.00658\,.$$

Hence

$$\ln\gamma_{\text{HCl}}=-0.2118$$

$$\log\gamma_{\text{HCl}}=-0.0920$$

$$\gamma_{\text{HCl}}=0.809\,.$$

Harned and Owen give the experimental value as $\gamma_{\text{HCl}}=0.797$.

Single ion activity coefficients for the solution. For the calculation of γ_{H} and γ_{Cl} the values of the parameters of HCl and SrCl$_2$ to be used in Eqs. (2.53) and (2.54) are:

$$\text{HCl}:\beta^{(0)}=0.1775\,;\qquad\beta^{(1)}=0.2945\qquad C^{\phi}=0.00080$$
$$\text{SrCl}_2:\beta^{(0)}=0.2858\,;\qquad\beta^{(1)}=1.667\,;\qquad C^{\phi}=-0.001305\,.$$

Neglecting the terms including only Θ and ψ

$$\ln \gamma_H = z_H^2 f^\gamma + m_{Cl}[2B_{HCl} + (2m_H z_H + 2m_{Sr} z_{Sr})C_{HCl}]$$

$$+ m_H m_{Cl}[z_H^2 B_{HCl}^1 + z_H C_{HCl}] + m_{Sr} m_{Cl}[z_{Cl}^2 B_{SrCl_2}^1 + z_{Cl} C_{SrCl_2}]$$

$$\ln \gamma_{Cl} = z_{Cl} f^\gamma + m_H[2B_{HCl} + 2m_{Cl} z_{Cl} C_{HCl}] + m_{Sr}[2B_{SrCl_2} + 2m_{Cl} z_{Cl} C_{SrCl_2}]$$

$$+ m_H m_{Cl}[z_{Cl}^2 B_{HCl}^1 + z_{Cl} C_{HCl}] + m_{Sr} m_{Cl}[z_{Cl}^2 B_{SrCl_2}^1 + z_{Cl} C_{SrCl_2}]$$

Calculation of the numerical values of the terms gives

$$\ln \gamma_H = -0.2799 + 0.0455 + 0.0003 + 0.0037 = -0.2304$$

$$\gamma_H = 0.7942$$

$$\ln \gamma_{Cl} = -0.2799 + 0.00758 + 0.0714 + 0.0003 + 0.0037 = -0.1969$$

$$\gamma_{Cl} = 0.8213$$

$$\gamma_{Cl} = \sqrt{\gamma_H \gamma_{Cl}} = 0.808 \, .$$

2.3.   Metal Complex Formation and Equilibrium Constants

2.3.1.   *Complex Formation*

The deviations from the Debye–Hückel limiting law which occur with solutions of strong electrolytes more concentrated than about 10^{-3} m indicate that in these the electrostatic forces of attraction between ions no longer predominate in determining the values of G^{ex}. Attempts such as that by N. Bjerrum to extend the range of validity of the limiting law had little success and empirical equations were developed to fit experimentally determined values of activity coefficients over ranges of concentration. In general these accepted the Debye–Hückel coefficient A, derived from Eq. (2.22), as taking account of long range electrostatic forces and added additional terms to allow for the effects of short range forces. It was always assumed that no chemical bonding due to electronic interactions between ions occurred and no discrete chemical species were formed. This assumption is necessary because there is at present no means of calculating what effect such interactions would have on the value of G^{ex} of a solution. That is to say it is not possible to calculate the value of the free energy of formation of a discrete chemical species formed by reaction between constituents in a solution. This is true whether the reaction is between ions and ions or between ions and neutral molecules. Such reactions are of fundamental importance to process chemists and hydrometallurgists. They are treated as chemical equilibria and described quantitatively in terms of experimentally determined equilibrium constants.

Consider a metal M of valency $z+$ in a solution which contains the anion L of valency 1. When ion interaction occurs and this model is used L is known as a ligand and the products as complexes. These are formed in a stepwise sequence and each step is controlled by an equilibrium constant.

$$M^{z+} + L^- = ML^{(z-1)+}$$

$$K_1 = \{ML^{(z-1)+}\}/\{M^{z+}\}\{L^-\}$$

$$ML^{(z-1)+} + L^- = ML_2^{(z-2)+}$$

$$K_2 = \{ML_2^{(z-2)+}\}/\{ML^{(z-1)+}\}\{L^-\}$$

$$ML_{n-1}^{(z-n+1)+} + L^- = ML_n^{(z-n)+}$$

$$K_n = \{ML_n^{(z-n)+}\}/\{ML_{n-1}^{(z-n+1)+}\}\{L^-\}.$$

The maximum number of ions of L which can form complexes with M^{z+} is n, and n is the coordination number of M^{z+}. A composite equilibrium constant, β, is defined as follows:

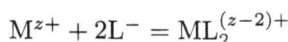

$$M^{z+} + 2L^- = ML_2^{(z-2)+}$$

$$\beta_2 = \{ML_2^{(z-2)+}\}/\{M^{z+}\}\{L^-\}^2$$

$$\beta_2 = K_1 K_2.$$

In general

$$\beta_m = K_1 K_2 K_3 \cdots K_m.$$

If the ligand is an uncharged molecule, ammonia for example, the equilibria are set out in the same way, but the charge on each complex ion is $z+$.

The factors which control the absolute and relative amounts of each species containing the metal, and the amount of free ligand present in the solution are: (i) the values of all the equilibrium constants; (ii) the total concentration of metal $[M_t]$ present in all forms; (iii) the total concentration of ligand $[L_t]$; (iv) the ratio of these two concentrations; and (v) the activity coefficient of each species taking part in the equilibrium. With a constant total concentration of metal in solution, as the total concentration of ligand is increased from zero, the complex ML first forms. Its concentration rises and then falls as ML_2 is produced; the concentration of this increases and again falls as the higher complexes form, Fig. 3.10.

The degree of formation of a complex ML_m is defined as

$$\alpha_{ML_m} = [ML_m]/[M_t]$$

but if polynuclear complexes are present, that is complexes which contain more than one metal atom in each ion or molecule, the concentration of each must be multiplied by the number of metal atoms contained in it.

The average ligand number is defined as

$$\bar{n}_L = ([L_t] - [L])/[M_t].$$

It is the concentration of ligand combined in complex species divided by the total concentration of metal, and is particularly important when equilibrium constants are being measured.

When writing the chemical equations representing equilibria it is conventional to ignore the solvation of the species taking part. In aqueous solutions the metal ions are strongly hydrated, and the ligand may in many cases be regarded as replacing a water molecule at a coordination position around the metal atom. For instance, the ion $Cu(NH_3)_4^{2+}$ has the four ammonia groups arranged at the four corners of a square with the copper atom at the centre, the molecule being planar. The ammonia groups can be regarded as successively replacing water molecules in the same positions.

Like all of the other di- and trivalent ions of the metals of the first transition series of the periodic table, the simple hydrated Cu^{2+} ion has six coordinated water molecules arranged octahedrally around it. Because of the Jahn–Teller effect, in the case of Cu^{2+} the octahedron is distorted and this is connected with the reason why the metal ion binds the fifth and sixth ligands, including water of hydration, only very weakly. Thus in the case of the ammonia complexes the successive equilibrium constants are:

$\log K_1$	$\log K_2$	$\log K_3$	$\log K_4$	$\log K_5$
4.15	3.50	2.89	2.13	−0.52 .

The $Cu(NH_3)_5^{2+}$ ion can be formed in strongly ammoniacal aqueous solutions but the sixth ammonia molecule can be added only in liquid ammonia.

The reason for the decrease in the values of K as the number of NH_3 groups already attached to the Cu^{2+} ion increases was considered by J. Bjerrum. In his terminology the logarithm of the ratio of two consecutive equilibrium constants is called the "total effect", $T_{(m-1),m}$. He subdivided this into two quantities, the "statistical effect" $S_{(m-1),m}$ and the "ligand effect" $L_{(m-1),m}$. The statistical effect was explained on the basis that the tendency for one ligand group L to be lost from the species ML_m is proportional to the value of m, while the tendency for that species to take up another ligand group is proportional to $(n - m)$. The n consecutive stability constants will then be proportional to

$$n/1, (n-1)/2, \cdots (n-m+1)/m, (n-m)/(m+1), \cdots 2/(n-1), 1/n$$

and the ratio between two consecutive constants due to the statistical causes alone is given by

$$K_m/K_{(m+1)} = (n-m+1)(m+1)/(n-m)m.$$

Hence

$$S_{m,(m+1)} = \log K_m - \log K_{(m+1)} = \log \frac{(n-m+1)(m+1)}{m(n-m)}.$$

This equation will apply if each ligand group occupies only one coordination position and if the n coordination positions around the metal ion are identical. The first four values of K for the Cu^{II}–NH_3 system are of the same order of magnitude and adjustment for the statistical factor gives "corrected" values which are even closer:

$$\log K_1(\text{corr}), 3.55; \quad \log K_2(\text{corr}), 3.32;$$

$$\log K_3(\text{corr}), 3.07; \quad \log K_4(\text{corr}), 2.73.$$

Thus most of the differences between the experimental values can be attributed to the statistical effect.

The ligand effect was itself subdivided by J. Bjerrum into an "electrostatic effect" and a "rest effect" (probably better described as the residual effect). The electrostatic effect is caused by any charges on the ligand and the metal-containing species. A ligand ion L^- will be attracted towards M^{2+} or ML^+ but repelled from ML_3^-. N. Bjerrum derived an equation to give the magnitude of the electrostatic effect but in view of the uncertainties in its use the rest effect is usually considered for cases of uncharged ligands only.

Some kinds of ligand can occupy two coordination positions and are known as bidentate ligands. Examples are ethylenediamine, $H_2N.CH_2.CH_2.NH_2$, (en), the carbonate ion, and many organic substances containing both a neutral coordinating group and an acidic group, for example the aminoacid glycine, $H_2N.CH_2.COOH$. This can attach itself to a metal ion by its acidic — O^-H^+ group, neutralising one positive charge, and also by forming a coordinate covalent bond with the metal, using the nitrogen atom. Thus it "grasps" the metal as a crab grasps an object, leading to the generic name of such substances, chelating compounds. Many reagents used in solvent extraction processes for metals are chelating extractants. The expression for the statistical effect for bidentate ligands has to be modified but this need not concern us.

2.3.2.   *Thermodynamics of Equilibrium Constants*

The value of the equilibrium constant for the formation of a metal complex species is a measure of its stability relative to the stabilities of the simple hydrated metal ion and the molecules or ions of the ligand in the solution. The strengths of the chemical bonds between metal and ligands plays a part in determining the stability of the complex but other factors are also involved.

The equilibrium constant is related to the standard free energy change of the reaction, ΔG°, by the van't Hoff isotherm

$$\Delta G^\circ = -RT \ln K$$

and this is controlled by the standard enthalpy and standard entropy changes, ΔH° and ΔS° respectively,

$$\Delta G^\circ = \Delta H^\circ - T\Delta S^\circ .$$

The standard enthalpy change is a measure of the change in heat content of the reactants and products when the complex is formed, and is determined by the kind of chemical bond formed between the metal ion and ligand. In the case of a monodentate singly charged ligand the value of ΔH° is usually between $+20$ and -20 kJ/mole for each step, but when strong covalent bonds are formed may be as large as -80 kJ/mole.

The standard entropy change on complex formation, unlike the enthalpy change, is sensitive to the structure of the environment of the complex. In aqueous electrolyte solutions $\Delta S°$ is usually positive. This rather unexpected fact is due to the disorganisation of the structure of the water around the complex. The positive entropy change resulting from this disorganisation is usually much greater than the negative entropy change due to conversion of the translational entropy of the separate metal ion and ligand to the vibrational and rotational contribution of the complex. If the ligand is negatively charged the neutralisation of charge on complex formation reduces the number of ions in the system, which affects the entropy change, as also does the net reduction in the number of coordinated water molecules. This results in the large positive entropy change, giving a more stable complex.

Association of a metal ion with an uncharged monodentate ligand does not reduce the number of ions present in the system and there is little reorientation of the water molecules. In this case a relatively small positive entropy change, or even a negative entropy change, occurs when the complex is formed. The value of $\Delta S°$ is generally the most important single factor controlling the stability of a complex compound. Standard entropies of formation of cations in aqueous solution tend to become more positive the higher the temperature, while the values for anions become more negative. Thus, in general, it seems that the higher the temperature, the more positive the entropy change on complex formation, and the more stable the complex.

At temperatures other than 25°C, the standard enthalpy change on reaction, ΔH_T^o, can be written

$$\Delta H_T^o = \Delta H_{298}^o + \int_{298}^{T} \left(\frac{\delta \Delta H°}{\delta T} \right)_p dT. \qquad (2.57)$$

Similarly, the standard entropy change can be written

$$\Delta S_T^o = \Delta S_{298}^o + \int_{298}^{T} \left(\frac{\delta \Delta S°}{\delta T} \right)_p dT. \qquad (2.58)$$

Thus, calculation of $\Delta G°$ at elevated temperatures requires knowledge of the way in which the standard enthalpy and entropy changes of the reaction vary with temperature. It is assumed that the pressure is unaltered, but at values other than 1 atm the effect of pressure on the values of ΔH_T^o and ΔS_T^o must also be considered.

The enthalpy change on reaction depends on the enthalpies of the substances taking part. The value of the heat content, or enthalpy, is defined as

$$H = U + PV$$

where U is the total internal energy, P the pressure, and V the volume of the system. Clearly, the heat content must be related to the quantity of heat required to raise the temperature of the system by one degree, that is the heat capacities C_ν at constant volume, and C_p at constant pressure. To raise the temperature by dT at constant volume, the quantity of heat required d_{q_ν} is

$$d_{q_\nu} = dU = C_\nu dT$$

since

$$C_\nu = (\delta U / \delta T)_\nu \,.$$

At constant pressure

$$d_{q_p} = dH = C_p dT$$

since

$$C_p = (\delta H / \delta T)_p \,.$$

These definitions of heat capacities lead to a relationship between the change in total heat capacity in a reaction and the variation of the heat of the reaction with temperature. If the initial state of the system is I and the final state is II and we write

$$\Delta C_\nu = (C_\nu)_{II} - (C_\nu)_I$$

$$\Delta U = U_{II} - U_I$$

$$\Delta H = H_{II} - H_I$$

then subtracting the relevant equations belonging to the final and initial states,

$$\Delta C_\nu = (\delta \Delta U / \delta T)_\nu \tag{2.59}$$

$$\Delta C_p = (\delta \Delta H / \delta T)_p \,. \tag{2.60}$$

These are Kirchhoff's laws. The heat capacity of 1 g of a substance or species is the specific heat, but for theoretical purposes the heat capacity of

1 g molecule is required and C_v and C_p denote these molar heat capacities. The latter is connected with the entropy change of a process as follows:

$$\Delta S = q/T$$

where q is the amount of heat absorbed. At constant pressure this amount of heat equals ΔH, so that

$$\Delta S = \Delta H/T.$$

Then, from (2.60)

$$\frac{\delta C_p}{T}\,\mathrm{d}T = \left(\frac{\delta \Delta H}{T\delta T}\right)_p \mathrm{d}T = \left(\frac{\delta \Delta S}{\delta T}\right)_p \mathrm{d}T.$$

It is now possible to substitute the measurable quantity ΔC_p into Eqs. (2.57) and (2.58) relating to the chemical reaction which was being considered. The values of ΔH_T° and ΔS_T° at any temperature become

$$\Delta H_T^\circ = \Delta H_{298}^\circ + \int_{298}^{T} \Delta C_p^\circ \,\mathrm{d}T \tag{2.61}$$

$$\Delta S_T^\circ = \Delta S_{298}^\circ + \int_{298}^{T} \frac{\Delta C_p^\circ \mathrm{d}T}{T}. \tag{2.62}$$

Substituting into the equation,

$$\Delta G^\circ = \Delta H^\circ - T\Delta S^\circ$$

gives

$$\Delta G_T^\circ = \Delta H_{298}^\circ + \int_{298}^{T} \Delta C_p^\circ \,\mathrm{d}T - T\Delta S_{298}^\circ - T\int_{298}^{T} \frac{\Delta C_p^\circ}{T}\,\mathrm{d}T. \tag{2.63}$$

The expression

$$\frac{1}{T}\int_{298}^{T} \Delta C_p^\circ \,\mathrm{d}T - \int_{0}^{T} \frac{\Delta C_p^\circ}{T}\,\mathrm{d}T$$

is known as the Gibbs free energy function, Δfef_T°, which may be written

$$\Delta fef_T^\circ = \frac{1}{T}\int_{298}^{T} \Delta C_p^\circ \,\mathrm{d}T - \int_{298}^{T} \frac{\Delta C_p^\circ}{T}\,\mathrm{d}T - \Delta S_{298}^\circ. \tag{2.64}$$

Substitution into (2.63) gives

$$\Delta G_T^{\circ} = \Delta H_{298}^{\circ} + T \Delta f e f_T^{\circ}. \tag{2.65}$$

and

$$\log K_T = -\frac{\Delta H_{298}^{\circ}}{2.303 RT} - \frac{\Delta f e f_T^{\circ}}{2.303 R}. \tag{2.66}$$

The equilibrium constant for the reaction can be obtained from (2.66) for any temperature provided the standard Gibbs free energy function of the reaction is known; that is provided ΔC_p°, the standard change in the heat capacity of reaction at constant pressure is known.

Although free energy functions are available for many reactions, they are not known for most equilibria in aqueous solutions. Thus it is necessary to make assumptions in order to estimate approximate values of $\log K_T$ for such reactions, which are those of interest in hydrometallurgical systems.

One such method of approximation is the van't Hoff isochore. The Gibbs–Helmholtz equation may be written

$$\Delta G^{\circ} - \Delta H = T(\delta \Delta G^{\circ}/\delta T)_p$$

and the alteration of the equilibrium constant with temperature may be obtained by substituting $-RT \ln K$ for ΔG°. This gives the van't Hoff isochore

$$\delta \ln K/\delta T = \Delta H/RT^2 \tag{2.67}$$

which in its integrated form is

$$\log K_T = -\frac{\Delta H}{2.303 R}\left(\frac{1}{T} - \frac{1}{298}\right) + \log K_{298}. \tag{2.68}$$

This equation need only be used if ΔH is assumed to be constant between the temperatures T and 298 K, since if the variation of ΔH with temperature is known, ΔC_p° can be calculated from (2.61). The assumption that ΔH can be replaced by ΔH_{298}° at all temperatures is equivalent to assuming that ΔC_p° is zero.

The values of ΔH°, ΔS°, and ΔC_p° are all obtained in the usual way for the reaction from the individual values for each product and reactant, which may all change with temperature. In the absence of sufficient heat capacity data it is better to assume that ΔC_p° is a constant than to assume that it is zero. Equation (2.64) then becomes

$$\Delta f e f_T^o = \frac{\Delta C_p^o}{T} \int_{298}^{T} dT - \Delta C_p^o \int_{298}^{T} \frac{dT}{T} - \Delta S_{298}^o \qquad (2.69)$$

and on integration

$$\Delta f e f_T^o = \frac{\Delta C_p^o}{T}(T - 298) - \Delta C_p^o(\ln T - \ln 298) - \Delta S_{298}^o. \qquad (2.70)$$

Combining this equation with (2.66) gives

$$\log K_T = -\frac{\Delta H_{298}^o}{2.303R}\left(\frac{1}{T} - \frac{1}{298}\right) + \frac{\Delta C_p^o}{R}$$

$$\times \left[\frac{1}{2.303}\left(\frac{298}{T} - 1\right) - \log\frac{298}{T}\right] + \log K_{298}. \qquad (2.71)$$

This is equivalent to assuming that ΔH^o for the reaction varies linearly with temperature.

Clearly, the errors in the values of $\log K_T$ introduced by assuming that $\Delta C_p^o = 0$ and using the van't Hoff isochore depend on the relative sizes of the enthalphy and heat capacity terms in (2.71). If ΔH_{298}^o is large the error is small, but it was pointed out above that the enthalphy change in many complex forming equilibria is small. The heat capacity changes in such reactions are generally of the order of $+200$ to -200 J/mole degree at $25°C$, and if the value of ΔH_{298}^o is of the order of 8 kJ the error introduced in the value of $\log K_T$ may be 1 or 2 units at elevated temperatures.

There is little information at present concerning values of ΔC_p^o, ΔH^o, and ΔS^o for complexing reactions. In a few cases values of the equilibrium constants are known over small ranges of temperature, and this means that ΔC_p^o may be calculated by assuming that ΔH^o is a linear function of temperature. The values obtained are, however, very sensitive to small errors in the equilibrium constants. Calculation of thermodynamic data for elevated temperatures is dealt with in Sec. 3.3.6.

Values of equilibrium constants depend on pressure and at constant temperature

$$(\delta \ln K / \delta P)_T = -\Delta V^o / RT \qquad (2.72)$$

where ΔV^o is the partial molal volume change for the reaction. The value of this and also the change in the volumes of the dissolved ions and molecules on passing from the standard state to the given solution at the higher pressure are small. Thus the effect of pressure on equilibrium constants in aqueous solutions is small below the critical pressure.

2.3.3.   *Selection of Values of Equilibrium Constants*

At the beginning of Sec. 2.3.2, it was stated that the equilibrium constant of a complex species refers to its formation from the simple hydrated metal ion and the ligand. In Sec. 2.3.1, the equations for the constants were shown in terms of activities of species taking part in the equilibrium reaction. In defining degree of formation of a complex and average ligand number, concentrations were employed. Methods of measuring equilibrium constants usually involve altering the ratio of ligand to metal ion concentration, with the latter held constant, and measuring the amount of ligand complexed or the amount of metal ion complexed at each ratio. The experiments are repeated at a number of total metal ion concentrations. Thus, during each series of measurements the total ionic strength of the solution will vary, and this will lead to changes in the activity coefficients of all the ions present in solution. The equilibrium constants will therefore change in a manner very difficult to allow for. Pitzer's method cannot be used because the interaction coefficients of the complex ion salts are not known.

In order to avoid this variation it is usual to have present in the solution a relatively high concentration, commonly 0.2–2 M, of another salt, the supporting electrolyte, so that changes of total ionic strength during the experiments are negligibly small. If equilibrium constants are measured at several ionic strengths it is possible to extrapolate the values to obtain figures for zero total ionic strength; that is for all activity coefficients equal to unity. The method used to carry out the extrapolation should always be considered in the light of the discussion in Sec. 2.2 before such constants are used uncritically in quantitative calculations.

The supporting electrolyte should be inert, that is its ions should not take part in any reaction which competes with the equilibria being studied. Alkali-metal perchlorates or nitrates are often employed, the assumption being that neither of these anions forms complexes to any significant extent with most metals. This may not always be the case however, and in particular if the ligand being studied forms relatively weak complexes with the metal, there may be a significant competing reaction. For example if the value of β_4 for the zinc chloride complex $ZnCl_4^{2-}$ is required in order to carry out calculations relating to the separation of zinc and nickel by a solvent extraction procedure it is necessary to consider whether, if a strong solution of sodium nitrate had been used as the supporting electrolyte an equilibrium of the type

$$ZnNO_3^+ + 4Cl^- = ZnCl_4^{2-} + NO_3^-$$

might have been involved. If $ZnNO_3^+$ has appreciable stability with respect to the simple hydrated Zn^{2+} ion the exchange reaction will have a significant effect on the measured value of β_4. The values of $\log \beta_3$ and $\log \beta_4$ at $I = 4$ and $25°C$ are $+1$ and -1 for Cl^-, and for nitrate $\log K_1$ under the same conditions is 0.11.[48]

There are two primary reference sources of collected data on stability constants. The first[46] and its supplement[47] were published by The Chemical Society (London), the information having been compiled under the auspices of the International Union of Pure and Applied Chemistry, and together cover the literature up to 1969. The intention was to cover all papers related to metal complex equilibria; heats, entropies and free energies. The policy of the Commission was to avoid decisions concerning the quality and reliability of the published work so that all published data are included. The second source[48] is in 4 volumes; inorganic complexes are dealt with in Volume 4, and this gives tables of free energies (K values) heats and entropies selected by two of the authors concerned with the first source according to criteria which are set out. This second source, Critical Stability Constants, is very convenient to use for many purposes since a single value of a constant at each of 1 to 3 ionic strengths (including zero) is given, together with ΔH and ΔS where appropriate and references to the literature are also included. It could be argued, however, that since thermodynamic calculations can now very easily be carried out by computer, and the effects of changing thermodynamic values used can be readily evaluated, it is useful to have at hand all of the experimentally determined information.

Before an equilibrium constant at zero ionic strength is used in calculations for solutions of moderate or high ionic strength the method used to extrapolate the experimental data to $I = 0$ must be known so that the reverse calculation does not lead to errors. In practice it is usually possible to find values for a system which were measured in solutions not dissimiliar to those for which the calculations are required. Thus for calculations concerning Ni–NH_3 species present in ammoniacal ammonium sulphate solutions used in the

Sherritt–Gordon process for producing nickel metal powder, values measured in 2 molar ammonium nitrate solution would be preferable to these in 1 or 0.1 molar ammonium nitrate. Calculations are then in concentration terms and activity coefficients have to be assumed in order to pass to equations in thermodynamic terms.

2.4.   Complex Formation or Ion Association?

The implication of the section heading is that the two means of dealing with the excess Gibbs energy of a solution should not be mixed when dealing with the values for a system over a range of concentrations. Pitzer developed his equation on the basis that interaction between ions did not result in the formation of discrete new species. Thus in a dilute solution of $CoCl_2$ the value of G^{ex}, and consequently γ_{CoCl_2}, is regarded as being determined by the extent of the interaction between Co^{2+} and Cl^- ions and this is described by the values of $\beta^{(0)}$, $\beta^{(1)}$ and C^ϕ in his Eqs. (2.34) and (2.39)–(2.41). Such a solution is pale pink and contains simple hydrated Co^{2+} ions. If HCl or a highly soluble metal chloride such as LiCl or $CaCl_2$ is added to it, when the concentration of Cl^- reaches about 6 M the colour becomes blue because of the formation of the species $CoCl_4^{2-}$ which has the tetrahedral (sp^3) configuration of Cl^- ions around the Co^{2+} ion, covalently bound. It seems unlikely that Pitzer's equations could be used to fit activity data for the $CoCl_2$–HCl system to high chloride concentrations without the use of many parameters. Because these high concentrations are required to form the species no value of β_4 for $CoCl_4^{2-}$ has been measured and no value of ΔG° of formation of it has been published.[50]

Many 2-valent metal sulphates have activity coefficients which are almost identical over the range 10^{-3} to 2 m, being about 0.7 and 0.035 respectively at those two concentrations. The low values of $\gamma_\pm$ are usually interpreted as being due to the presence of ion association species such as $NiSO_4(aq)$, $CoSO_4(aq)$, $Zn(SO_4)(aq)$, etc. However the activity data for these 2:2 electrolytes can be fitted by adding an additional term to Pitzer's equation.[20] Thus the behaviour of the metal salt–H_2O systems can be described by interaction coefficients for M^{2+}, SO_4^{2-}, or by equilibrium constants which provide thermodynamic data for the species $MSO_4(aq)$, both ΔG° and ΔS° if equilibrium constants have been measured over a suitable range of temperature. If the data for M^{2+}, SO_4^{2-} and $MSO_4(aq)$ are used together with the interaction coefficients for

M^{2+} SO_4^{2-} serious errors may arise because two independent models for a single set of observations are being combined.

The thermodynamics of transition metals in aqueous solutions is almost always considered in terms of ion association models and equilibrium constants for discrete species. These constants are usually derived from measurements of the amount of ligand or of metal complexed at different ligand–metal ratios using a number of total metal ion concentrations. The measurements are interpreted by postulating the presence in solution of certain species. Thus in the Cu^{2+}–Cl^-–H_2O system the following species are considered to be present in the solution under various conditions of pH and chloride concentration: Cu^{2+}, H^+, OH^-, Cl^-, CuO_2^{2-}(aq), $HCuO_2^-$(aq), $Cu(OH)^+$(aq), $Cu(OH)_2$(aq), $Cu(OH)_4^{2-}$(aq), $Cu_2(OH)_2^{2+}$(aq), $CuCl^+$(aq), $CuCl_2$(aq), $CuCl_3^-$, $CuCl_4^{2-}$. The experimental measurements are fitted if it is supposed that these species are present under some conditions and that they have the equilibrium constants assigned to them. Additional evidence is required before it may be positively assumed that a species does actually exist in a solution. Such evidence may be based on spectra of some species, ultrasound, electrical measurements, etc.

The ion interaction model for this system would require the cation–anion interaction Cu^{2+}–OH^- and Cu^{2+}–Cl^-, also possibly the anion–anion interaction OH^-–Cl^- and the triple ion interaction Cu^{2+}–OH^-–Cl^-. The advantage to the hydrometallurgist of the ion association model is that it is possible to easily produce speciation diagrams which indicate the conditions under which a particular species is formed. These are described in Sec. 3.11.

Chapter 3

Thermodynamics of Reactions

3.1. Introduction

As is usual when dealing with chemical processes, thermodynamics applied to hydrometallurgy is used to calculate conditions of chemical equilibrium between species. These species may be solids, liquids, gases or solutes. Different kinds of information may be required in different circumstances and it can be set out in several different ways. Three of the most used methods are (i) predominance area diagrams; (ii) speciation diagrams and (iii) determination of phase equilibria in concentrated solutions.

Predominance area diagrams show the conditions in which solids, liquids, gases and solute species are dominant in a system which in most cases of interest to hydrometallurgists involves water. The system may be simple, for example $Fe-H_2O$, or more complex, $Fe-S-H_2O$, or $Cu-Fe-S-H_2O$. The variables used are pH (minus log hydrogen ion activity) and oxidation potential E (volts on the hydrogen scale, see Sec. 3.3.3). These diagrams are analogous to the Pourbaix diagrams[49] used particularly in corrosion studies. Alternatively if the system involves a gas which reacts with and dissolves in water, CO_2 for example, pH and log partial pressure of the gas could be used as the variables. Such diagrams indicate the chemistry of the system at a particular temperature and set of activities of solute species, and where applicable fugacities of gaseous reactants.

It is frequently necessary to know the concentration, or at least the activity of a particular species in the solution and how the proportions of the different species present, which can be numerous in multicomponent solutions, change with changes in chemical conditions. Convenient forms of speciation diagrams show how the fraction of the total amount of a metal present which exists as a particular species changes as some condition is altered, for example pH, $\log E$, $\log\{Cl^-\}$ etc. In multicomponent solutions, containing a number of metals and anions for example, if the total activity of each of these is set the fraction or the activity of each of the species being considered, simple and complex, can be plotted. Such calculations are carried out using a computer as also is the drawing of predominance area diagrams. It is essential to understand the theoretical basis of these in order to use them properly.

Phase equilibria in concentrated solutions are not treated so as to produce diagrams. Instead, a model of the aqueous Gibbs energy is used to calculate solubilities in a system. In all of these methods of displaying information concerning equilibria, the Gibbs energy is a minimum when thermodynamic equilibrium is achieved. In a real system, however, kinetic factors may prevent this being reached.

3.2.   Standard Free Energies of Formation

The standard free energy of formation of a substance is the energy change which takes place when it is formed from the elements of which it is composed. For purposes of calculation the convention is adopted that the heat content and internal energy of all elements are in their standard states, that is in their stable form at ordinary temperature and, until recently, 1 atmosphere pressure, is zero. For example the standard state of hydrogen is gaseous H_2 and for this the free energy ΔG° and enthalpy ΔH° of formation are defined as zero. Hydrogen is an equilibrium mixture of two molecular species, ortho- and para-hydrogen; in the latter the spins of the two hydrogen nuclei are opposed and in the former additive. The standard state of the element is the diatomic molecule in the equilibrium mixture of the two molecular species. The entropy S of hydrogen gas is not zero, it is 130.684 J mol^{-1} K^{-1} at 298.15 K and 0.1 MPa (1 bar). It should be noted that the values of the properties apply for the pure substances under a pressure of 0.1 MPa (1 bar), not 1 atmosphere. This alteration does not noticeably affect changes in values of ΔG° and ΔS° for processes except where the number of moles for gas changes during the process.

Metallic tin has three crystalline forms, α- or grey tin, β- or white Sn, and γ-Sn. The equilibrium transition temperatures are α–β 13.2°C, β–γ 161°C. The standard state is white Sn and grey tin has ΔG° 0.13 kJ mol^{-1}. In the case of phosphorus the reference phase has been selected to be the crystalline white form because the more stable phases have not been well defined thermochemically. ΔG° for the red triclinic form is given as -12.1 kJ mol^{-1}.

The free energy of formation of an ion in an aqueous solution is derived from the elements which form it. As an example the value of ΔH° of formation of Na$^+$(aq) depends on the values of ΔH° for the following steps:

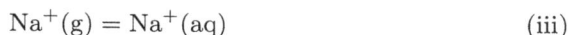

$$\text{Na(s)} = \text{Na(g)} \tag{i}$$

$$\text{Na(g)} = \text{Na}^+(\text{g}) + e(\text{g}) \tag{ii}$$

$$\text{Na}^+(\text{g}) = \text{Na}^+(\text{aq}) \tag{iii}$$

ΔH° for step (i) involves destruction of the crystal of sodium metal with formation of single gaseous atoms which do not interact with one another. The value of this sublimation energy is controlled by the lattice energy of sodium metal.

ΔH° for step (ii) is the ionization potential of sodium.

ΔH° for step (iii) is the heat of hydration of the gaseous sodium ion.

Thus ΔH° for the overall change

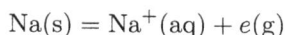

$$\text{Na(s)} = \text{Na}^+(\text{aq}) + e(\text{g})$$

which is the heat of formation of Na$^+$(aq), is the heat which produces from the element in its standard state one mole of hydrated sodium ions in water with no other interactions occurring between those ions and any other species present. As shown in Chapter 2, that means that the solution containing the sodium ions is infinitely dilute. The tabulated data for Na$^+$(aq) gives values of ΔH°, ΔG°, S° and heat capacity C_p at 298.15 K and 1 bar. Since these are related for a substance by

$$\Delta G^\circ = \Delta H^\circ - T\Delta S^\circ$$

the values of ΔG° and ΔS° also relate to an infinitely dilute solution and the free energy change caused by interactions between ions in more concentrated solutions gives rise to the excess free energy in the solution G^{ex} dealt with in Chapter 2.

Reaction (ii) above is a half-cell reaction and indicates that the electron is present as a gas, e.g. in a plasma. In chemical reactions at low temperatures the sodium atom is oxidised to Na^+ by an oxidising agent which is itself reduced by accepting the electron. This pair of coupled reactions is known as a full reaction, because no other chemical process is necessary to complete the electron transfer reaction, or a redox reaction because it involves both reduction and oxidation.

Standard free energies of formation of substances are used to analyse quantitatively the thermodynamics of chemical reactions which proceed to equilibrium. If the stoichiometric equation for such a reaction is

$$A + 2B = 3C + D$$

then the standard free energy change for it is

$$\Delta G_r^o = 3\Delta G_C^o + \Delta G_D^o - \Delta G_A^o - 2\Delta G_B^o .$$

The values of ΔG^o for the reactants and products are those considered above and so refer to 1 mole of them in their standard states. Thus for a pure solid or liquid reactant, which has unit activity, its value of ΔG^o is used in the calculation of ΔG_r^o. In the case of an impure reactant, which can be regarded as a solution of the reacting solute in a solvent which does not take part in the reaction, the activity of the reactant in its solution is used in the calculation. In many cases of interest in hydrometallurgy a number of different reactions can occur in an apparently simple system, giving rise to various products according to the prevailing conditions. It is very convenient to set out diagramatically the ranges of conditions within which these species are thermodynamically stable. These are known as predominance area diagrams.

3.3.   Predominance Area Diagram for the Fe–H_2O System, 298.15 K

3.3.1.   *Reactions Involving Dissolved Species and Solids, Nonredox*

The basic principles of the thermodynamic treatment of reactions of interest in hydrometallurgy will be dealt with using the iron water system. The first step is to decide which species are to be considered and to list their free energies of formation. This is done in Table 3.1.

Table 3.1.

Species	Fe^{2+}	Fe^{3+}	$Fe(OH)_3$	Fe_2O_3	Fe_3O_4	$Fe_{0.947}O$	$H_2O(l)$	OH^-
$\Delta G°$ kJ	-78.90	-4.7	-696.5	-742.2	-1015.4	-245.12	-237.129	-157.244

Fe^{2+} and Fe^{3+} are the simple hydrated ions; $Fe(OH)_3$ is precipitated crystalline solid; Fe_2O_3 is hematite; Fe_3O_4 magnetite; $FeO_{0.947}O$ wustite; $H_2O(l)$ liquid water and OH^- simple hydrated hydroxide ion.

When ferric oxide is in contact with pure water it will dissolve to some extent.

$$Fe_2O_3 + 3H_2O = 2Fe^{3+} + 6OH^- . \tag{3.1}$$

The equilibrium constant for the reaction is given by

$$K = \{Fe^{3+}\}^2\{OH^-\}^6/\{Fe_2O_3\}\{H_2O\}^3$$

but since the activities of pure ferric oxide and pure liquid water are unity

$$K = \{Fe^{3+}\}^2\{OH^-\}^6 .$$

As the solid is only very slightly soluble, the activity coefficients of both kinds of ion are unity and the stoichiometric equilibrium constant is related to the solubility product K_{so}, the subscript "so" indicating that both the ferric and hydroxide ions are uncomplexed in solution.

The standard free energy of the reaction is

$$\Delta G° = 2\Delta G°_{Fe^{3+}} + 6\Delta G°_{OH^-} - \Delta G°_{Fe_2O_3} - 3\Delta G°_{H_2O}$$

$$= (2 \times -4.7) + (6 \times -157.24) - (-742.2) - (3 \times -237.13)$$

$$= +500.75 \text{ kJ} .$$

This is related to the equilibrium constant by the equation

$$\Delta G° = -RT \ln K$$

where R is the molar gas constant $= 8.3143$ J mol^{-1} K^{-1} and T is the absolute temperature. Thus at $25°C$

$$500.75 \times 10^3 = -8.3143 \times 298.15 \times 2.303 \log K$$

$$\log K = -87.71 = \log\{Fe^{3+}\}^2\{OH^-\}^6 .$$

The solubility product $K_{so} = \{Fe^{3+}\}\{OH^-\}^3$ so that

$$\log K_{so} = -43.86.$$

This value refers to ferric and hydroxide ions in equilibrium with the solid phase hematite, α-Fe_2O_3, whereas experimental determinations of the solubility product will normally be carried out with precipitated ferric hydroxide. The nature of this will be determined by the conditions under which precipitation was carried out and the solid will usually be amorphous. Using the value of ΔG° for crystalline $Fe(OH)_3$ given in Table 3.1 for the reaction

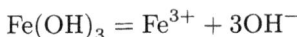

$$Fe(OH)_3 = Fe^{3+} + 3OH^-$$

$$\Delta G^\circ = \Delta G^\circ_{Fe^{3+}} + 3\Delta G^\circ_{OH^-} - \Delta G^\circ_{Fe(OH)_3} = +220.07 \text{ kJ}.$$

At 25°C

$$220.07 \times 10^3 = -8.314 \times 298.15 \times 2.303 \log K = -5708.7 \log K$$

or, using ΔG° in kJ

$$220.07 = -5.709 \log K$$

$$\log K = -38.55$$

and this is the solubility product of the thermochemically well defined crystalline $Fe(OH)_3$.

Experimentally determined values of $\log K_{so}$ for Fe^{3+} and the ligand OH^- have been tabulated.[46] A value for the solid phase identified as α-Fe_2O_3 is -42.7. For unspecified solids figures given are: -36.35; -37.7 to -39.2; -37.50; -39.43. All of the experimental data had been corrected for infinite dilution. The practical implication of the calculations for hematite and $Fe(OH)_3$ and comparison with experimental results is that the amount of iron remaining in a solution after the bulk of it has been precipitated as an oxide or hydroxide depends on the nature of the solid phase in equilibrium with the solution.

Effect of pH.   From the solubility product of α-Fe_2O_3 it is seen that

$$\log\{Fe^{3+}\} = -43.86 - 3\log\{OH^-\}$$

$$= -1.86 - 3pH \tag{3.2}$$

since

$$\log\{H^+\} + \log\{OH^-\} = -14$$

and

$$pH = -\log\{H^+\}.$$

The solubility of hematite in water is therefore dependent on pH and Eq. (3.1) represents the reaction

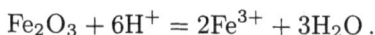

$$Fe_2O_3 + 6H^+ = 2Fe^{3+} + 3H_2O.$$

3.3.2. *Reactions Involving Oxidation and Reduction*

In a chemical reaction in which a substance or species is oxidised, another is reduced. It is convenient here to separate the reaction into its two parts, one involving donation of an electron and the other involving acceptance of the electron. In the Fe–H_2O system two kinds of reaction involving the ions Fe^{3+} and Fe^{2+} and metallic Fe are significant. One is an equilibrium such as

$$Fe^{2+} + 2e = Fe$$

which gives rise to a single ion electrode potential and the other is a redox equilibrium between the ions in solution

$$Fe^{3+} + e = Fe^{2+}.$$

The convention used in this book is that a half-cell reaction is written with the oxidised state on the left and the reduced state on the right of the equals or equilibrium symbol.

The value of the potential E is given by the equation

$$E = E^\circ - \frac{RT}{zF} \ln \frac{\{\text{Reduced state}\}}{\{\text{Oxidised state}\}} \tag{3.3}$$

where F is the Faraday constant $= 96\,487$ JV^{-1} mol^{-1}. For a single ion electrode the activity of the metal is unity, so that in the case of a metal M of valency z

$$E_{M^{z+},M} = E^\circ_{M^{z+},M} + \frac{RT}{zF} \ln a_{M^{z+}} \tag{3.4}$$

where $a_{M^{z+}}$ is the activity of the metal ions in solution. E° is the standard electrode potential, corresponding to the value of E when $a_{M^{z+}}$ is unity. In the case of an anion the free element of unit activity is the oxidised material, so that

$$E_{A^-} = E^\circ_{A^-} - \frac{RT}{zF} \ln a_{A^-} \, . \tag{3.5}$$

In a redox reaction two ions are involved and neither has necessarily to be at unit activity. The value of E is then controlled by the ratio of the activities of the two ions, and E° corresponds to the value when their activities are equal.

The value of E is controlled by the free energy change which occurs during the reaction as written according to the equations

$$-\Delta G = zEF$$

$$-\Delta G^\circ = zE^\circ F = RT \ln K$$

where K is the equilibrium constant for the reaction. Thus E° can be calculated from the standard free energy change.

For the reaction

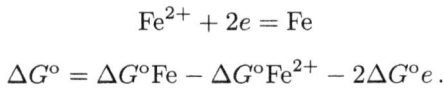

$$\mathrm{Fe}^{2+} + 2e = \mathrm{Fe}$$

$$\Delta G^\circ = \Delta G^\circ \mathrm{Fe} - \Delta G^\circ \mathrm{Fe}^{2+} - 2\Delta G^\circ e \, .$$

Suppose for the moment that $\Delta G^\circ e = 0$, then

$$\Delta G^\circ = 0 - (-78.90) - 0 = +78.90 \text{ kJ} \, .$$

As two electrons are involved $z = 2$, and F, the Faraday constant, is 96.487 kJV^{-1} mol^{-1}.

Thus

$$E^\circ = -78.90/192.974 = -0.409 \text{ V} \, .$$

In the alternative sign convention the chemical equation is written

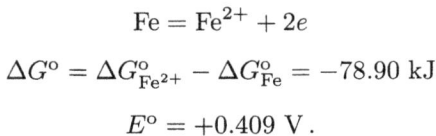

$$\mathrm{Fe} = \mathrm{Fe}^{2+} + 2e$$

$$\Delta G^\circ = \Delta G^\circ_{\mathrm{Fe}^{2+}} - \Delta G^\circ_{\mathrm{Fe}} = -78.90 \text{ kJ}$$

$$E^\circ = +0.409 \text{ V} \, .$$

There is no thermodynamic difference between the two conventions, it is necessary only to specify which way the reaction is written.

$$E^o_{Fe^{2+},Fe} = -0.409 \text{ V}$$

$$E^o_{Fe,Fe^{2+}} = +0.409 \text{ V}.$$

To avoid unnecessary confusion over signs of numbers in the equations used in calculating data for preparing predominance area diagrams it is desirable to use only one convention.

For the redox reaction

$$Fe^{3+} + e = Fe^{2+}$$

$$\Delta G^o = -78.90 + 4.7 = -74.2 \text{ kJ}$$

$$E^o = +74.2/96.487 = +0.769 \text{ V}.$$

Thus the activities of Fe^{2+} and Fe^{3+} are equal at 0.769 V.

3.3.3. *The Hydrogen Scale*

The methods outlined above enable electrode potentials to be correlated with free energy changes occurring during reactions. It is not possible to measure the potential of a single electrode, only the potential of a cell made up of two or more electrodes. Thus only algebraic differences between electrode potentials can be measured and it is necessary to define a potential of one electrode under specified conditions. This has been done by defining the activity normal hydrogen electrode as a perfectly reversible electrode with hydrogen gas at fugacity 0.1 MPa in a solution of hydrogen ions at unit activity. For all practical purposes this corresponds to a platinum electrode coated with platinum black, continually bathed in pure hydrogen gas at 0.1 MPa pressure, dipping into a solution in which hydrogen ions are at unit activity and which contains no other substance which affects the potential of the platinum electrode. The potential between the normal hydrogen electrode and the solution is considered to be zero at all temperatures.

The reaction at the hydrogen electrode is

$$2H^+ + 2e = H_2.$$

Since $E^o_{H^+,1/2H_2} = 0$, it follows that ΔG^o for the reaction $= 0$. Since the standard state of hydrogen is gaseous H_2 the free energy of formation of $H^+(aq)$ has now been defined as zero at all temperatures, as also has $S^o_{H^+}$. This definition provides the link between thermodynamic data for substances in aqueous solution and for those in the solid, liquid and gaseous states. It does however cause difficulties when dealing with the thermodynamics of aqueous solutions at temperatures significantly higher than $25°C$ (Sec. 3.6.1). For this reason, an alternative approach to the problem of defining one electrode potential may be put forward in future, using calculated values of the potential of the hydrogen electrode at different temperatures.

The value calculated above from free energy data of $E^o_{Fe^{2+},Fe} = -0.409$ V is the standard potential on the hydrogen scale because the reaction can be written as

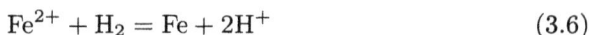

$$Fe^{2+} + H_2 = Fe + 2H^+ \tag{3.6}$$

without changing the value of ΔG^o or E^o as long as the fugacity of hydrogen gas is 1 or, for all practical purposes, when the partial pressure of hydrogen is 1 atm. Since Reaction (3.6) has been balanced by having one H_2 molecule provide the 2 electrons required to reduce Fe^{2+} to Fe, forming $2H^+$ by the oxidation reaction, the free energy of formation of electrons does not appear in the calculation of ΔG^o for Reaction (3.6) and so may be taken as zero as was assumed above. For half-cell reactions at temperatures significantly higher than $25°C$ this is not the case, see Sec. 3.6.1.

3.3.4.   *The pe Scale*

In thermochemical calculations by computer it may be convenient to set out a redox equilibrium for a reaction written as a reduction involving one electron rather than in the conventional way used above and in this book subsequent to this section.

Consider the redox equilibrium

$$Ox + ze = Red.$$

The equilibrium constant can be written so as to include the activity of the electrons

$$K = \frac{\{Red\}}{\{Ox\}\{e\}^z}.$$

Thus

$$\{e\} = \left[\frac{1}{K}\frac{\{\text{Red}\}}{\{\text{Ox}\}}\right]^{1/z} \tag{i}$$

$$\ln\{e\} = \frac{1}{z}\left[-\ln K + \ln\frac{\{\text{Red}\}}{\{\text{Ox}\}}\right] \tag{ii}$$

$$zE^\circ F = RT\ln K \tag{iii}$$

$$E = E^\circ - \frac{RT}{zF}\ln\frac{\{\text{Red}\}}{\{\text{Ox}\}} = \frac{RT}{zF}\ln K - \frac{RT}{zF}\ln\frac{\{\text{Red}\}}{\{\text{Ox}\}}$$

$$E = \frac{RT}{zF}\left[\ln K - \ln\frac{\{\text{Red}\}}{\{\text{Ox}\}}\right]$$

$$E = -\frac{RT}{F}\ln\{e\} = -\frac{2.303RT}{F}\log\{e\} = \frac{2.303RT}{F}\text{pe}. \tag{iv}$$

At 25°C, $E = 0.05917$ pe.

The equilibrium activity of the electron for any redox couple is defined formally and mathematically by Eq. (i) even though the electron is not considered to be an individual species and its concentration is certainly not defined. The pe scale is derived from (ii) where pe $= -\log\{e\}$. Electron activities can be expressed in either of the scales, pe or E, the latter based on (iii). The relationship between them is shown in (iv).

The parameter pe provides a nondimensional scale, like pH, whereas E is conventionally measured in volts. A few examples of equations for chemical reactions written for the pe scale[51] are given here together with their equivalent written for the E scale. Where E° does not equal 0.05917 pe° different values for ΔG° of formation of Fe^{2+}(aq) have been used in the calculations.

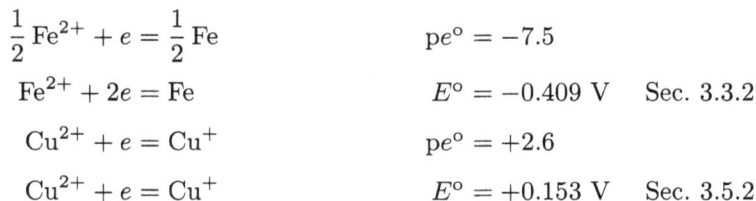

$\frac{1}{2}Fe^{2+} + e = \frac{1}{2}Fe$	pe° $= -7.5$
$Fe^{2+} + 2e = Fe$	$E^\circ = -0.409$ V Sec. 3.3.2
$Cu^{2+} + e = Cu^+$	pe° $= +2.6$
$Cu^{2+} + e = Cu^+$	$E^\circ = +0.153$ V Sec. 3.5.2

$$\frac{1}{2}\,Cu^{2+} + e = \frac{1}{2}\,Cu \qquad\qquad pe^o = +5.7$$

$$Cu^{2+} + 2e = Cu \qquad\qquad E^o = +0.339\ V \qquad Sec.\ 3.5.2$$

$$\frac{1}{2}\,Fe_3O_4 + 4H^+ + e = \frac{3}{2}\,Fe^{2+} + 2H_2O \qquad pe^o = +16.6$$

$$Fe_3O_4 + 8H^+ + 2e = 3Fe^{2+} + 4H_2O \qquad E^o = +0.880\ V \qquad Sec.\ 3.3.6$$

$$\frac{1}{8}\,SO_4^{2-} + \frac{5}{4}\,H^+ + e = \frac{1}{8}\,H_2S(aq) + \frac{1}{2}H_2O \quad pe^o = +5.13$$

$$SO_4^{2-} + 10H^+ + 8e = H_2S(aq) + 4H_2O \qquad E^o = +0.303\ V \qquad Sec.\ 3.4.3$$

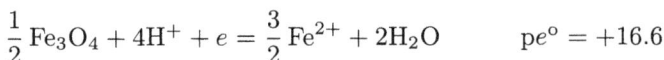

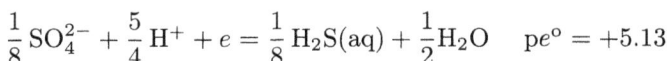

3.3.5. *Equilibria between Solids*

In the iron system the solids considered are Fe, $Fe(OH)_3(cr)$, Fe_2O_3, Fe_3O_4 and $Fe_{0.947}O$. The regions of stability of all except $Fe(OH)_3$ (cr) will be determined. That of the hydroxide will correspond with that of hematite with the boundary lines shifted slightly because of the different values of ΔG^o of formation of the two solids.

FeO–Fe

The oxidation of iron to wustite could be written, approximating the formula to FeO

$$2FeO = 2Fe + O_2$$

putting the oxidised form of the metal on the left. Since the reaction considered takes place in water, however, it is convenient to combine this equation with that representing the water equilibrium

$$4H^+ + O_2 + 4e = 2H_2O$$

to give

$$FeO + 2H^+ + 2e = Fe + H_2O\,. \qquad\qquad (3.7)$$

This shows the pH dependence of the reaction to form the oxide.

The standard free energy change of this reaction is

$$\Delta G^o = \Delta G^o_{Fe} + \Delta G^o_{H_2O} - \Delta G^o_{FeO} - 2\Delta G^o_{H^+}$$

$$= 0 - 237.129 + 245.12 - 0 = 7.99\ kJ$$

so that the standard potential is

$$E^\circ = -\Delta G/zF = -7.99/(2 \times 96.487) = -0.041 \text{ V}.$$

When reactants are not all at unit activity E for the reaction is given by

$$E = E^\circ - \frac{RT}{zF} \ln \frac{\{\text{Fe}\}\{\text{H}_2\text{O}\}}{\{\text{FeO}\}\{\text{H}^+\}^2} = E^\circ - \frac{0.0592}{2} \log \frac{1}{\{\text{H}^+\}^2}$$

$$= E^\circ - 0.0592 \text{ pH} = -0.041 - 0.0592 \text{ pH}. \qquad (3.8)$$

Fe_3O_4–Fe

The reaction is written

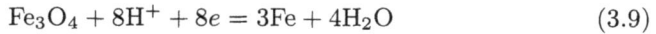

$$\text{Fe}_3\text{O}_4 + 8\text{H}^+ + 8e = 3\text{Fe} + 4\text{H}_2\text{O} \qquad (3.9)$$

$$\Delta G^\circ = (4 \times -237.129) - (-1015.4) = 66.9 \text{ kJ}$$

$$E^\circ = -66.9/(8 \times 96.487) = -0.0867 \text{ V}$$

$$E = E^\circ - \frac{0.0592}{8} \log \frac{1}{\{\text{H}^+\}^8}$$

$$= -0.0867 - 0.0592 \text{ pH}. \qquad (3.10)$$

Half-cell reactions such as (3.9) can be set up as follows when uncertainty as to the valency of a species may arise, Fe in Fe_3O_4 for example. Write the species for which the equation is being written on either side of the equals sign

$$\text{Fe}_3\text{O}_4 + \qquad = 3\text{Fe} + 4\text{O}^{2-}.$$

Fe_3O_4 contains 4 oxygen atoms and each is always taken as having charge -2 except as mentioned below. These oxygen atoms form $4\text{H}_2\text{O}$ which can be inserted on the right of the equation and the 8H^+ necessary to form the water on the left. The four oxygen atoms in Fe_3O_4 had a total charge of -8, neutralised by the 3Fe ions. The Fe metal in the equation has no charge so 8 electrons are required to achieve charge balance. In hydrogen peroxide and other peroxides oxygen is present as the O_2^{2-} ion so the above argument would not be valid. Similarly in sulphide systems the sulphide ion is taken as having 2 negative charges, except in polysulphides and in solids such as pyrite, FeS_2, which contains S_2^{2-} ions.

Fe_3O_4–FeO

The reaction is

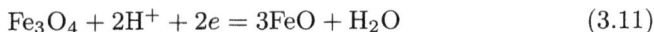

$$Fe_3O_4 + 2H^+ + 2e = 3FeO + H_2O \tag{3.11}$$

$$\Delta G^\circ = (3 \times -245.12) + (-237.129) - (-1015.4) = 42.9 \text{ kJ}$$

$$E^\circ = -42.9/(2 \times 96.487) = -0.222 \text{ V}$$

$$E = E^\circ - \frac{0.0592}{2} \log \frac{1}{\{H^+\}^2}$$

$$= -0.222 - 0.0592 \text{ pH}. \tag{3.12}$$

Fe_2O_3–Fe_3O_4

$$3Fe_2O_3 + 2H^+ + 2e = 2Fe_3O_4 + H_2O \tag{3.13}$$

$$\Delta G^\circ = (2 \times -1015.4) + (-237.129) - (3 \times -742.2) = -41.3 \text{ kJ}$$

$$E^\circ = 41.3/(2 \times 96.487) = 0.214 \text{ V}$$

$$E = E^\circ - \frac{0.0592}{2} \log \frac{1}{\{H^+\}^2}$$

$$= 0.214 - 0.0592 \text{ pH}. \tag{3.14}$$

Fe_2O_3–FeO

$$Fe_2O_3 + 2H^+ + 2e = 2FeO + H_2O \tag{3.15}$$

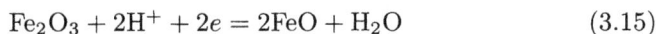

$$\Delta G^\circ = (2 \times -245.12) + (-237.129) - (-742.2) = -14.8 \text{ kJ}$$

$$E^\circ = 14.8/(2 \times 96.487) = 0.0725 \text{ V}$$

$$E = 0.0725 - 0.0592 \text{ pH}. \tag{3.16}$$

3.3.6.   *Equilibria between Ions and Solids*

$Fe\text{–}Fe^{2+}$

$$Fe^{2+} + 2e = Fe. \tag{3.17}$$

It was shown in Sec. 3.3.2 that $E^{\circ}_{Fe^{2+},Fe} = -0.409$ V.
    From (3.4)

$$E = -0.409 + \frac{RT}{2F}\ln\{Fe^{2+}\}$$

$$= -0.409 + 0.0296\log\{Fe^{2+}\}. \tag{3.18}$$

$Fe^{2+}\text{–}FeO$

$$FeO + 2H^{+} = Fe^{2+} + H_2O. \tag{3.19}$$

This is not a redox reaction since no electrons are involved, iron being Fe^{II} in the solid and the hydrated ion.

$$\Delta G^{\circ} = (-78.90) + (-237.129) - (-245.12) = -70.91 \text{ kJ}$$

$$\Delta G^{\circ} = -RT\ln K = -5.7089\log K \ (\text{using } R \text{ in kJ})$$

$$\log K = +12.42 = \log\{Fe^{2+}\} - 2\log\{H^{+}\} = \log\{Fe^{2+}\} + 2\text{ pH}. \tag{3.20}$$

$Fe_3O_4\text{–}Fe^{2+}$

$$Fe_3O_4 + 8H^{+} + 2e = 3Fe^{2+} + 4H_2O \tag{3.21}$$

$$E^{\circ} = -[-169.8/(2 \times 96.487)] = 0.880 \text{ V}$$

$$E = 0.880 - \frac{RT}{2F}\ln\frac{\{Fe^{2+}\}^3}{\{H^{+}\}^8}$$

$$= 0.880 - 0.0888\log\{Fe^{2+}\} - 0.237\text{ pH}. \tag{3.22}$$

Fe$_2$O$_3$–Fe^{2+}

$$Fe_2O_3 + 6H^+ + 2e = 2Fe^{2+} + 3H_2O \qquad (3.23)$$

$$E^\circ = 127.0/(2 \times 96.487) = 0.658 \text{ V}$$

$$E = 0.658 - 0.0592 \log\{Fe^{2+}\} - 0.178 \text{ pH}. \qquad (3.24)$$

3.3.7.   Equilibria between Ions

The only example in this system has been dealt with in Sec. 3.3.2.

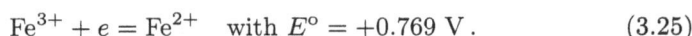

$$Fe^{3+} + e = Fe^{2+} \quad \text{with } E^\circ = +0.769 \text{ V}. \qquad (3.25)$$

3.3.8.   Representation of Equilibria

Stability limits of water.   The chemical reactions and the equations of the lines representing the equilibria in the Fe–H$_2$O system at 298.15 K have been prepared in Eqs. (3.7) to (3.25) and all of the equilibria are written in terms of pH and redox potential on the hydrogen scale. Thus pH and E are used conventionally as the x and y axes respectively. The ranges 0–14 for pH and -1.0 to $+1.0$ volt for E are usually sufficient but can be extended if desired.

Water itself is stable only within a limited region of this range of potential, oxygen or hydrogen being evolved when the limits are exceeded. The upper limit of water stability can be calculated using the reaction

$$O_2 + 4H^+ + 4e = 2H_2O \qquad (3.26)$$

$$\Delta G^\circ = -474.26 \text{ kJ}$$

$$E^\circ = +474.26/(4 \times 96.487) = 1.229 \text{ V}$$

$$E = 1.229 - \frac{0.0592}{4} \log \frac{1}{\{H^+\}^4} - \frac{0.0592}{4} \log \frac{1}{P_{O_2}}. \qquad (3.27)$$

At unit fugacity of oxygen, that is for practical purposes 1 atm partial pressure of oxygen gas, the last term disappears so that

$$E = 1.229 - 0.0592 \text{ pH}. \qquad (3.28)$$

The lower limit of water stability is obtained similarly by using the half-cell reaction

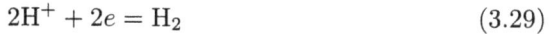

$$2H^+ + 2e = H_2 \tag{3.29}$$

since no other material is involved in the reaction. The presence of water is implied in the term H^+, which is derived from the water itself.

The value of $E^\circ = 0$. Thus

$$E = 0 - \frac{0.0592}{2} \log P_{H_2} - 0.0592 \text{ pH}. \tag{3.30}$$

At unit fugacity of hydrogen

$$E = -0.0592 \text{ pH}. \tag{3.31}$$

The stability limits of water can be plotted in the frame of Fig. 3.1. The upper limit given by (3.28) indicates that at pH = 10, $E = 0.637$ V; at pH = 0, $E = 1.229$ V. After drawing the line the predominant species are indicated on the two sides of it. For the lower stability limit at pH = 0, $E = 0$ V; at pH = 10, $E = -0.592$ V.

Equilibria involving iron species. It is convenient to construct the diagram systematically and here it is begun at the bottom left, i.e. the most acidic and reducing region. The data necessary for plotting the lines are first set out.

Fe–FeO

   Equation (3.8). At pH = 0, $E = -0.041$; at pH = 10, $E = -0.633$ V.

Fe–Fe₃O₄

   Equation (3.10). At pH = 0, $E = -0.0867$; at pH = 10, $E = -0.678$ V.

FeO–Fe₃O₄

   Equation (3.12). At pH = 0, $E = -0.222$; at pH = 10, $E = -0.814$ V.

Fe₃O₄–Fe₂O₃

   Equation (3.14). At pH = 0, $E = 0.214$; at pH = 10, $E = -0.378$ V.

FeO–Fe$_2$O$_3$

Equation (3.16). At pH $= 0$, $E = 0.0725$; at pH $= 10$, $E = -0.519$ V.

Fe–Fe^{2+}

Equation (3.18). The diagram will be constructed for activities of ionic species containing iron (Fe^{2+} and Fe^{3+}) $= 10^{-6}$.

$$\text{For } \{\text{Fe}^{2+}\} = 10^{-6}, \quad E = -0.587 \text{ V}.$$

Note that this value is independent of pH.

FeO–Fe^{2+}

Equation (3.20)

$$\text{For } \{\text{Fe}^{2+}\} = 10^{-6}, \quad \text{pH} = 9.21.$$

Note that this is a solubility product equation, independent of E.

Fe$_3$O$_4$–Fe^{2+}

Equation (3.22)

$$\text{For } \{\text{Fe}^{2+}\} = 10^{-6}, \quad E = 1.413 - 0.237 \text{ pH}$$

At pH $= 0$, $E = 1.413$; at pH $= 10$, $E = -0.957$ V.

Fe$_2$O$_3$–Fe^{2+}

Equation (3.24)

$$\text{For } \{\text{Fe}^{2+}\} = 10^{-6}, \quad E = 1.013 - 0.178 \text{ pH}$$

At pH $= 0$, $E = 1.013$; at pH $= 10$, $E = -0.767$ V.

Fe$_2$O$_3$–Fe^{3+}

Equation (3.2)

$$\text{For } \{\text{Fe}^{3+}\} = 10^{-6}, \quad \text{pH} = 1.38.$$

$Fe^{3+}-Fe^{2+}$

Equation (3.25)

$${Fe^{3+}} = {Fe^{2+}} = 10^{-6}, \quad E^\circ = 0.769 \text{ V}.$$

3.3.9. *Discussion of the Diagram, Fig. 3.1*

The most reduced solid in the system is iron metal and it must therefore be predominant in the area below any of the oxides formed from it by an

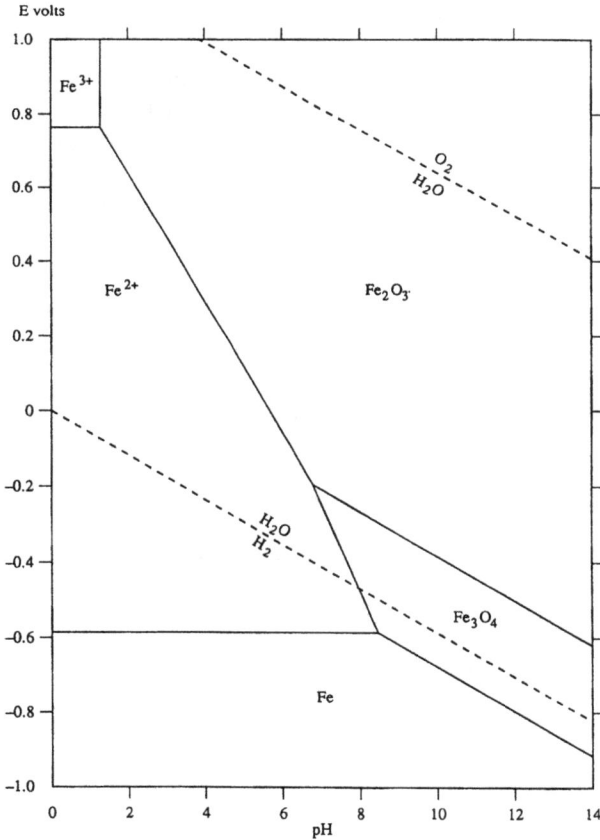

Fig. 3.1. *E*-pH diagram, Fe–H$_2$O system, 298.15 K, { } Fe ionic species 10^{-6}.

oxidation reaction. The line forming the boundary between Fe and Fe_3O_4 lies at more negative values of E than does the Fe–FeO boundary. Therefore in aqueous systems iron metal is thermodynamically unstable because it lies outside the limits of stability of water and when it reacts with the water it produces Fe_3O_4, not wustite. In the more acid region of the diagram iron metal produces Fe^{2+} ions when it is oxidised and here also lies outside the region of stability of water. Iron dissolves in acidified water with evolution of hydrogen. Iron (II) ions are oxidised to iron (III) ions in solutions of low pH but only at rather high values of E, that is under quite strongly oxidising conditions. As a consequence, a solution containing Fe^{3+} ions is strongly oxidising and is used in hydrometallurgical processes to oxidise some metals present in ores at a low valency to make them soluble, U^{IV} to U^{VI} for example. In such a leaching system in which Fe^{3+} is reduced to Fe^{2+}, the value of E of the solution can be measured using a redox electrode. When it reaches too low a value it can be raised by addition of another oxidising agent such as pyrolusite, MnO_2, or sodium chlorate, $NaClO_3$, which reoxidises Fe^{2+} to Fe^{3+}. Solutions containing ferric salts are also used to oxidise some metal sulphides and predominance area diagrams of sulphur-containing systems are of use in understanding the behaviour of such leaching systems.

It is important to realise that a totally misleading diagram can result if one, or more, important species is not taken into account in preparing it. For example if the existence of Fe^{2+} is ignored the horizontal line separating the Fe^{3+} and Fe^{2+} areas disappears and the line separating the Fe^{3+} and Fe_2O_3 areas extends down to the boundary between Fe_2O_3 and Fe_3O_4 and then changes slope to separate the Fe^{3+} and Fe_3O_4 areas. When constructing a diagram for the first time all possible equilibria should be considered, including Fe^{3+}–Fe_3O_4, in order to determine which are of no significance. This in fact does not appear.

3.4.   The Sulphur–Water System, 298.15 K

3.4.1.   *Data Used*

It has been shown[52] that the only species in this system which are thermodynamically stable at appreciable activities are elemental sulphur, $H_2S(aq)$, $HS^-(aq)$, $S^{2-}(aq)$, $SO_4^{2-}(aq)$ and HSO_4^-. Other sulphur acids or their salts are known but are metastable. They include sulphurous, H_2SO_3; thiosulphuric,

$H_2S_2O_3$; dithionous, $H_2S_2O_4$; disulphurous, $H_2S_2O_5$; dithionic, $H_2S_2O_6$; and polythionic, $H_2S_nO_6$ acids. In addition monosulphides of the alkali metals, M_2S, and hydrogen sulphides MSH form polysulphides of general formula M_2S_n which contain unbranched chain anions $(S_n)^{2-}$. Thermodynamic data for many of these have been tabulated[50] and the interrelationships between the metastable compounds can be examined by drawing predominance area diagrams in which sulphuric acid and the ions derived from it are not considered, see Sec. 5.3.1 and Fig. 5.10.

The free energy data (298.15 K) used for producing the equilibrium diagram Fig. 3.2, are given in Table 3.2.

Table 3.2.

Species	S^{2-}	HS^-	$H_2S(aq)$	SO_4^{2-}	HSO_4^-	$H_2O(l)$
ΔG° kJ	+111.4	+12.08	−27.83	−744.53	−755.91	−237.129

The value of ΔG° for S^{2-} is not from Ref. 50; because of their importance in the hydrometallurgy of sulphides the dissociation constants of $H_2S(aq)$ are discussed in Sec. 3.8.2.

3.4.2. *Acid Dissociation Constants*

The species H_2SO_4 is not present in solutions which are represented by predominance area diagrams.

$$HSO_4^- = H^+ + SO_4^{2-}$$
$$\log K_2 = -1.99 = \log\{SO_4^{2-}\} - \log\{HSO_4^-\} - pH \tag{1}$$

$$H_2S(aq) = H^+ + HS^-$$
$$\log K_1 = -6.99 = \log\{HS^-\} - \log\{H_2S(aq)\} - pH \tag{2}$$

$$HS^- = H^+ + S^{2-}$$
$$\log K_2 = -17.39 = \log\{S^{2-}\} - \log\{HS^-\} - pH . \tag{3}$$

3.4.3. *Redox Equilibria*

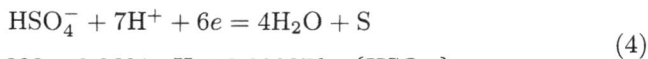

$$HSO_4^- + 7H^+ + 6e = 4H_2O + S$$
$$E = 0.333 - 0.0691 \, pH + 0.00987 \log\{HSO_4^-\} \tag{4}$$

$$SO_4^{2-} + 8H^+ + 6e = 4H_2O + S \tag{5}$$

$$E = 0.352 - 0.0789 \text{ pH} + 0.00987 \log\{SO_4^{2-}\}$$

$$S + 2H^+ + 2e = H_2S(aq) \tag{6}$$

$$E = 0.144 - 0.0592 \text{ pH} - 0.0296 \log\{H_2S(aq)\}$$

$$S + H^+ + 2e = HS^- \tag{7}$$

$$E = -0.0626 - 0.0296 \text{ pH} - 0.0296 \log\{HS^-\}$$

$$SO_4^{2-} + 9H^+ + 8e = HS^- + 4H_2O \tag{8}$$

$$E = 0.249 - 0.0666 \text{ pH} - 0.00740 \log(\{HS^-\}/\{SO_4^{2-}\})$$

$$SO_4^{2-} + 10H^+ + 8e = H_2S(aq) + 4H_2O \tag{9}$$

$$E = -0.300 - 0.0740 \text{ pH} - 0.00740 \log(\{H_2S(aq)\}/\{SO_4^{2-}\}).$$

3.4.4.   *Discussion of the Diagram, Fig. 3.2*

The equilibrium diagram for the cases where the activity of each sulphur-containing species in solution is 0.1 or 10^{-4} is given in Fig. 3.2. Under these conditions there is a region of stability of sulphur. That is reduction of a solution 0.1 M in sulphate ions at a pH between 1.96 and about 7.7 will produce sulphur. At a pH below 1.96, the equilibrium is between HSO_4^- and sulphur. As the potential is lowered to more negative values sulphur should produce H_2S in solution at pH values below 6.99 and HS^- at higher pH. In fact of course the ratios of HSO_4^- to SO_4^{2-} and of $H_2S(aq)$ to HS^- change progressively on passing through the pH values corresponding to their pK values. This is shown by curved lines in the corrosion diagrams of Poubaix[49] where the total activity of an ionic substance is kept constant as it changes species, i.e. $H_2S(aq)$ to HS^-; HSO_4^- to SO_4^{2-}; Fe^{3+} to Fe^{2+} etc. When the activity of the two ions in equilibrium is made equal as in Fig. 3.2, where three equilibrium lines meet they do so at a point, which provides a check on the accuracy of calculations and of drawing.

At pH values above about 7.7, there is direct transformation between SO_4^{2-} and HS^- in 0.1 M solutions. This higher pH limit for the stability of sulphur is not greatly affected by higher activities of SO_4^{2-} and HS^-, rising only to about pH 8.4 for 1 M solutions of these ions. The existence of a region of stability of sulphur does depend on the values at lower activities of the sulphur acids in

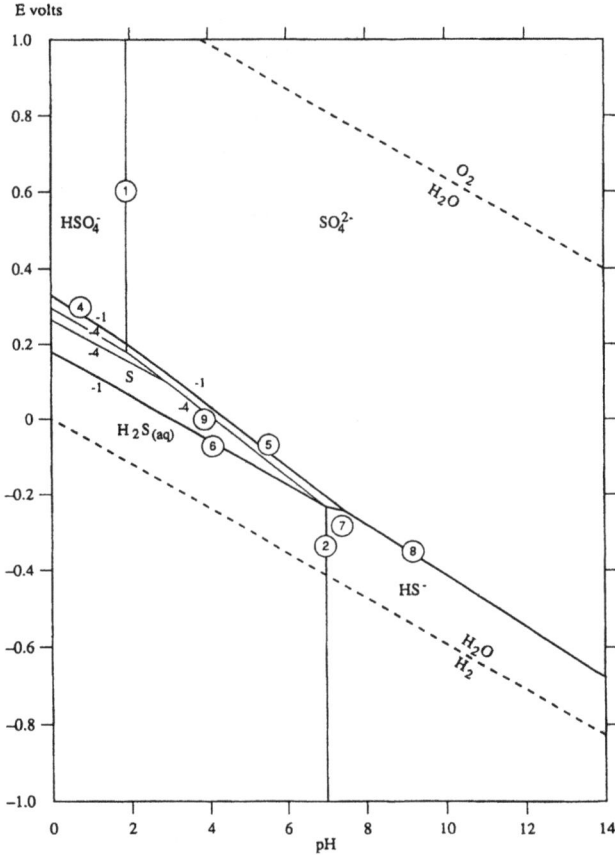

Fig. 3.2. E-pH diagram, S–H_2O system, 298.15 K, { } S ionic species and H_2S(aq) 10^{-1} and 10^{-4}, lines for both values are given and identified.

solution, however. As {H_2S(aq)} decreases E for Eq. (8) rises to more positive values at a given pH. As {SO_4^{2-}} or {HSO_4^-} decreases the values of E for Eqs. (4) and (5) fall to less positive values at a given pH. Thus the area of the sulphur region shrinks as can be seen from the lines for 0.1 and 10^{-4} M species in Fig. 3.2. Eventually the sulphur region disappears and there is an equilibrium line between H_2S(aq) and SO_4^{2-} or HSO_4^-.

The diagram indicates regions of thermodynamic stability for species. However, the rate of attainment of equilibrium between some species may be slow.

For example sulphur is quite stable in pure oxygenated water although it is oxidised to sulphate by certain microorganisms. Oxidation of dissolved sulphides to sulphate can take place fairly readily, but it is difficult to reduce sulphate to sulphide. Again, however, some microorganisms carry out this reduction and use it as a source of energy. The constraints which can be applied by kinetic factors to the attainment of chemical equilibrium can be severe but they can frequently be overcome by the use of high (by aqueous system standards) temperatures.

This discussion has been carried out in terms of molarities of reactants since these are used in practice and trends in behaviour are of interest here rather than accurate numerical values based on activities.

3.5.   The Copper – Sulphur – Water System, 298.15 K

3.5.1.   *Data Used*

The species considered and data used for sulphur are given in Table 3.2. The data for the species containing copper which are considered are given in Table 3.3.

Table 3.3.

Species	Cu^+	Cu^{2+}	CuO	Cu_2O	CuO_2^{2-}	$HCuO_2^-$	Cu_2S	CuS
$\Delta G°$ kJ	49.98	65.49	−129.7	−146.0	−183.6	−258.5	−86.2	−53.6

3.5.2.   *Equilibria*

$$Cu^{2+} + H_2O = CuO + 2H^+$$
$$-2 \text{ pH} - \log\{Cu^{2+}\} = -7.346 \tag{1}$$

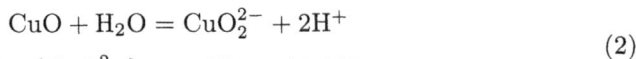

$$CuO + H_2O = CuO_2^{2-} + 2H^+$$
$$\log\{CuO_2^{2-}\} - 2 \text{ pH} = -32.095 \tag{2}$$

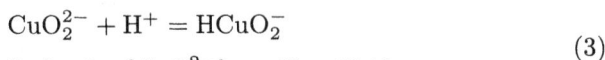

$$CuO_2^{2-} + H^+ = HCuO_2^-$$
$$\log\{HCuO_2^-\} - \log\{CuO_2^{2-}\} + \text{pH} = 13.12 \tag{3}$$

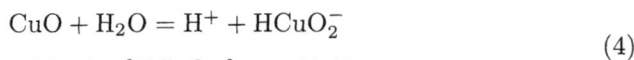

$$CuO + H_2O = H^+ + HCuO_2^-$$
$$-\text{pH} + \log\{HCuO_2^-\} = -18.98 \tag{4}$$

$$2\text{CuO} + 2\text{H}^+ + 2e = \text{Cu}_2\text{O} + \text{H}_2\text{O}$$
$$E = +0.641 - 0.0592\,\text{pH} \tag{5}$$

$$\text{Cu}_2\text{O} + 2\text{H}^+ + 2e = 2\text{Cu} + \text{H}_2\text{O}$$
$$E = +0.472 - 0.0592\,\text{pH} \tag{6}$$

$$\text{Cu}^{2+} + 2e = \text{Cu}$$
$$E = +0.339 + 0.0296\,\log\{\text{Cu}^{2+}\} \tag{7}$$

$$\text{Cu}^+ + e = \text{Cu}$$
$$E = +0.518 + 0.0592\,\log\{\text{Cu}^+\} \tag{8}$$

$$\text{Cu}^{2+} + e = \text{Cu}^+$$
$$E = +0.161 - 0.0592\,\log\{\text{Cu}^+\} + 0.0592\,\log\{\text{Cu}^{2+}\} \tag{9}$$

$$2\text{Cu}^{2+} + \text{H}_2\text{O} + 2e = \text{Cu}_2\text{O} + 2\text{H}^+$$
$$E = +0.207 + 0.0592\,\text{pH} + 0.0592\,\log\{\text{Cu}^{2+}\} \tag{10}$$

$$2\text{Cu}^{2+} + \text{HSO}_4^- + 7\text{H}^+ + 10e = \text{Cu}_2\text{S} + 4\text{H}_2\text{O}$$
$$E = +0.424 + 0.0118\,\log\{\text{Cu}^{2+}\} + 0.00592\,\log\{\text{HSO}_4^-\} - 0.0414\,\text{pH} \tag{11}$$

$$2\text{Cu}^{2+} + \text{SO}_4^{2-} + 8\text{H}^+ + 10e = \text{Cu}_2\text{S} + 4\text{H}_2\text{O}$$
$$E = +0.437 + 0.0118\,\log\{\text{Cu}^{2+}\} + 0.00592\,\log\{\text{SO}_4^{2-}\} - 0.0473\,\text{pH} \tag{12}$$

$$2\text{Cu}^{2+} + \text{H}_2\text{S(aq)} + 2e = \text{Cu}_2\text{S} + 2\text{H}^+$$
$$E = +0.981 + 0.0592\,\log\{\text{Cu}^{2+}\} + 0.0296\,\log\{\text{H}_2\text{S(aq)}\} + 0.0592\,\text{pH} \tag{13}$$

$$2\text{Cu}^{2+} + \text{HS}^- + 2e = \text{Cu}_2\text{S} + \text{H}^+$$
$$E = +1.188 + 0.0592\,\log\{\text{Cu}^{2+}\} + 0.0296\,\log\{\text{HS}^-\} + 0.0296\,\text{pH} \tag{14}$$

$$2\text{CuS} + 2\text{H}^+ + 2e = \text{Cu}_2\text{S} + \text{H}_2\text{S(aq)}$$
$$E = +0.071 - 0.0296\,\log\{\text{H}_2\text{S(aq)}\} - 0.0592\,\text{pH} \tag{15}$$

$$2CuS + H^+ + 2e = Cu_2S + HS^-$$
$$E = -0.171 - 0.0296 \log\{H_2S(aq)\} - 0.0296 \text{ pH}$$
(16)

$$Cu_2S + SO_4^{2-} + 8H^+ + 6e = 2CuS + 4H_2O$$
$$E = +0.389 + 0.00987 \log\{SO_4^{2-}\} - 0.0789 \text{ pH}$$
(17)

$$Cu_2S + HSO_4^- + 7H^+ + 6e = 2CuS + 4H_2O$$
$$E = +0.369 + 0.00987 \log\{HSO_4^-\} - 0.0691 \text{ pH}$$
(18)

$$2Cu + HSO_4^- + 7H^+ + 6e = Cu_2S + 4H_2O$$
$$E = +0.482 + 0.00987 \log\{HSO_4^-\} - 0.0691 \text{ pH}$$
(19)

$$2Cu + SO_4^{2-} + 8H^+ + 6e = Cu_2S + 4H_2O$$
$$E = +0.501 + 0.00987 \log\{SO_4^{2-}\} - 0.0789 \text{ pH}$$
(20)

$$Cu_2S + 2H^+ + 2e = H_2S(aq) + 2Cu$$
$$E = -0.303 - 0.0296 \log\{H_2S(aq)\} - 0.0592 \text{ pH}$$
(21)

$$2HCuO_2^- + 4H^+ + 2e = Cu_2O + 3H_2O$$
$$E = +1.764 + 0.0592 \log\{HCuO_2^-\} - 0.1184 \text{ pH}$$
(22)

$$2CuO_2^{2-} + 6H^+ + 2e = Cu_2O + 3H_2O$$
$$E = +2.540 + 0.0592 \log\{CuO_2^{2-}\} - 0.1776 \text{ pH}$$
(23)

$$CuO_2^{2-} + 4H^+ + 2e = Cu + 2H_2O$$
$$E = +1.506 + 0.0296 \log\{CuO_2^{2-}\} - 0.1184 \text{ pH}$$
(24)

$$Cu^{2+} + HSO_4^- + 7H^+ + 8e = CuS + 4H_2O$$
$$E = +0.404 + 0.0074 \log\{Cu^{2+}\} + 0.0074 \log\{HSO_4^-\} - 0.0518 \text{ pH}.$$
(25)

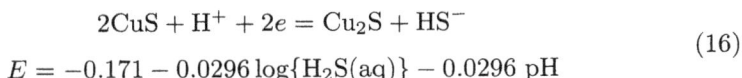

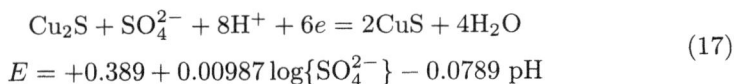

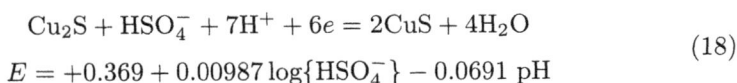

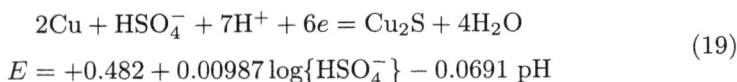

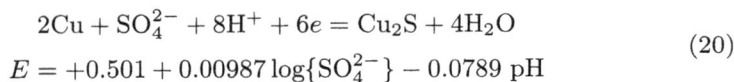

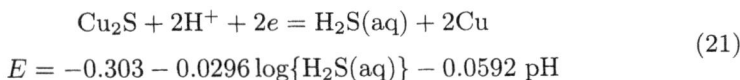

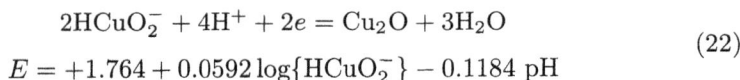

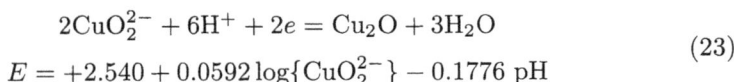

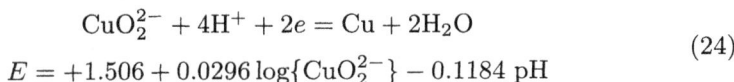

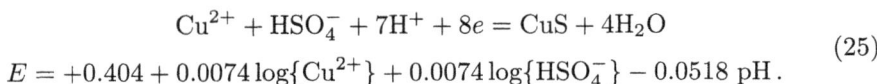

3.5.3.   *Discussion of the Diagram, Fig. 3.3*

The diagram, Fig. 3.3, is drawn for activities of sulphur-containing species in solution 0.1 and of copper-containing ions in solution 10^{-3} and 10^{-6} M.

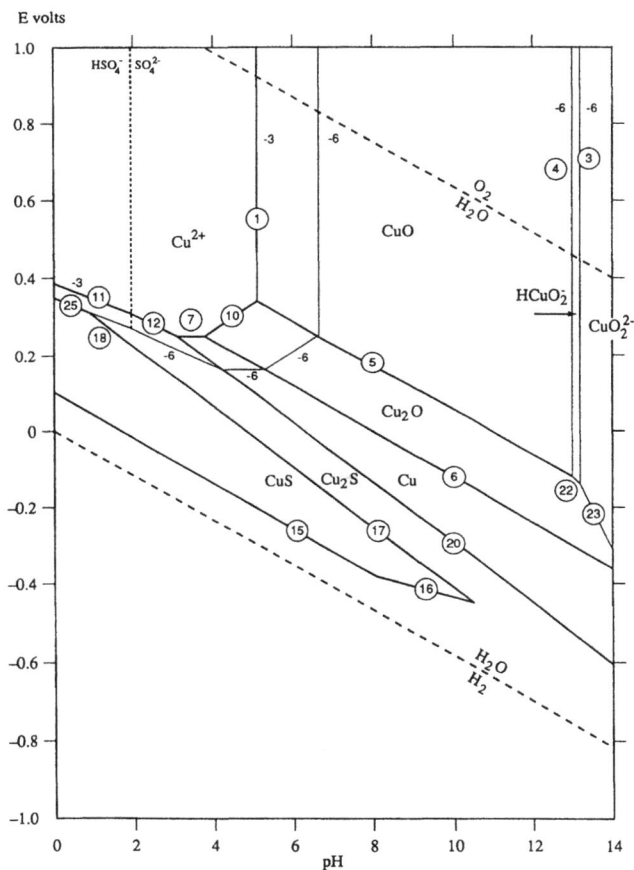

Fig. 3.3. *E*-pH diagram, Cu–S–H$_2$O system, 298.15 K, { } S ionic species and H$_2$S(aq) 10^{-1}, { } Cu ionic species 10^{-3} and 10^{-6}.

Thus it is drawn for a system in which excess sulphur is present. The diagram for the Cu–H$_2$O system having the same values of activities has the areas of stability of the sulphides absent and the boundaries between Cu^{2+} and Cu metal, equilibrium 7, extended to the axis where pH = 0. That is, CuS and Cu$_2$S occur in the region of the diagram in which the stable phase is the metal if sulphide is absent, with some extension into the Cu^{2+} region. The equilibrium between Cu^{2+} and H$_2$S to precipitate sulphide is considered further in Sec. 3.8.3. There is no region of stability of the Cu$^+$ ion in this diagram, thus at 298.15 K {Cu^+} is less than 10^{-6} M. Copper oxides are in equilibrium with low activities of ions derived from H_2CuO_2 at pH values of about 13 and above.

If the region of stability of elemental sulphur with {sulphur species} 10^{-1} shown in Fig. 3.2 were plotted on Fig. 3.3, it would lie entirely within the region of stability of CuS. Thus sulphur can coexist with CuS at 298.15 K in thermo-dynamic equilibrium, but not with Cu_2S. The diagram indicates clearly that under acidic oxidising conditions the minerals chalcocite, Cu_2S, and covellite, CuS, can in theory be leached to produce a solution containing Cu^{2+}, but not Cu^+, while the sulphur content is oxidised to sulphur or, at higher values of E, to sulphate. The requirement for this reaction to proceed is that the value of E in the solution is maintained at such a high value that essentially no H_2S or ions derived from it are present in the solution. The boundary between HSO_4^- and SO_4^{2-} is shown in Fig. 3.3 because the equilibrium line between Cu_2S and Cu^{2+}, equilibria 11 and 12, changes slope at that pH value as also does the line between Cu_2S and CuS, equilibria 17 and 18, although it is barely noticeable on the diagram.

3.6. Thermodynamics of Electrolyte Solutions at Elevated Temperatures

3.6.1. *Construction of Predominance Area Diagrams for Elevated Temperatures*

Construction of predominance area diagrams for a temperature substan-tially different from 298.15 K requires that ΔG° for each equilibrium reaction must be calculated using free energies of formation of the species at that temperature. These can be obtained by using tabulated data for $\Delta G^\circ_{298.15}$, $\Delta S^\circ_{298.15}$ and the heat capacities of the species, Sec. 2.3.2.

The most useful function for predicting the thermodynamic properties of electrolyte solutions at temperatures above 298 K is the molal heat capacity, C_p°, as a function of temperature. C_p° for a substance can usually be assumed to be constant over a temperature range of about 100 K if no phase change occurs. However the value of C_p° of a species in aqueous solution usually changes rapidly with temperature and a smaller interval of temperature is desirable.

The equations used to calculate ΔG° for species taking part in reactions substantially above 298.15 K are

$$\Delta H_T^o = \Delta H_{298.15}^o + \int_{298.15}^{T} \Delta C_p^o \, dT \qquad (2.54)$$

$$\Delta S_T^o = \Delta S_{298.15}^o + \int_{298.15}^{T} \frac{\Delta C_p^o dT}{T} \qquad (2.55)$$

and

$$\Delta G_T^o = \Delta H_T^o - T \Delta S_T^o.$$

If the value of C_p^o for a species changes significantly between 298.15 and T and if the change is not extreme then the average value between the two temperatures, $C_p^o/_{298.15}^T$ may be taken for the calculations and leads to

$$\Delta G_T^o = \Delta G_{298}^o - \Delta S^o(T - 298.15) + \Delta C_p^o/_{298}^T(T - 298.15)$$

$$- T\Delta C_p^o/_{298}^T \ln \frac{T}{298.15}. \qquad (3.32)$$

Values of $C_p^o/_{298}^{373}$ and of $C_p^o/_{298}^{423}$ are given in the Appendix for all of the species requiring consideration in preparing predominance area diagrams for the systems S–H_2O; Cu–S–H_2O and Fe–S–H_2O at 298.15 K together with the values of ΔG^o for each species at 423 K, calculated using Eq. (3.32) which for T reference $= 298$ K and $T = 423$ becomes

$$\Delta G_{423}^o = \Delta G_{298}^o - (125 \times \Delta S_{298}^o) + (\Delta C_p^o/_{298}^{423} \times 125)$$

$$- 423\Delta C_p^o/_{298}^{423} \times 0.3503.$$

It was pointed out by Ferreria[53] that if half-cell equations such as those used in Secs. 3.3–3.5 are used to draw predominance area diagrams for temperatures much higher than 298 K, incorrect results are obtained even if values of ΔG^o which have been calculated for the species at those temperatures are used. He had found no published diagram which had been prepared correctly. He attributed the difficulty to the fact that in a half-cell reaction the free energy of formation of the electrons which are included in the equations are not taken into account. In Sec. 3.3.2, when dealing with the equilibrium

$$Fe^{2+} + 2e = Fe$$

$$\Delta G^o = \Delta G_{Fe}^o - \Delta G_{Fe^{2+}}^o - 2\Delta G_e^o$$

the value of ΔG_e° was assumed temporarily to be zero. In Sec. 3.3.3, it was pointed out that the value of $E_{Fe^{2+},Fe}^{\circ} = -0.409$ V, calculated from ΔG° for the reaction, is the standard potential on the hydrogen scale because the reaction can be written

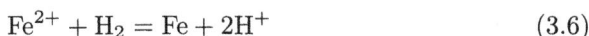

$$Fe^{2+} + H_2 = Fe + 2H^+ \qquad (3.6)$$

with $p_{H2} = 1$ atm. The electrons required to reduce Fe^{2+} have been supplied by H_2 which is itself reduced to $2H^+$. Ferreira proposed therefore that in calculations for diagrams for elevated temperatures full cell reactions including the hydrogen should be used. This is now the generally accepted practice.

The potentials calculated are on the hydrogen scale. In defining this ΔG° and ΔS° of formation for the H^+ ion are taken as zero at all temperatures and E° of the normal hydrogen electrode is considered to be zero at all temperatures. However, H_2 has entropy $S^{\circ}298.15 = 130.684$ kJ mol^{-1} K^{-1} and both H^+ and H_2 have heat capacities. By using thermodynamic data for H_2 and H^+ corrected to the elevated temperature being considered, correct diagrams are obtained. It should be noted that this results in the hydrogen scale being different at different temperatures.

3.6.2.   *The S–H_2O System,* **423 K**

The free energy data used are given in Table 3.4, taken from Appendix B

Table 3.4.

Species	S	HS$^-$	H$_2$S(aq)	SO$_4^{2-}$	HSO$_4^-$	H$_2$O(l)	H$^+$	H$_2$	O$_2$
ΔG° kJ, 423 K	-4.60	10.23	-48.9	-736.9	-770.6	-247.63	-3.20	-17.01	-26.45

In view of the uncertainty at present concerning the values of C_p° and S° of the sulphide ion S^{2-} and the fact that it does not appear on the diagram for 298.15 K, it is not considered here. The equations are numbered as in Sec. 3.4.2, but are written here as full cell reactions

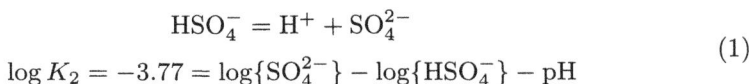

$$HSO_4^- = H^+ + SO_4^{2-}$$
$$\log K_2 = -3.77 = \log\{SO_4^{2-}\} - \log\{HSO_4^-\} - pH \qquad (1)$$

$$H_2S(aq) = H^+ + HS^-$$

$$\log K_1 = -6.91 = \log\{HS^-\} - \log\{H_2S(aq)\} - pH \tag{2}$$

$$HSO_4^- + H^+ + 3H_2 = 4H_2O + S$$

$$E_{423} = +0.294 - 0.0979\,pH + 0.0140\log\{HSO_4^-\} \tag{3}$$

$$SO_4^{2-} + 2H^+ + 3H_2 = 4H_2O + S$$

$$E_{423} = +0.347 - 0.1119\,pH + 0.0140\log\{HSO_4^-\} \tag{4}$$

$$S + H_2 = H_2S(aq)$$

$$E_{423} = +0.141 - 0.0420\log\{H_2S(aq)\} - 0.0839\,pH \tag{5}$$

$$S + H_2 = HS^- + H^+$$

$$E_{423} = -0.148 - 0.0420\log\{HS^-\} - 0.0420\,pH \tag{6}$$

$$SO_4^{2-} + H^+ + 4H_2 = HS^- + 4H_2O$$

$$E_{423} = +0.223 - 0.0944\,pH - 0.0105\log(\{HS^-\}/\{SO_4^{2-}\}) \tag{7}$$

$$SO_4^{2-} + 2H^+ + 4H_2 = H_2S(aq) + 4H_2O$$

$$E_{423} = +0.295 - 0.1049\,pH - 0.0105\log(\{H_2S(aq)\}/\{SO_4^{2-}\}) \tag{8}$$

$$O_2 + 2H_2 = 2H_2O$$

$$E_{423} = 1.127 - 0.0839\,pH + 0.0210\log p_{O_2} \tag{9}$$

$$2H^+ + H_2 = H_2 + 2H^+$$

$$E_{423} = 0 - \frac{0.0839}{2}\log p_{H_2} - 0.0839\,pH\,. \tag{10}$$

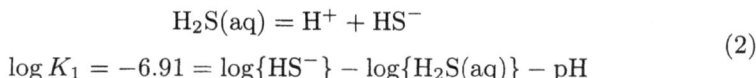

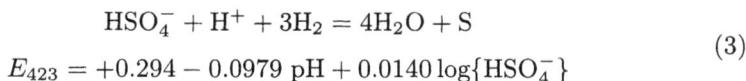

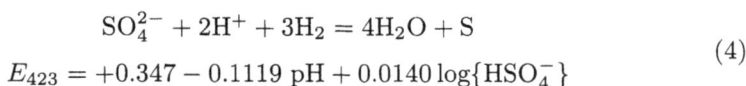

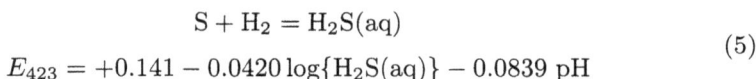

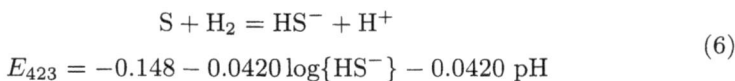

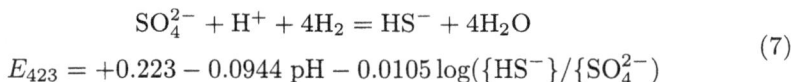

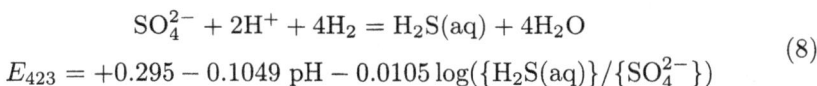

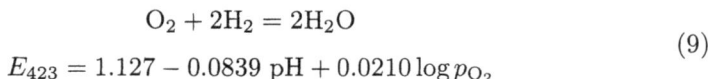

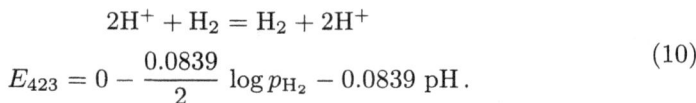

Notes on the Equations

At 423 K $\log K = -\Delta G^\circ/8.099$

$$\frac{2.303RT}{F} = 0.0839\,.$$

In this brief discussion of the derivation of the numerical equations for the reactions the half-cell reactions should be referred to also (Secs. 3.4.2 and

3.4.3). In Reaction (5), 6 electrons are shown in the half-cell equation. In the full cell equation these are derived from $3H_2$. This equivalence of 2 electrons from one H_2 molecule always holds. Oxidation of one H_2 produces $2H^+$ so that whereas in the half-cell equation for (4) $7H^+$ are involved, in the full cell equation only 1 H^+ is shown. However the $6H^+$ from the $3H_2$ molecules also take part in the reaction, giving the correct total of $7H^+$.

Thus in calculating $\Delta G°$ for a reaction the full cell equation is used. For Reaction (4) as an example

$$\Delta G° = (4 \times -247.63) + (-4.60) - (-736.9) - (2 \times -3.20)$$

$$- (3 \times -17.01) = -200.79$$

$$E_{423}° = +200.79/(96.487 \times z) = +2.081/z = +0.347 \text{ V}.$$

From Eq. (3.3)

$$E_{423} = 0.347 - \frac{RT}{zF} \ln \frac{\{\text{Reduced state}\}}{\{\text{Oxidised state}\}}$$

$$= +0.347 + 0.0140 \log\{SO_4^{2-}\} + 0.0140 \log\{H^+\}^8$$

since $z = 6$, $\{S\} = 1$, $\{H_2O\} = 1$ and $\{H_2\} = 1$ at 1 atmosphere pressure.

Equations (10) and (11) give the upper and lower stability limits for water. For the lower limiting line $E_{423}°$ is zero since (10) is the hydrogen electrode reaction. Thus E at pH $= 0$ is zero if p_{H_2} is 1 atm. The slope of the line is -0.0839 instead of -0.0592 at 273 K.

Discussion of the Diagram, Fig. 3.4

The diagram is drawn for activities of ionic species of sulphur and of $H_2S(aq)$ 10^{-1}. Two major points of interest become apparent when the diagrams for 298 K (Fig. 3.2) and 423 K are compared. The first is that at the higher temperature the area of predominance of HSO_4^- over SO_4^{2-} extends to higher pH values than at 298 K. The second is that at 423 K the area of stability of elemental sulphur is much smaller than at 298 K. Equilibrium (7), giving the boundary between HS^- and S is unimportant even at $\{10^{-1}\}$ and 298 K and the area of the sulphur region is controlled at higher temperatures and lower activities of sulphur species by equilibria (4), HSO_4^-–S; (5), SO_4^{2-}–S and (6) $H_2S(aq)$–S; Fig. 3.5. At a particular temperature, as the activity of HSO_4^- or SO_4^{2-} is lowered, the values of E for the equilibria fall to lower values.

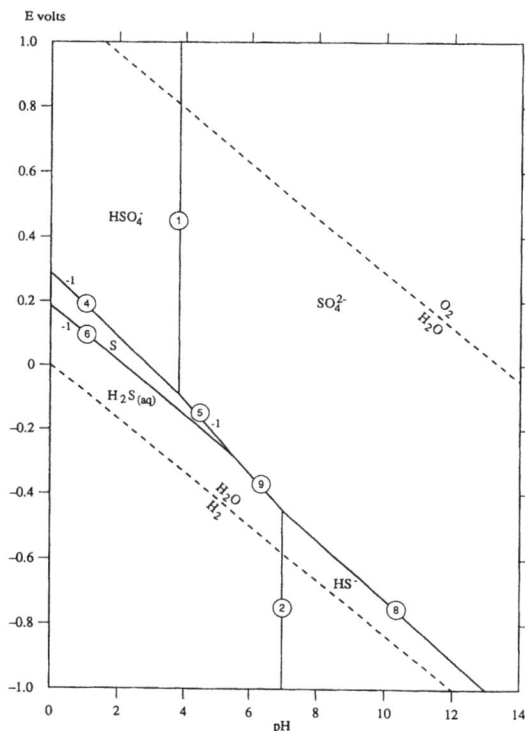

Fig. 3.4. E-pH diagram, S–H_2O system, 423 K, { } S ionic species and H_2S(aq) 10^{-1}.

However, as the activity of H_2S(aq) falls, the value of E rises. This is shown in Fig. 3.5 for 298 K and 423 K. At a particular pH and set of activities of the dissolved species, when E for the H_2S(aq)–S line (6) rises above that for SO_4^{2-} or HSO_4^- sulphur is not thermodynamically stable under those conditions. Whereas the ratio of HSO_4^- and SO_4^{2-} is controlled by the pH of the solution the activity of total sulphate and of H_2S(aq) can be controlled independently.

3.6.3.   *The Cu–S–H_2O System, 423 K*

The free energy data used are given in Table 3.5. The species S^{2-} is not considered.

   The E-pH diagram is given in Fig. 3.6 for activities of ionic sulphur species and H_2S(aq) 10^{-1}, and of ionic copper species 10^{-3}.

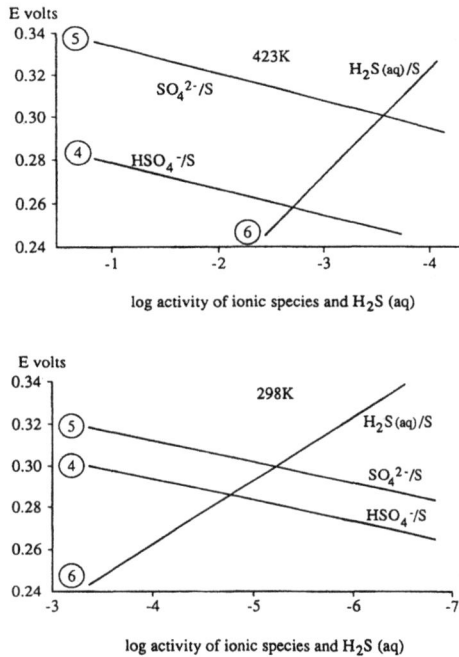

Fig. 3.5.   Effect of log activity of ionic species of sulphur and of H_2(aq) on the equilibria which control the area of stability of elemental sulphur at 298 K and 423 K.

Table 3.5.

Species	Cu	Cu$^+$	Cu^{2+}	CuO	Cu$_2$O	HCuO$_2^-$	CuO$_2^{2-}$	CuS	Cu$_2$S
ΔG° kJ, 423 K	-4.72	39.76	71.54	-136.4	-159.3	-252	-156	-63.03	-103.4

Species	S	HS$^-$	H$_2$S(aq)	SO$_4^{2-}$	HSO$_4^-$	H$_2$O(l)	H$^+$	H$_2$
ΔG° kJ, 423 K	-4.60	10.23	-48.9	-736.9	-770.6	-247.63	-3.20	-17.01

Half-cell reactions are given in Sec. 3.5.2. The full cell equations given here are numbered to correspond

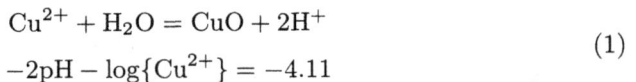

$$Cu^{2+} + H_2O = CuO + 2H^+$$
$$-2pH - \log\{Cu^{2+}\} = -4.11$$

(1)

$$CuO + H_2O = CuO_2^{2-} + 2H^+$$
$$\log\{CuO_2^{2-}\} - 2pH = -33.79 \tag{2}$$

$$CuO_2^{2-} + H^+ = HCuO_2^-$$
$$\log\{HCuO_2^-\} - \log\{CuO_2^{2-}\} + pH = 11.46 \tag{3}$$

$$CuO + H_2O = H^+ + HCuO_2^-$$
$$-pH + \log\{HCuO_2^-\} = -15.91 \tag{4}$$

$$2CuO + H_2 = Cu_2O + H_2O$$
$$E_{423} = +0.602 - 0.0839\ pH \tag{5}$$

$$Cu_2O + H_2 = 2Cu + H_2O$$
$$E_{423} = +0.419 - 0.0839\ pH \tag{6}$$

$$Cu^{2+} + H_2 = Cu + 2H^+$$
$$E_{423} = +0.340 + 0.0420\log\{Cu^{2+}\} \tag{7}$$

$$Cu^+ + \frac{1}{2}H_2 = Cu + H^+$$
$$E_{423} = +0.406 + 0.0839\log\{Cu^+\} \tag{8}$$

$$Cu^{2+} + \frac{1}{2}H_2 = Cu^+ + H^+$$
$$E_{423} = +0.274 - 0.0839\log\{Cu^+\} + 0.0839\log\{Cu^{2+}\} \tag{9}$$

$$2Cu^{2+} + H_2O + H_2 = Cu_2O + 4H^+$$
$$E_{423} = +0.262 + 0.0839\log\{Cu^{2+}\} + 0.0839\ pH \tag{10}$$

$$2Cu^{2+} + HSO_4^- + 5H_2 = Cu_2S + 4H_2O + 3H^+$$
$$E_{423} = +0.405 + 0.0168\log\{Cu^{2+}\} + 0.00839\log\{HSO_4^-\} - 0.0587\ pH \tag{11}$$

$$2Cu^{2+} + SO_4^{2-} + 5H_2 = Cu_2S + 4H_2O + 2H^+$$
$$E_{423} = +0.437 + 0.0168\log\{Cu^{2+}\} + 0.00839\log\{SO_4^{2-}\} - 0.0671\ pH \tag{12}$$

$$2Cu^{2+} + H_2S(aq) + H_2 = Cu_2S + 4H^+$$
$$E_{423} = +0.982 + 0.0839\log\{Cu^{2+}\} + 0.0420\log\{H_2S(aq)\} + 0.0839\,pH$$
(13)

$$2Cu^{2+} + HS^- + H_2 = Cu_2S + 3H^+$$
$$E_{423} = +1.292 + 0.0839\log\{Cu^{2+}\} + 0.0420\log\{HS^-\} + 0.0420\,pH$$
(14)

$$2CuS + H_2 = Cu_2S + H_2S(aq)$$
$$E_{423} = +0.048 - 0.0420\log\{H_2S(aq)\} - 0.0839\,pH$$
(15)

$$2CuS + H_2 = Cu_2S + HS^- + H^+$$
$$E_{423} = -0.242 - 0.0420\log\{HS^-\} - 0.0420\,pH$$
(16)

$$Cu_2S + SO_4^{2-} + 2H^+ + 3H_2 = 2CuS + 4H_2O$$
$$E_{423} = +0.378 + 0.01398\log\{SO_4^{2-}\} - 0.1119\,pH$$
(17)

$$Cu_2S + HSO_4^- + H^+ + 3H_2 = 2CuS + 4H_2O$$
$$E_{423} = +0.320 + 0.01398\log\{HSO_4^-\} - 0.0979\,pH$$
(18)

$$2Cu + HSO_4^- + H^+ + 3H_2 = Cu_2S + 4H_2O$$
$$E_{423} = +0.449 + 0.01398\log\{HSO_4^-\} - 0.0979\,pH$$
(19)

$$2Cu + SO_4^{2-} + 2H^+ + 3H_2 = Cu_2S + 4H_2O$$
$$E_{423} = +0.501 + 0.01398\log\{SO_4^{2-}\} - 0.1119\,pH$$
(20)

$$Cu_2S + H_2 = H_2S(aq) + 2Cu$$
$$E_{423} = -0.322 - 0.0420\log\{H_2S(aq)\} - 0.0839\,pH$$
(21)

$$2HCuO_2^- + 2H^+ + H_2 = Cu_2O + 3H_2O$$
$$E_{423} = +1.942 + 0.0839\log\{HCuO_2^-\} - 0.1678\,pH$$
(22)

$$2CuO_2^{2-} + 4H^+ + H_2 = Cu_2O + 3H_2O$$
$$E_{423} = +2.904 + 0.0839\log\{CuO_2^{2-}\} - 0.2517\,pH$$
(23)

$$CuO_2^{2-} + 2H^+ + H_2 = Cu + 2H_2O$$

$$E_{423} = +1.661 + 0.0420 \log\{CuO_2^2\} - 0.1678 \text{ pH} \tag{24}$$

$$Cu^{2+} + HSO_4^- + 4H_2 = CuS + 4H_2O + H^+$$

$$E_{423} = +0.375 + 0.01049 \log\{Cu^{2+}\} + 0.01049 \log\{HSO_4^-\} - 0.0734 \text{ pH} \tag{25}$$

$$2Cu^+ + H_2O = Cu_2O + 2H^+$$

$$-2 \text{ pH} - 2\log\{Cu^+\} = -0.298 \tag{26}$$

$$2Cu^+ + HSO_4^- + 4H_2 = Cu_2S + 4H_2O + H^+$$

$$E_{423} = +0.438 + 0.01049 \log\{HSO_4^-\} + 0.02098 \log\{Cu^+\} - 0.0734 \text{ pH} \tag{27}$$

$$(2Cu^+ + HSO_4^- + 7H^+ + 8e = Cu_2S + H_2O).$$

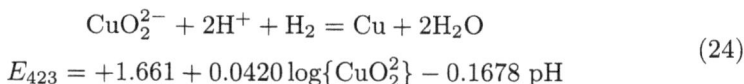

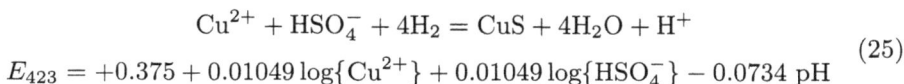

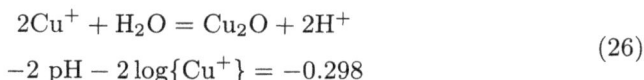

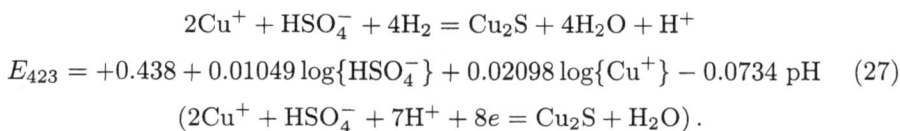

Discussion of the Diagram, Fig. 3.6

The diagram is drawn for activities of sulphur-containing ions 0.1 and of copper-containing ions 10^{-3} M. At this activity $HCuO_2^-$ is in equilibrium with CuO at pH 12.91 but CuO_2^{2-} does not appear.

The most important difference between this diagram for 423 K and that for 298 K (Fig. 3.3) is that whereas the Cu^+ ion does not appear at $\{Cu^+\} = 10^{-6}$ at the lower temperature, it has a region of stability with $\{Cu^+\} = 10^{-3}$ at 423 K. The Cu^+ region of the diagram is expanded in Fig. 3.7 which also shows as broken lines the equilibria between HSO_4^-, sulphur and $H_2S(aq)$ at 423 K. The Cu^+ area is bounded by four equilibrium lines; with Cu^{2+}, Cu_2O, copper metal and Cu_2S. The last of these, Reaction (27), forms Cu_2S by reduction of the ion HSO_4^-. If the activity of sulphur-containing ions is very low in the system or, alternatively, reduction of HSO_4^- is very slow, the Cu^+ ion predominance area will be bounded at the top and bottom by the Cu^{2+} ion and copper metal areas and on the right by the solubility product of Cu_2O. The pH limits for activities of Cu^+, 10^{-2}, 10^{-1} and 1 are 2.15, 1.15 and 0.15. Whether or not the Cu^+ ion has any area of predominance at all depends on whether copper metal on oxidation forms Cu^{2+} ions or Cu^+ ions at a particular activity as the redox potential is raised. At 298 K with $\{Cu^+\} = \{Cu^{2+}\} = 10^{-6}$ Cu^{2+} forms at a lower value of E than does Cu^+. At 423 K, Cu^+ forms first with activities 10^{-3}.

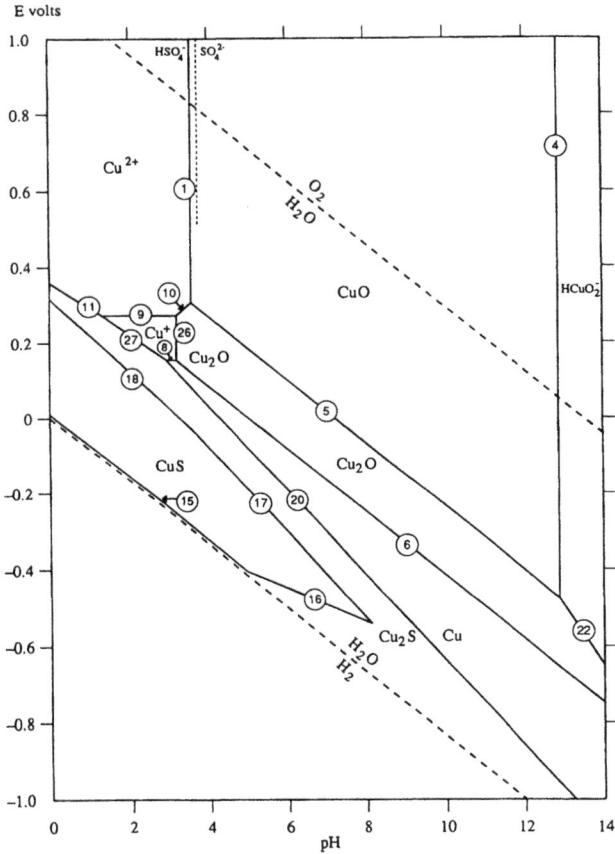

Fig. 3.6. E-pH diagram, Cu–S–H$_2$O system, 423 K, { } S ionic species and H$_2$S(aq) 10^{-1}, { } Cu ionic species 10^{-3}.

It is conventional in predominance area diagrams to set the activities of ions containing a particular element at the same value. In practice this is not a common situation. It is of interest to know the thermodynamically limiting activity of Cu$^+$ in the presence of a given activity of Cu^{2+} at 423 K. This can be calculated by setting the values of E for Reactions (7) and (8) equal. Thus

$$0.340 + 0.0420 \log\{Cu^{2+}\} = 0.406 + 0.0839 \log\{Cu^+\}$$

$$0.0420 \log\{Cu^{2+}\} - 0.0839 \log\{Cu^+\} = 0.066\,.$$

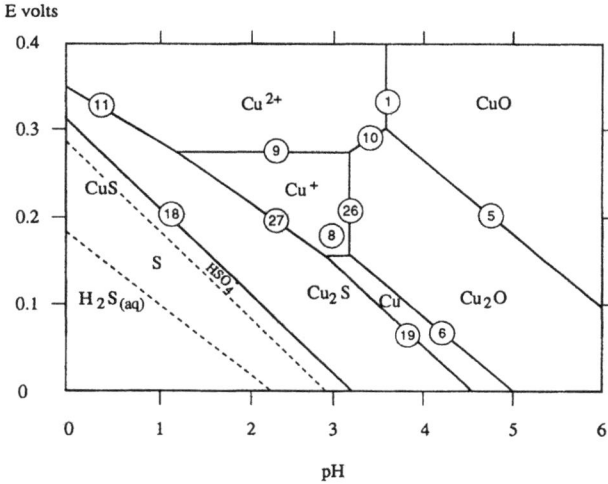

Fig. 3.7.   Expansion of scale of area of Fig. 3.6. in which the species Cu^+ is stable.

This gives

$\log\{Cu^{2+}\}$	-3	-2	-1	0
$\log\{Cu^+\}$	-2.29	-1.79	-1.29	-0.79
$\{Cu^+\}$	5.13×10^{-3}	1.62×10^{-2}	5.13×10^{-2}	1.62×10^{-1}.

Plotting $\log\{Cu^+\}$ against $\log\{Cu^{2+}\}$ gives a straight line of slope 0.5 because

$$\log\{Cu^{2+}\} - 2\log\{Cu^+\} = 0.066/0.042 = 1.57.$$

The intercept value of $\log\{Cu^{2+}\}$ at $\log\{Cu^+\} = 0$ is 1.57.
   A similar calculation for 298 K gives

$$0.339 + 0.0296\log\{Cu^{2+}\} = 0.518 + 0.0592\log\{Cu^+\}$$

$$0.0296\log\{Cu^{2+}\} - 0.0592\log\{Cu^+\} = 0.179$$

$\log\{Cu^{2+}\}$	-3	-2	-1	0
$\log\{Cu^+\}$	-4.52	-4.02	-3.52	-3.02
$\{Cu^+\}$	3.02×10^{-5}	9.56×10^{-5}	3.02×10^{-4}	9.56×10^{-4}

$$\log\{Cu^{2+}\} - 2\log\{Cu^+\} = 0.179/0.0296 = 6.04\,.$$

These calculations indicate that in a solution in which cuprous ions are not strongly complexed, sulphate solution for example, at 298 K the maximum Cu^+ ion activity in a solution of cupric sulphate up to 1 activity molal, will be about 10^{-3}. Since $\gamma_\pm$ for $CuSO_4$ is 0.062 in a 0.5 molal solution and 0.15 in 0.1 molal solution (Table 3.8), the achievable value is much lower than this. At 423 K, however free cuprous ions must be considered to be potentially available as a reactant species.

The main difference between the areas of predominance of the copper sulphides at 298 K and 423 K is that CuS is stable only in solutions with pH below about 8 at 423 K while at 298 K it extends to pH about 11. It should be noted that at both temperatures Cu_2S is formed by reduction of HSO_4^-;[11] SO_4^{2-},[12,20] or CuS.[15,16] CuS is formed from Cu_2S with SO_4^{2-} or HSO_4^- providing additional sulphide[17,18] and, at 298 K by reduction of HSO_4^- and precipitation of Cu^{2+}. It is very well known however that copper, like many other elements, is precipitated from process liquors by the addition of sulphide, often in the form of NaHS. This reaction is indicated on the copper diagrams since the sulphides are stable in the $H_2S(aq)$ and HS^- areas and CuS can coexist with elemental sulphur although Cu_2S cannot. The diagrams show that if a cupric sulphate solution having $\{Cu^{2+}\}10^{-3}$ or 10^{-6} and $\{SO_4^{2-}\}$ or $\{HSO_4^-\}10^{-1}$ is subjected to a progressively more reducing environment, a sulphide of copper is produced by reduction of the sulphate at a higher value of E than is required to reduce the sulphate to sulphur or to $H_2S(aq)$ or HS^-. This is due to the thermodynamic stability of the sulphides and their low solubility products, Sec. 3.8.3.

3.6.4.   *The Fe–S–H_2O System, 423 K*

A diagram for this system is given in Fig. 3.8, reproduced from that drawn by Ferreira.[53] Equilibria such as those involving pyrite have not been used above and as an example that between Fe_3O_4, SO_4^{2-} and FeS_2 is given in both half-cell and full cell form:

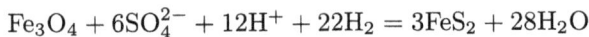

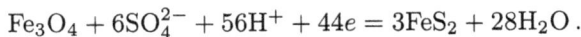

$$Fe_3O_4 + 6SO_4^{2-} + 12H^+ + 22H_2 = 3FeS_2 + 28H_2O$$

$$Fe_3O_4 + 6SO_4^{2-} + 56H^+ + 44e = 3FeS_2 + 28H_2O\,.$$

Pyrite, FeS_2, contains the S_2^{2-} group (Sec. 1.2.4).

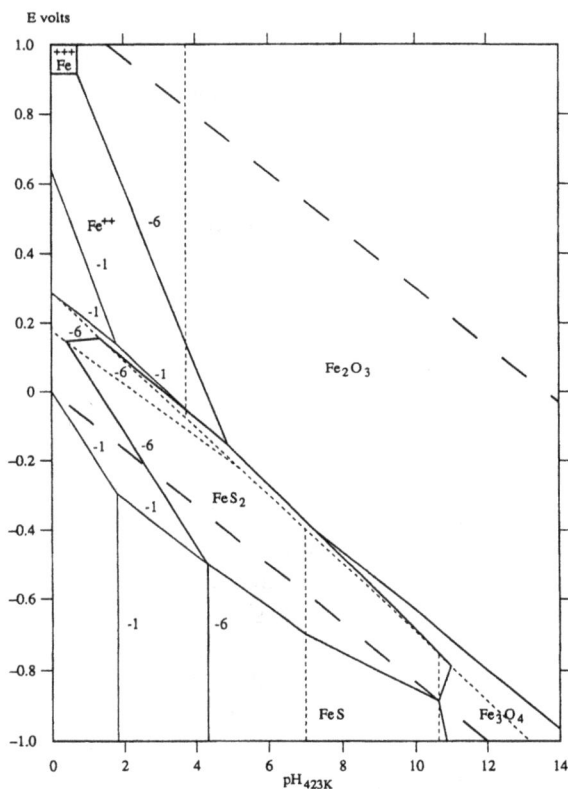

Fig. 3.8. E-pH diagram, Fe–S–H_2O system, 423 K, { } S ionic species and H_2S(aq) 10^{-1}, { } Fe ionic species 10^{-1} and 10^{-6}. Finely broken lines relate to the S–H_2O system, see Fig. 3.4.

The data used to produce Fig. 3.8 are given in Appendix B.

3.6.5.   *Estimation of Entropies and Heat Capacities of Ionic Species*

Free energies of formation of substances and species involved in reactions at temperatures substantially different from 298 K are calculated using Eq. (3.32) in Sec. 3.6.1. The last two terms involve ΔC_p^o and tend to cancel out one another so that at temperatures close to 298 K the change in ΔG^o for a reaction is not very sensitive to ΔC_p^o. When dealing with a heterogeneous equilibrium between pure substances at a high temperature, Eq. (3.32) can be used to estimate the change in ΔG° over a thousand degrees or more because the heat capacities for such substances are small, change slowly and predictably with temperature and so can be estimated accurately if necessary. This is not the case for species in aqueous solution, particularly for electrolytes. Values of C_p^o for these can be very large, negative numbers which change very rapidly with temperature. Most solutes for which data are available have a negative heat capacity at 298 K, rising to a maximum between about 320 and 373 K before falling rapidly at higher temperatures, particularly at low molalities. At 573 K, the value may be more negative than -1000 JK^{-1} mol^{-1}. Because of the existence of the maximum in the curve, estimation of the change in ΔG° for a solute may be quite successful if the value of C_p^o at 298 K is used, at least up to 423 K. However, use of the mean value between 298 K and the temperature for which the calculation is being carried out is preferable, if data are available.

Entropies of ionic and other species and substances are also required if Eq. (3.32) is to be used. Ionic entropies at high temperatures can be used to derive partial molal heat capacities for almost any species and during the 1950's some success was achieved in relating partial molal ionic entropies of species to structural parameters such as charge, size, mass and geometry.[54] Although these correlations are not soundly based on theory they permit the estimation of entropies for a wide variety of ions at 25°C with sufficient accuracy to give rise to only small errors in free energy calculations.

Criss and Cobble[55,56] developed their Correspondence Principle after noting that experimentally derived entropy data for higher temperatures indicated that the functional dependencies appeared to be the same as those for the correlations observed at 25°C. They suggested that it is not necessary to know what these dependencies are, but that the entropy of an ion which has no internal degrees of freedom at temperature T is given by some function of the ionic charge z, the dielectric constant (relative permittivity) of the solvent ε_r, the mass m and radius r of the ion and any other variable considered to be of significance

$$S^o_{(T)} = f(z, \varepsilon, r, m, T, \text{etc.}).$$

Some of these functions will be largely independent of the nature of the ion and will depend only on the choice of standard state, solvent and temperature. The equation for $S^\circ(T)$ was, therefore written

$$S^o_{(T)} = g(T, \text{solvent}) + f(\text{ion}) + [g(T, \text{solvent}) \times f(\text{ion})]. \tag{3.33}$$

The third, cross term was included to allow for the possibility that some of the variables may be interdependent. It was assumed that Eq. (3.33) can be expanded about a reference entropy $S^o_{(T_1)}$ at temperature T_1 , so that

$$S^o_{(T_2)} = a_{(T_2)} + b_{(T_2)}S^o_{(T_1)} + c_{(T_2)}(S^o_{(T_1)})^2 + d_{(T_2)}(S^o_{(T_1)})^3 + \cdots . \qquad (3.34)$$

This is the general correspondence relationship and it was shown that the first two terms of the equation were adequate to express the entropy data known in 1964 for ions in water up to 200°C.

Ions were assigned to one of four classes; simple cations, simple anions including OH^-, oxyanions XO_n^{m-}, and acid oxyanions $XO_n(OH)_1^{m-}$. A set of ionic entropies was constructed for each temperature chosen, 60°C, 80°C and 150°C, by fixing the entropy of $H^+(aq)$ at each temperature. This set a standard state for each temperature. The entropy values assigned to $H^+(aq)$ were as follows.

Temperature°C	25	60	100	150
S^o J mol^{-1} deg^{-1}	-20.92	-10.46	8.37	27.20 .

They were chosen by using the first two terms of Eq. (3.34) with experimentally determined values of the entropies of a particular class of ions to obtain a linear relationship between $S°298.15$ and $S^o_{(T_2)}$. With other values linear relationships were not obtained. Because the value assigned at 25°C lies within the range of values (-8.8 to -26.4 J mol^{-1} deg^{-1}) suggested by a number of workers for the "absolute" ionic entropy of $H^+(aq)$, Criss and Cobble proposed that the values of ionic entropy they used should be regarded as being on the "absolute" scale. Their scale had to be distinguished from that used in thermodynamic calculations.

In use the correspondence principle can be described by the general relationship

$$S^o_{T_2}(\text{abs}) = a_{T_2} + b_{T_2}S_{298}(\text{abs}) \qquad (3.35)$$

where a_{T_2} and b_{T_2} are constants dependent on the class of ions; cations, anions, oxyanions and acid oxyanions, and on the temperature considered. $S^o_{298}(\text{abs})$ refers to the ionic partial molal entropies on the "absolute" scale which is in fact not absolute, considering the selected standard state at 298 K so that

$$S^o_{298}(\text{abs}) = S^o_{298}(\text{conventional}) - 20.92z \qquad (3.36)$$

where z is the ionic charge. The values of the constants a and b in Eq. (3.35) are given[55,56] in Table 3.6. Note that the values relate to entropies in calories mol^{-1} deg^{-1} and to use them values of S_{298}^o in J mol^{-1} deg^{-1} should be converted to calories (1 cal = 4.184 J exactly) and $S_{(T_2)}^o$ calculated reconverted to joules.

Table 3.6. Entropy constants for Eq. (3.35) in cal mol^{-1} deg^{-1}.

Temp °C	Simple cations		Simple anions and OH$^-$		Oxyanions		Acid oxyanions XO$_n$(OH)$_1^{m-}$	
	a_T	b_T	a_T	b_T	a_T	b_T	a_T	b_T
25	0	1.000	0	1.000	0	1.00	0	1.000
60	3.9	0.955	−5.1	0.969	−14.0	1.217	−13.5	1.380
100	10.3	0.876	−13.0	1.000	−31.0	1.476	−30.3	1.894
150	16.2	0.792	−21.3	0.989	−46.4	1.687		

The average value of the heat capacity between two temperatures T and 298, $C_p^o/_{298}^T$ used in Eq. (3.32) can be calculated if the values of C_p^o, S^o and ΔG^o at 298 K are known and also S^o at temperature T. Since S_T^o can now be calculated using the correspondence principle the average values of heat capacities of ions can be calculated between 25°C and 60°C, 100°C and 150°C.

$$C_p^o/_{298}^T = \frac{S_T^o - S_{298}^o}{\ln(T/298)}. \tag{3.37}$$

From Eqs. (3.35) and (3.36), it follows that

$$C_p^o/_{298}^T = \frac{a_T - S_{298}^o[1.000 - b_T]}{\ln(T/298)} \tag{3.38}$$

which can be rewritten

$$C_p^o/_{298}^T = \alpha_T + \beta_T S_{298}^o \tag{3.39}$$

where

$$\alpha_T = a_T/\ln(T/298)$$

and

$$\beta_T = -[1.000 - b_T]/\ln(T/298).$$

It is emphasised that the values of S^o used in Eqs. (3.35) and (3.37)–(3.39) are on the "absolute" scale and before use in thermodynamic calculations must be converted to the conventional scale using (3.36) with the value of S^o for $H^+(aq)$ at the temperature being considered. The heat capacity constants α_T and β_T corresponding to the entropy constants in Table 3.6 are given in Table 3.7.

Table 3.7. Heat capacity constants for Eq. (3.39) in cal mol^{-1} deg^{-1}.

Temp °C	Simple cations		Simple anions and OH$^-$		Oxyanions		Acid oxyanions		H^{+*}
	α_T	β_T	α_T	β_T	α_T	β_T	α_T	β_T	$C_p^o/_{298}^T$
60	35	−0.41	−46	−0.28	−127	1.96	−122	3.44	23
100	46	−0.55	−58	0.000	−138	2.24	−135	3.97	31
150	46	−0.59	−61	−0.03	−133	2.27	–	–	33

*Entries in the H^+ column refer to the average value of the heat capacity for $H^+(aq)$ as derived from fixing the entropy of $H^+(aq)$ at each temperature (cal mol^{-1} deg^{-1}).

A table of ionic partial molal heat capacities has been given[55,56] for temperature increments 25–60°C, 25–100°C, 25–150°C and 25–200°C. Values of entropy constants and of heat-capacity parameters were also included for 200°C in the tables given above as 3.6 and 3.7. It is unlikely that these can be regarded as providing reliable guidance to the behaviour of reacting aqueous systems at 200°C because of the rapid increase in values of C_p^o for electrolytes in aqueous solutions in this temperature range, referred to above.

3.7. Precipitation of Metals from Solution by Reduction

3.7.1. *Introduction*

Predominance area diagrams show the conditions of E and pH at which metallic elements are stable. A line respresenting an equilibrium between a metal and one of its ions shows the potential at which there is thermodynamic equilibrium between the metal and the ion present in solution at the given activity. In order to reduce the ion to form metal, the potential must be more negative, or less positive, than that value in order to provide a driving force. In the case of iron the metal is stable in an aqueous environment only at values of E outside the region of stability of water, Fig. 3.1. Thus, writing the full

cell reaction

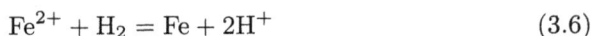

$$Fe^{2+} + H_2 = Fe + 2H^+ \tag{3.6}$$

the metal cannot be formed by driving it from left to right because the effort to do so will cause water to decompose with evolution of hydrogen. Zinc is even more electronegative with $E^o_{Zn^{2+}}$, $Zn = -0.762$ volt compared with $E^o_{Fe^{2+},Fe} = -0.409$ volt, yet zinc metal can be produced by electrolysis of zinc sulphate solution. Application of the necessary electrical potential across the cathode and anode does not cause decomposition of the water because hydrogen gas does not form nuclei on a zinc surface at that potential and so the gas is not evolved. This is a case of a reaction, electrolytic dissociation of water, not occurring as thermodynamics predicts it should, because of a kinetic factor, absence of a mechanism for forming hydrogen nuclei on a zinc metal surface. Lack of, or slow attainment of equilibrium in the sulphur–water system has been referred to previously in Sec. 3.4.4. The value of $E^o_{Cu^{2+},Cu}$ is $+0.339$ volt, Sec. 3.5 Eq. (7), and the metal is extensively produced by electrowinning from solutions and by electrorefining copper anodes made from metal produced in smelters.

3.7.2.   *Deposition of Metals by Electrolysis*

When electrowinning copper from acidic solutions an inert electrode is used and the overall reaction at the anode is evolution of oxygen. At 1 atm pressure the reversible electrode potential is

$$E_a = 1.229 - 0.0592 \, pH \,. \tag{3.28}$$

The reversible potential for reduction of Cu^{2+} to metal at the cathode is given by Reaction (7) in Sec. 3.5.2

$$E_c = +0.339 + 0.0296 \log\{Cu^{2+}\} \,.$$

When these reactions are used to produce copper they are driven by the application of a voltage, E_{cell}, across the electrodes, the rate at which the metal deposits on the cathode depending on the current passing which in turn is controlled by the magnitude of E_{cell}. This can be broken down as follows

$$E_{cell} = (E_a - E_c) + (E_{oa} - E_{oc}) + IR + E_R \,.$$

$(E_a - E_c)$ is the voltage at which deposition and redissolution of copper are in equilibrium. The term $(E_{oa} - E_{oc})$ expresses the overpotentials at the anode and cathode due to concentration polarization at the electrodes. These are the potentials necesary to overcome gradients in the concentrations of the ions reacting at the electrode surfaces and being removed from the solution. Concentration polarization increases as the rate of metal deposition is increased by raising E_{cell} and in consequence the current density (amps metre^{-2} of electrode surface) and depends also on the rate and direction of movement of the electrolyte with respect to the electrode surface. E_{oa} and E_{oc} depend on the composition of the electrolyte. Other factors which may affect the magnitude of this term are sometimes referred to as activation overvoltages and attributed to some slow step in the electrode reactions. IR is a measure of the electrical resistance to the passage of the current through the electrolyte and E_R is the additional potential required to produce the required rate of reaction. Included in E_R is a small potential required to overcome contact resistance at anode and cathode connections to the busbar carrying the electric power, and losses in the leads. In copper electrowinning the terms in the equation for E_{cell} are

$$E_{cell} = (1.19 - 0.30) + (0.5 - 0.05) + 0.5 + 0.1 = 1.94 \text{ V}.$$

When copper or nickel metal anodes are electrorefined $(E_a - E_c)$ is approximately zero. In commercial practice the anodic overvoltage in the case of nickel is about $+0.3$ V, the cathodic overvoltage about -0.2 V giving $(E_{oa} - E_{oc}) = +0.5$ V. $(IR + E_R) = +0.9$ V so that the cell voltage for electrorefining of nickel metal is around 1.4 V.

Nickel is also produced by electrowinning from nickel sulphide anodes. These are essentially Ni_3S_2 cast from a liquid matte and cooled under carefully controlled conditions to obtain satisfactory mechanical strength. Since the nickel and metal impurities are present in the solid as ions the principal reaction at the anode is oxidation of S^{2-} in the solid sulphide to elemental sulphur. $\Delta G°$ for $Ni^{2+} = -45.6$ and for $Ni_3S_2 = -197.1$ kJ mol^{-1} so that the half-cell reaction

$$2S + 3Ni^{2+} + 6e = Ni_3S_2$$

has $E° = +0.104$ V. The principal reaction at the cathode is reduction of Ni^{2+} to metal; $E°_{Ni^{2+},Ni} = -0.236$ so that $(E_a° - E_c°) = 0.34$ V. In commercial practice the anodic overvoltage is $+1.1$ V and the cathodic overvoltage -0.2 V giving $(E_{oa} - E_{oc}) = 1.3$ V. $(IR + E_R) = +0.9$ V so that the working voltage is around 2.54 V. Contact resistance between the busbars carrying the electric power and the electrodes requires an additional voltage and actual tank values for freshly introduced anodes are about 2.8 V. As the anodes corrode the anode sludge forms a porous coating which eventually doubles their bulk and the tank voltage must be gradually increased to about 4 V near the end of the life of a 5 cm thick electrode.

Clearly although the thermodynamics of the reactions which occur at the electrodes in electrowinning or electrorefining a metal are of importance a number of additional factors greatly affect the cell voltage required. The most important economic requirement is high current efficiency coupled with high cathode quality.

Some metals can also be precipitated from solution by a cementation reaction, using a more electropositive metal as the reducing agent. This kind of reaction, using iron, is also employed commercially.

3.7.3. *Precipitation of Metals by Hydrogen Gas*

The reaction

$$Cu^{2+} + H_2 = Cu + 2H^+$$

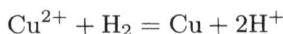

should take place at 25°C with $p_{H_2} = 1$ atm but it does not do so. Metallic copper can be produced by reaction of cupric sulphate solution with hydrogen but elevated temperatures and pressures are required. The reaction is not used commercially because it is self-nucleating, that is fresh metal nuclei are produced throughout the period during which the reduction is taking place and therefore the particle size of the product cannot be controlled. Some metals are produced commerically by reduction of aqueous solutions of their salts by hydrogen however, particularly nickel and cobalt.

In such a process the concentration of metal in the feed solution is typically about 1 M and reduction is continued until the concentration is about 0.1 M or less. As the concentration falls so also does the thermodynamic activity and in consequence the driving force necessary increases. Predominance area diagrams are not convenient for representing such changes and the kind of diagram usually used is shown in Fig. 3.9. This has been drawn for the sulphates of the metals using the mean activity coefficients given in Table 3.8.[7,8]

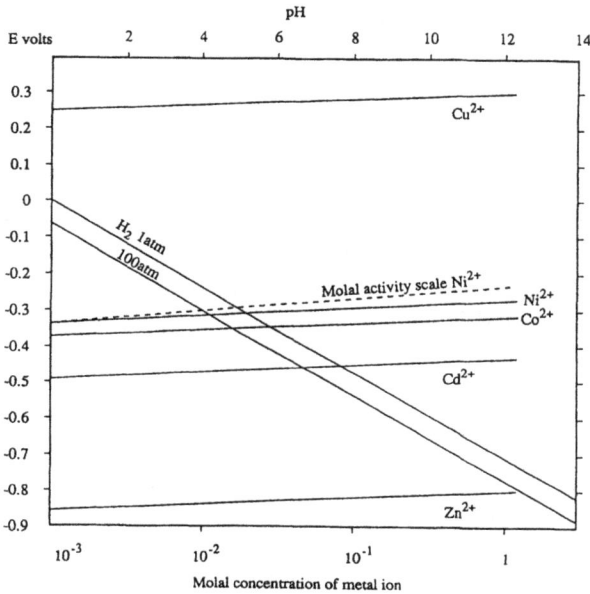

Fig. 3.9.   Effect of molal concentration of metal ions on the values of E for their reduction to metal. The broken line for Ni^{2+} was drawn using the concentration scale to show the values of the activity of Ni^{2+} in the $NiSO_4$ solution. Activity coefficients used were those of the metal sulphates.

Table 3.8. Mean activity coefficients in solutions.

Molality	0.001	0.005	0.01	0.05	0.10	0.30	0.50	1.0	2.0
$NiSO_4$	–	–	–	–	(0.150)	0.084	0.063	0.043	0.034
$CuSO_4$	0.74	0.53	0.41	0.21	(0.150)	0.083	0.062	–	–
$ZnSO_4$	0.700	0.477	0.387	0.202	(0.150)	0.083	0.063	0.043	0.035
$CdSO_4$	0.699	0.476	0.383	–	(0.150)	0.082	0.061	0.041	0.032

Activity coefficients are small in solutions of practical importance here and in view of the uncertainties about the absolute accuracy of the measured values it is sufficient to consider that the values are identical for all of these salts. The degree of dissociation of the salts, and also of cobalt sulphate, as measured by the electrical conductance of their solutions shows that the latter can be included in Fig. 3.9. The data used to draw the solid line drawn for Ni^{2+} is given in Table 3.9.

Table 3.9. $E^\circ_{Ni^{2+},Ni} = -0.236$ volt.

Concentration m	10^{-3}	10^{-2}	10^{-1}	0.3	0.5	1.0
Activity	0.0007	0.004	0.015	0.025	0.032	0.043
E volt	-0.329	-0.307	-0.290	-0.283	-0.280	-0.276

The broken line for Ni^{2+} was drawn taking the concentration as being identical with the activity. The decrease in activity coefficient as concentration increases partly offsets the rise in E as the concentration increases. The lines are almost straight and have been drawn so in the figure.

The lines for the reaction of hydrogen [hydrogen electrode reaction, Sec. 3.3.8 Eq. (3.29)] have been drawn over the pH range 0 to 14 since this is the available experimental range. The pH and metal concentration scales are independent of one another and the points at which lines relating E and pH cross those relating E and $\{M^{2+}\}$ have no significance. Hydrogen gas can reduce a solution containing a metal ion when the potential of the hydrogen reaction is more negative than that of the metal reaction; that is when the metal line lies above the hydrogen line in Fig. 3.9. By far the most important factor controlling the relative positions of the hydrogen and metal lines is the pH of the solution. Increasing the hydrogen partial pressure lowers the whole E-pH hydrogen line, and increasing the metal-ion concentration raises the value of E on the metal line, thereby making reduction to metal easier. A higher metal ion concentration and hydrogen partial pressure, particularly the latter, have far greater effects on the kinetics of the reduction than on the thermodynamics (Sec. 4.4).

A diagram similar to Fig. 3.9 was used by Schaufelberger[57] while working on the process now known as the Sherritt Gordon reduction process for nickel and cobalt. The correction for activity coefficients was made by Meddings and Mackiw[58] who pointed out that the hydrogen pressure necessary to reduce a metal ion in solution at a given concentration, at a particular pH can be calculated from the information incoporated in Fig. 3.9. Thus for Zn^{2+} in 0.01 m zinc sulphate solution $E = -0.833$ V. At pH 6, for the hydrogen reaction, $E = -0.355$ volt. The difference, 0.478 volt, must be made up by increasing the hydrogen pressure, using the term $0.0296 \log p_{H_2}$ in Eq. (3.30), giving $\log p_{H_2} = 16.148$. Thus if a solution of zinc sulphate at pH 6 is reduced to give a minimum zinc concentration of 0.01 m the hydrogen pressure necessary is 1.406×10^{16} atm. Zinc is not produced by such a process. However, the necessary overpotential can be applied in this case by electrolysis.

Considering now the reduction of Ni^{2+} by hydrogen, if the initial concentration of nickel in the solution is 1 m complete reduction to metal will produce 2 moles of H^+. The reaction will cease when the pH reaches a value at which E for the two half-cell reactions become equal.

$$E_{Ni^{2+},Ni} = E^{\circ}_{Ni^{2+},Ni} + 0.0296 \log\{Ni^{2+}\} = E_{H^+,1/2H_2}$$

$$= -0.0296 p_{H_2} - 0.0592 \text{ pH}.$$

Thus at 1 atm of hydrogen

$$-0.236 + 0.0296 \log\{Ni^{2+}\} = -0.0592 \text{ pH}.$$

It can be seen from Fig. 3.9 that at pH 6 the H_2 line for 1 atm lies below the nickel line so that reduction will occur at 25°C, if kinetic considerations allow equilibrium to be established. However, at pH 4, the hydrogen line lies above the nickel concentration line to 1 m. At pH 5, $\log\{Ni^{2+}\} = -2.027$ giving a nickel concentration slightly below 10^{-1} m as can be seen from the activity-concentration $-E$-value data given above in Table 3.9. It is possible, therefore, to reduce a 1 m solution of nickel sulphate by hydrogen gas, if thermodynamic equilibrium can be achieved, at 25°C and a hydrogen partial pressure of 1 atmosphere until the concentration has fallen to about 0.1 m as long as the pH of the solution is kept above about 4. That is, the almost 2 moles of H^+ produced per mole of nickel reduced must be eliminated in some way. In the first plant in which nickel was produced commercially by this reaction the hydrogen ions were neutralised by addition of anhydrous ammonia.

A much more elegant means of achieving the same results was devised by the same workers; they added ammonia to the nickel sulphate solution to form nickel ammines and the reaction as now used commerically can be written

$$Ni(NH_3)_2^{2+} + H_2 = Ni + 2NH_4^+. \tag{3.40}$$

The fact that a complex species rather than the simple hydrated metal ion is reduced to form the metal alters the free energy change of the reaction by an amount which is measured by the equilibrium constant of the metal–ligand reaction to form the complex ion.

When considering how the formation of complexes of a metal influences the potential of the metal ion–metal electrode it is usual to deal with the case in which a large excess of the ligand L is present. For simplicity in writing equations it is assumed here that L is uncharged. The reaction which forms the complex is then

$$M^{z+} + nL = ML_n^{z+}$$

and the overall equilibrium constant β_n is given by

$$\beta_n = \{ML_n^{z+}\}/\{M^{z+}\}\{L\}^n \tag{3.41}$$

n being the maximum coordination number of the metal ion with ligand L.

In order to calculate the electrode potential it is assumed that only the uncomplexed ion M^{z+} is reduced to metal and the effect of complex formation is to lower the activity of the simple metal ion which is calculated using (3.41) and the value of β_n taken from tables. This is done as follows: since the complex is stable, and for simplicity all activity coefficients are taken as unity, $\{ML_n^{z+}\}$ may be taken as the total concentration of the metal present in solution. That is, because β_n is large, $\{M^{z+}\}$ is so small that this approximation introduces no significant error. $\{L\}$ is the concentration of free, uncomplexed ligand and may be calculated from the total concentrations of ligand and of metal in the solution and the value of n. The value of E° for the complex ion ML_n^{z+}, referring to its reduction to metal and free ligand, is obtained by setting $\{ML_n^{z+}\}$, that is the total concentration of metal, equal to 1 m and $\{L\}$ also to 1 m. Thus

$$E^\circ_{ML_n^{z+},M} = E^\circ_{M^{z+},M} + \frac{2.303RT}{zF} \log \frac{1}{\beta_n}. \tag{3.42}$$

The argument used in this thermodynamic calculation is that the free energy change on reduction of a complex metal ion to metal is equal and opposite in sign to that on oxidation of the metal from its standard state to M^{z+} and reaction of this with n ligand molecules to form ML_n^{z+}. The argument is an assumption that Hess's law is correct. It does not imply that direct reduction of a complex ion does not take place. Many complex ions are reduced in steps, the metal being stable in a lower valency state. In a few cases stable complexes in which the metal is formally zerovalent can be produced by reduction under suitable conditions. There is no reason why a complex ion should not be able to accept electrons and then dissociate because the product is unstable.

It is emphasised that the method of calculating electrode potentials given above is valid only if ML_n^{z+} is the sole complex ion containing the metal present

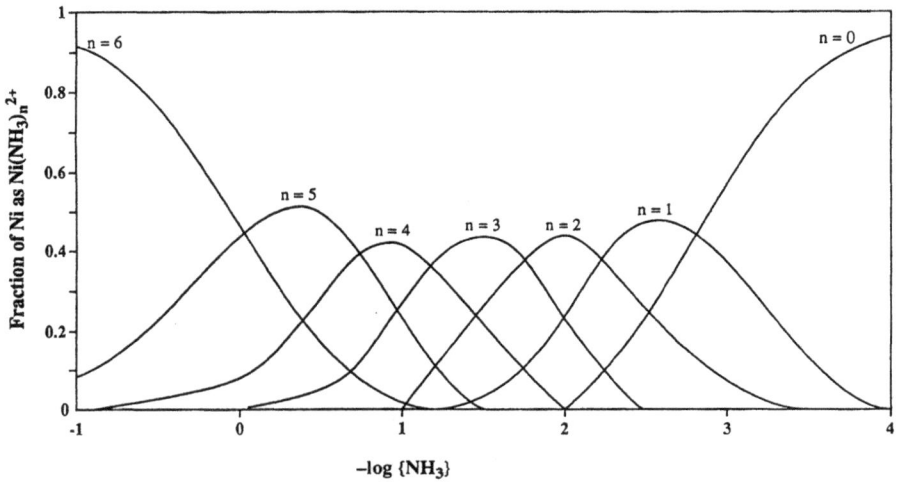

Fig. 3.10.   Sequential formation of the complexes of NH_3 with Ni^{2+} as the activity of NH_3 is increased (from right to left).

in the solution. If lower complexes such as $ML_{(n-1)}^{z+}$ are present it is not sufficient to calculate $\{M^{z+}\}$ using only the value of β_n. Even if a suitable correction were put in by utilising the successive equilibrium constants, liberation of L, as metal is deposited will continuously alter the ratio of ligand to total metal in solution and so change the proportions of the different species present.

This is shown in Fig. 3.10 for the nickel–ammonia system. The successive equilibrium constants have been determined under various conditions of total ionic strength by a number of workers.[46] In order to obtain information relevant to the industrial process appropriate values must be selected. Nickel is produced by reduction of an ammoniacal ammonium sulphate solution having a free ammonia to nickel ratio of 2.1 to 1, and $[Ni^{2+}]$ about 40 gl^{-1} in 2.5 to 3 M ammonium sulphate solution. Values obtained by measurements at 30°C in 2 M ammonium nitrate as base electrolyte[59] are used here.

		1	2	3	4	5	6
	n						
Nickel	$\log K_n$	2.80	2.24	1.73	1.19	0.75	0.03
Cobalt	$\log K_n$	2.11	1.63	1.05	0.76	0.18	−0.62

These practical constants refer to concentrations of reactants and products in equations such as

$$K_n = [\text{Ni(NH}_3)_n^{2+}]/[\text{Ni(NH}_3)_{(n-1)}^{2+}][\text{NH}_3] .$$

As the ratio of free ammonia to nickel (in all forms) concentration increases, the intermediate complex ammines form, increase in concentration and then decrease as the higher ammines are produced, Fig. 3.10. The points drawn were calculated from $\bar{n}$ values and tabulated.[59]

The equilibrium constants were measured at 30°C but will be assumed here to hold at 298.15 K. Suppose also at present that all activity coefficients are equal to unity. Then for the first equilibrium

$$\text{Ni}^{2+} + \text{NH}_3 = \text{NiNH}_3^{2+}$$

$$\Delta G^\circ = 2.303RT \log K_1 = -5.7089 \log K_1 = -15.98 \text{ kJ at } 298.15 \text{ K} .$$

(3.43)

The free energy of formation of Ni^{2+} is -45.6 kJ mol^{-1} and of $\text{NH}_3(\text{aq})$ -26.50 kJ mol^{-1} so that the free energy of formation of the $\text{Ni(NH}_3)^{2+}$ ion is -88.08 kJ mol^{-1}. When the complex is reduced to nickel the reaction is

$$\text{Ni(NH}_3)^{2+} + 2e = \text{Ni} + \text{NH}_3$$

$$\Delta G^\circ = -26.50 + 88.08 = 61.58 \text{ kJ}$$

(3.44)

and the standard electrode potential $E^\circ_{3.44}$ is -0.319 V. The complex ions are more difficult to reduce to metal than the simple hydrated ion Ni^{2+} because they have larger free energies of formation.

It is not necessary to calculate the free energies of formation of the individual ammines in order to calculate E° for them. Thus for the first complex

$$-\Delta G^\circ_{(3.44)} = \Delta G^\circ_{\text{Ni}^{2+}} + \Delta G^\circ_{(3.43)} = -45.6 - 2.303RT \log K_1$$

and in the general case of the ion $\text{Ni(NH}_3)_n^{2+}$ $\log K_1$ is replaced by $\log \beta_n$. The values for the individual complex ions are

n	1	2	3	4	5	6
$-E^\circ$ volt	0.319	0.385	0.436	0.472	0.494	0.495 .

A solution not having a sufficiently large ammonia to nickel ratio to produce only the hexammine will contain several of the complexes and the potential required to reduce each individual species will depend on its activity. It can be seen from Fig. 3.9. however that the effect of this is very much less than the change in potential required for reduction due to the formation of the ammine complexes. The species with $n = 4$ is, in thermodynamic terms, becoming quite difficult to reduce by reduction with hydrogen gas. That is why in the process using this reaction for the production of nickel metal, the ratio of free ammonia to total nickel in solution is 2.1 to 1. All of the H^+ produced is neutralised by the ammonia formed by the complex on reduction so that the pH cannot fall, but the excess quantity of ammonia is not sufficient to make the thermodynamics of reduction significantly less favourable.

The reaction between hydrogen and a solution of nickel ammines does not occur below a temperature of approximately 180°C and even at and above that value requires a catalytically reactive surface. This is commonly the case in hydrogenation reactions. At such temperatures the nickel solution precipitates basic salts due to high temperature hydrolysis and this is prevented by having present a high concentration of ammonium sulphate to increase the ionic strength of the solution. Using the values for the equilibrium constants measured at 30°C in 2 M ammonium nitrate, given above, the values of K at 213°C, calculated using the van't Hoff isochore, are as follows:

n	1	2	3	4	5	6
Nickel $\log K_n$	1.71	1.27	0.509	0.343	−0.056	−1.16
Cobalt $\log K_n$	1.55	1.07	0.49	0.199	−0.38	−1.18

This procedure had to be used since no heat capacity data for the complex species are available.

3.8.   Precipitation of Compounds from Solution

3.8.1.   *Introduction*

Precipitation of oxides and hydroxides and the thermodynamic basis of the solubility product have been dealt with in Sec. 3.3.1. Precipitation of many other classes of solids is also of great importance in hydrometallurgical processes. For example a metal value such as copper or nickel in a liquor produced by leaching is often precipitated as a sulphide, giving a concentrate for further treatment. Impurities present in a solution are frequently removed by precipitation and although hydroxides are of great importance for this purpose arsenates, phosphates and many other compounds are used also. All of these solids are salts of an acid and the equilibrium values of their solubilities depend on the pH of the solution. Thus calculations based on solubility products can be carried out to establish optimum conditions for removal of impurities from solution while not losing valuable metal, or for high recovery of metal values in a precipitate containing acceptable levels of impurities, or many other desired results. The appropriate solubility relationships can be conveniently represented as diagrams[60] but for some purposes it is more accurate to calculate for specific conditions and requirements, using the equations which are used to produce the diagrams.

Equilibria between metal ions and sulphides of the metals are of particular importance because they control both leaching of sulphide minerals and precipitation of sulphides from process liquors. In Sec. 3.5.2 the equilibria which are set out for producing a predominance area diagram for the copper–sulphur–water system include reactions which form CuS and Cu_2S by a variety of reactions, but not by that which gives the solubility product of CuS

$$CuS = Cu^{2+} + S^{2-} \quad K_{so} = \{Cu^{2+}\}\{S^{2-}\}\,.$$

The reason for this is that the second acid dissociation constant which is accepted in this work is so small that the ion S^{2-} does not appear in solution at an activity of significance for preparing such diagrams at pH 14 or less. That fact is not relevant when calculating chemical equilibria and rejection of the previously accepted value of K_{a2} requires justification.

3.8.2.   *The Second Dissociation Constant of $H_2S(aq)$*

When solubility products and solubilities of sparingly soluble metal sulphides are being considered it is necessary to know $\Delta G°$ of formation for S^{2-} which is calculated from measurements of the second acid dissociation constant of $H_2S(aq)$, $K_{a2} = \{S^{2-}\}\{H^+\}/\{HS^-\}$. This has in the past been taken as having the value 10^{-14} so that $\{S^{2-}\} = \{HS^-\}$ at pH 14 and S^{2-} has been included in a predominance area diagram if desired by extending the pH scale above 14.

In 1967, it was reported that both dissociation constants for $H_2S(aq)$ had been measured over the temperature range 20–90°C using a spectrophotometric method. An absorbance band at 230 nm was assigned to the HS^- ion and one at 360 nm to the S^{2-} ion.[61] A large range of values for K_{a2} was obtained even at 25°C and this was attributed to rapid oxidation of the alkaline sulphide solution even if only traces of oxygen were present. This can give a range of products from sulphate to elemental sulphur which dissolves to form polysulphides by association with remaining sulphide. Giggenbach in 1971[62] suggested that the band at 360 nm was due to the presence of such polysulphides and found that the band was not observable when oxygen was completely excluded from the system. His measurements gave a value of pK_{a2} of approximately 17.1. It was suggested also[63] that the usual procedure for dealing with equilibria involving sulphides by using activities of S^{2-} ion is not satisfactory because this ion appears in solution only at such high pH values, and therefore hydroxide concentration, that the activity coefficients are not determinable and so an alternative equilibrium should be employed. This argument is fallaceous. It is difficult to measure K_{a2} because the species S^{2-} is present in solution at a concentration which can be measured only in very strongly alkaline solutions. Once the second dissocation constant has been measured accurately, the value of $\Delta G°$ of formation of S^{2-} can be used for any solution at an appropriate temperature.

In spite of the evidence suggesting that K_{a2} for $H_2S(aq)$ was much smaller than had been previously reported the earlier value continued to be accepted after review of the evidence, and included in compilations of data. In 1983, the difficulty of calculating values of K_{a2} from experimental data obtained using solutions having sodium hydroxide concentrations up to 22 M was again pointed out[64] when Raman spectroscopy was used to identify the presence of the HS^- ion from its peak at 2570 cm^{-1} (stretch). The concentration of HS^- was determined by comparing the ratio of the observed intensity of this peak to that of an internal Raman scattering standard, 0.10 M sodium perchlorate. Standard solutions containing known concentrations of HS^- and 0.1 M sodium perchlorate were used to correlate the observed ratio with HS^- concentrations. The value of pK_{a2} reported was 17 ± 1.

A fundamentally different method of obtaining the value made use of the fact that KHS is very much more soluble in concentrated solutions of potassium hydroxide than is NaHS in sodium hydroxide solutions. Thus significantly higher concentrations of S^{2-} can be obtained using the potassium salts than in the sodium system, making measurement of its concentration or activity potentially more accurate. The classical method of measuring acid dissociation constants is by measuring hydrogen ion activities. However the hydrogen electrode is poisoned by the presence of sulphide and conventional glass electrodes for pH measurement cannot be used to measure $\{H^+\}$ at the very high pH values necessary when K_{a2} is very small. A major reason is that they respond also to the activities of alkali metal ions in the solution. The useful range of glass pH electrodes has now been extended to pH 17.6 in aqueous solutions and these have been used to measure pK_{a2} for $H_2S(aq)$.[65,66] A low alkali error glass electrode together with a saturated calomel electrode was used to determine $\{H^+\}$ by means of calibration data, and $\{K^+\}$ using a glass cationic selective electrode with a saturated calomel electrode. The value obtained was $pK_{a2} = 17.6 \pm 0.3$.

Since the value obtained by the potentiometric method is consistent with that obtained using Raman spectroscopy to measure concentrations of HS^- in solution, and there is convincing evidence that the much lower values accepted previously were in error because of oxidation of some sulphide, the value of pK_{a2} used here for $H_2S(aq)$ is 17.4 giving ΔG° of formation for $S^{2-}(aq) = 111.4$ kJ at 25°C.

3.8.3. *Precipitation of Sparingly Soluble Metal Sulphides*

For a salt M_aX_z, the solubility product is given by

$$K_{so} = \{M^{z+}\}^a \{X^{a-}\}^z$$

and for a sparingly soluble salt activities may be taken as being equal to concentrations. In a hydrometallurgical process precipitation of a compound is frequently used to remove an ion from a process liquor by addition of a reagent, nickel and cobalt as a mixed sulphide concentrate in the treatment of a lateritic nickel ore for example or, in a more subtle manner, copper as sulphide from a nickel–cobalt leach liquor produced by treatment of a copper–nickel matte. In such a case it is useful to calculate the conditions under which the precipitation should be carried out in order to achieve the desired result.

When a simple salt such as a metal sulphide is precipitated the calculation is based on a chemical reaction such as:

$$Cu^{2+} + S^{2-} = CuS$$

or

$$2Ag^+ + S^{2-} = Ag^2S.$$

In each case the activity of the metal-ion and of S^{2-} is involved. When precipitation is almost complete $\{S^{2-}\} = [S^{2-}]$ unless a large excess of the source of S^{2-}, usually H_2S gas or NaHS solution, is added. It is assumed here that activities equal concentrations. Whichever of these reagents is used the equilibria involving $\{S^{2-}\}$ are those of the acid $H_2S(aq)$ at pH values below 7, which are usually those of importance. The equilibria are:

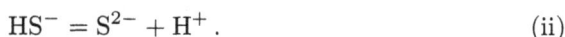

$$H_2S(aq) = HS^- + H^+ \tag{i}$$

$$HS^- = S^{2-} + H^+. \tag{ii}$$

The relevant thermodynamic data are:

Species	$H_2S(aq)$	HS^-	S^{2-}
ΔG° kJ	-27.83	12.08	111.4

These give

$$\log K_{a1} \quad \text{for (i)} = -6.99$$

$$\log K_{a2} \quad \text{for (ii)} = -17.4$$

so that when calculating solubility products of metal sulphides in solutions having pH values below 7 the relevant equilibrium is:

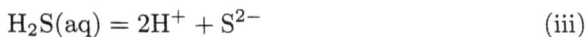

$$H_2S(aq) = 2H^+ + S^{2-} \tag{iii}$$

$$\Delta G^\circ = 139.23 \text{ kJ}$$

$$\log \beta_{12} \quad \text{for (iii)} = -24.39.$$

For the overall equilibrium between S^{2-} and total sulphide in solution, written as $[H_2S]$

$$\log[S^{2-}] = -24.4 + 2\text{pH} + \log[H_2S].$$

Precipitation of CuS

The equation for the solubility product of CuS is

$$K_{so} = \{Cu^{2+}\}\{S^{2-}\}$$

$$\Delta G^\circ = 65.49 + 111.4 + 53.6 = 230.5 \text{ kJ}$$

$$\log K_{so} = -40.4 .$$

Figure 3.3 was drawn for activities of sulphur species 10^{-1} and $\{Cu^{2+}\} = 10^{-3}$ and 10^{-6}. At pH 1

$$\log\{S^{2-}\} = -24.4 + 2 - 1 = -23.4$$

so that $\log\{Cu^{2+}\} = -17$. Thus the reaction between Cu^{2+} and S^{2-} ions is not relevant to a predominance area diagram. If the total sulphide concentration in solution is 10^{-6} M, at pH 1 $\log\{S^{2-}\} = -28.4$ and $\log\{Cu^{2+}\} = -12$

Precipitation of NiS

For crystalline NiS $\Delta G^\circ = -79.5$ kJ and for $Ni^{2+} \Delta G^\circ = -45.6$ kJ so that $\log K_{so}$ for NiS $= -25.45$. In a solution with total activity of sulphide species 10^{-1}, at pH 1 $\log\{S^{2-}\} = -23.4$ and the equilibrium activity of Ni^{2+} is 10^{-2}.

If the total activity of sulphide species is 10^{-2} at pH 1 $\log\{S^{2-}\} = -24.2$ and the equilibrum value of $\log\{Ni^{2+}\} = -1.2$. At pH 2, $\log\{S^{2-}\} = -22.2$ and $\log\{Ni^{2+}\} = -3.2$.

Thus in a solution containing 10^{-2} M total sulphide in equilibrium with crystalline NiS the nickel concentration should be approximately 0.06 M at pH 1 and 0.0006 M at pH 2. Activity coefficients have been taken as 1.

In the Moa Bay process for treating laterite ore the leach liquor containing nickel and cobalt sulphates at pH 2.4 is treated with H_2S gas at 1.03 MPa pressure (150 lb in^{-2} g) at 120°C. In order to obtain satisfactory yields of the sulphides it is necessary to recycle large amounts of precipitated nickel–cobalt sulphide to act as "seed".

3.8.4.   *Nucleation and Crystal Growth*

If in the laboratory, H_2S gas is passed into a pure solution of nickel sulphate made slightly acidic with sulphuric or hydrochloric acid no precipitation of NiS occurs even after long standing. If the solution has been made alkaline with ammonium hydroxide (no precipitation of $Ni(OH)_2$ occurs because complex ammines form) complete precipitation of nickel sulphide takes place rapidly. The solid is insoluble in cold dilute sulphuric or hydrochloric acid. In the acidic solution in which no NiS has precipitated the equilibrium solubility of the compound has been exceeded and the solution is supersaturated with respect to it.

Supersaturation is commonly found in precipitation reactions and occurs because there is an activation energy barrier which has to be overcome before a nucleus of solid can be formed on which more of the material can deposit. This is considered further below. It is most commonly observed with substances which have appreciable solubility and there is a limit to the degree of supersaturation which can occur for a particular substance. Before this limit is reached and homogeneous nucleation occurs, producing nuclei of the solid substance itself, it may deposit on small particles of other solids present in the liquid. This is heterogeneous nucleation and is responsible in many cases for the precipitation of solids having appreciable solubility. Once some particles of the solid have formed supersaturation cannot persist for long. As a reagent which is causing the precipitation continues to be added to the solution the question is whether it causes more solid to deposit on existing particles or whether new nuclei are formed. If the rate of addition is so slow that solid has time to deposit on particles formed by heterogeneous nucleation then the limiting degree of supersaturation may not be reached, even locally at the point of addition, and relatively few large crystalline particles will be formed. If, however, homogeneous nucleation does occur either because of rapid addition of reagent, or because the solid is very insoluble so that addition of a very little reagent causes the degree of supersaturation to be outside the metastable region of concentration, then fresh particles will be formed continuously during addition of reagent. This causes the solid to have a very small ultimate particle size. Such particles have a large surface free energy and they often tend to reduce this by forming aggregates in which they are "welded" together in some way. Because of their high surface free energy, small particles of a compound have a higher solubility in a given solution than do larger particles so that in a slurry, small particles tend to disappear and large particles to grow by "Ostwald ripening".

If the solid being precipitated has almost zero solubility then it is practically impossible to add the reagent so that no new nuclei are formed and growth of crystals due to the presence of particles of different size cannot occur. If they do not aggregate then a sol will form. Sols are basically colloidal suspensions and are stable only under certain conditions. In practically all normal industrial environments they flocculate and form gels or gelatinous precipitates, depending on the concentration of the solid. The physical characteristics of solids of small or very small particle size in suspension in aqueous solutions depend on the electrostatic charges on their surfaces, due to adsorption of ions or of organic flocculating agents or dispersants. This behaviour lies in the realm of colloid chemistry and is not considered here. A brief review including consideration of particle morphology is available.[67]

Nucleation Theory

The classical theory concerning the factors which control homogeneous nucleation in supersaturated solutions is outlined briefly here.[68] The objective is to decide what controls the critical size of the embryo and the rate at which new nuclei are formed.

The free energy barrier to nucleation of a given sized crystalline assembly of ions or atoms (embryo) is, at a given temperature and pressure

$$\Delta G = A\sigma + V\Delta G_v \qquad \text{(i)}$$

where A is the total interfacial area of the embryo, V is the total volume of the embryo, σ the average interfacial free energy based on A and ΔG_v the excess free energy per unit volume of the crystalline phase in the embryo over that in solution. Conventionally the embryo, or nucleus, has been considered to be a spherical assembly of ions or atoms but in this treatment it is supposed to be a very small crystal of the substance with respect to which the solution is supersaturated. Based on experimental data for supersaturated solutions of sodium chloride and ammonium sulphate the properties of nucleation embryos have been calculated,[68] Table 3.10 below.

Let y be a characteristic dimension of the embryo defined so that the total surface area for each type of surface and the total volume of the crystalline embryo are given by

$$A_i = k_i y^2 \qquad \text{(ii)}$$

$$V = ly^3 \qquad \text{(iii)}$$

where k_i and l are geometrical constants dependent on the morphology of the crystal being considered but independent of y. By definition

$$A = \sum_i A_i = \sum_i k_i y^2 \qquad \text{(iv)}$$

Table 3.10. Properties of nucleation embryos.

	NaCl in H_2O	$(NH_4)_2SO_4$ in H_2O
Critical supersaturation		
wt. fraction	0.395 ± 0.015	0.798 ± 0.012
molality	11.2	29.9
supersaturation ratio S	3.37	7.11
Critical nucleus size (Å)		
spherical (diameter)	15.7	16.4
cubic (side)	12.6	–
orthorhombic (diagonal)	–	23.9
Interfacial free energy (erg cm^{-2})		
spherical	88	80
cubic	71	–
orthorhombic	–	64

and

$$\sigma = \frac{\sum_i A_i \sigma_i}{A} = \frac{\sum_i k_i \sigma_i}{\sum_i k_i} . \tag{v}$$

Substituting (iii) and (iv) into (i) gives

$$\Delta G = \left(\sum_i k_i \right) \sigma y^2 + l \Delta G_v y^3 . \tag{vi}$$

The supersaturated solution is in a metastable condition and a free energy barrier must be overcome before crystal nucleation can occur. According to this theory the critical size of the nucleus is that at which ΔG reaches a maximum with respect to y, that is $(\delta G/\delta y) = 0$. Once the embryo has reached this size nucleation has occurred and the crystal grows spontaneously. This critical size is given by

$$y_i = -\frac{2\alpha\sigma}{3\Delta G_v} \tag{vii}$$

where $\alpha = \sum_i k_i/l$, the geometric shape factor. The minus sign results from the fact that ΔG_v is always negative. Substitution of this initial size into (vi) gives the critical free energy change for nucleation, that is the maximum free energy

$$\Delta G_i = \frac{4(\sum_i k_i)\alpha^2 \sigma^3}{27\Delta G_v^2}. \qquad \text{(viii)}$$

If the solute in the saturated solution is chosen as the reference state

$$\Delta G_v = \frac{\rho_{\text{cryst}}}{W_s} \left(\mu^{\text{sat}} - \mu^{\text{supersat}} \right) \qquad \text{(ix)}$$

where ρ_{cryst} is the density of the crystalline phase, W_s is the molecular weight of the solute, μ^{sat} is the chemical potential of the solute in a solution saturated with respect to the crystalline solid considered, and μ^{supersat} is the chemical potential of the solute in the supersaturated solution.

The definition of the mean activity of the electrolyte $a_{\pm}$ permits Eq. (ix) to be rewritten as

$$\Delta G_v = \frac{v\rho RT}{W_s} \ln \frac{a_{\pm}^{\text{sat}}}{a_{\pm}^{\text{supersat}}} \qquad \text{(x)}$$

where v is the number of ions produced by 1 mole of the electrolyte. If the initial supersaturation ratio S is defined as the activity ratio Eq. (x) becomes

$$\Delta G_v = -\frac{v\rho RT}{W_s} \ln S. \qquad \text{(xi)}$$

The rate of nucleation is written as

$$J = K \exp \left(-\frac{\Delta G_c}{kT} \right) \qquad \text{(xii)}$$

where ΔG_c is the value of ΔG_i given by (viii) and k is the Boltzmann constant. The value of the preexponential factor K is not known but theoretical estimates range from 10^{24} to 10^{36} cm^{-3} s^{-1}. An intermediate value, 10^{30} cm^{-3} s^{-1} has been commonly used.

From measurements of concentrations of sodium chloride and ammonium sulphate in levitated micron size droplets, in contact with no surfaces, in which crystallisation occurred from supersaturated solutions the properties of nucleation embryos were calculated using the general theory outlined above and some other estimates and assumptions.[68] They are of interest irrespective of their absolute accuracy and are shown in Table 3.10. The critical nucleus sizes were based on two models, a spherical particle and a crystal; a cubic structure with side 5.64 Å for NaCl and an orthorhombic unit cell with $a = 5.99$,

$b = 10.64$, $c = 7.80$ Å for $(NH_4)_2SO_4$. It was proposed that the critical nucleus of NaCl is a block of 8 unit cells having sides 12.6 Å. That of $(NH_4)_2\,SO_4$ has 4 unit cells, 2 above 2 so that the b dimension is increased slightly to 11.4 Å while the $2a$ length is 12.8 and the $2c$ length is 16.7 Å.

3.8.5. *Precipitation of Arsenates*

Solubility products of metal arsenates are calculated using the activity of the AsO_4^{3-} ion. Arsenic acid is tribasic and the solubility of the metal arsenate depends on which of the anions is formed at the pH being considered. The relevant thermodynamic data for the case of copper (II) arsenate is given in Table 3.11.

Table 3.11. Thermodynamic properties for the copper-arsenate system.

Species	$Cu_3(AsO_4)_2$	Cu^{2+}	$H_3AsO_4(aq)$	$H_2AsO_4^-$	$HAsO_4^{2-}$	AsO_4^{3-}
$\Delta G°$ kJ	-1300.7	65.49	-766.0	-753.17	-714.60	-648.41

The equilibrium constants for the acid are

$$H^+ + H_2AsO_4^- = H_3AsO_4 \quad \log K_{13} = 2.24$$

$$H^+ + HAsO_4^{2-} = H_2AsO_4^- \quad \log K_{12} = 6.76$$

$$H^+ + AsO_4^{3-} = HAsO_4^{2-} \quad \log K_1 = 11.59$$

Thus, at pH values below 2.24, the species H_3AsO_4 is considered to be the stable species; between 2.24 and 6.76 $H_2AsO_4^-$; etc.

The solubility product of Cu^{II} arsenate is given by

$$Cu_3(AsO_4)_2 = 3Cu^{2+} + 2AsO_4^{3-}$$

for which $\Delta G° = 200.35$ kJ giving $\log K_{so} = -35.1$.

In the pH range 0 to 2.24 the equilibrium is

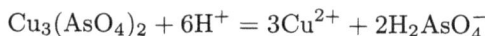

$$Cu_3(AsO_4)_2 + 6H^+ = 3Cu^{2+} + 2H_2AsO_4^-$$

for which $\Delta G° = -9.17$ kJ giving $\log K$ for the reaction $= +1.606$. Thus within this pH range the relationship between copper and arsenic concentrations in solution, and pH is, if activity coefficients are unity

$$3\log[Cu^{2+}] + 2\log[H_2AsO_4^-] = 1.606 - 6\ pH.$$

Each of these concentrations and the pH can be changed independently in a process.

In the pH range 2.24–6.76, the equilibrium is

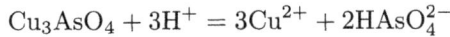

$$Cu_3AsO_4 + 3H^+ = 3Cu^{2+} + 2HAsO_4^{2-}$$

for which $\Delta G° = +67.97$ giving $\log K = -11.91$ so that

$$3\log[Cu^{2+}] + 2\log[HAsO_4^{2-}] = -11.91 - 3\ pH.$$

Similar calculations give equations for the pH ranges 6.76–11.59 and above 11.59.

3.8.6.   *Precipitation of Iron as Oxides*

Iron is very frequently present in hydrometallurgical process liquors and its concentration has to be controlled. If the amount is greater than about $5\ gl^{-1}$ precipitation of $Fe(OH)_3$ by raising the pH usually gives a gelatinous precipitate which is difficult to filter or settle, in some cases impossible. In the electrolytic zinc process considerable amounts of iron dissolve together with the zinc when the calcine from the sulphide concentrate is dissolved in sulphuric acid. This iron originates from zinc ferrite and in the early days of the process care was taken to leach the calcine under conditions such that the ferrite was not attacked because of the difficulty in removing the iron from solution. This led to significant loss of zinc in leach residues and strenuous efforts were made to find ways of producing easily filterable precipitates which could be used to reject dissolved iron from the circuit. Four such methods were developed, three of which are used by companies operating the electrolytic zinc process. These are the goethite proces, hematite process and jarosite process. The fourth is the magnetite process.

Precipitation of "Goethite"

The physical chemistry of iron precipitation by these methods has been reviewed in depth.[69] In the goethite process the precipitate formed consists of goethite, α-FeOOH; β-FeOOH; γ-FeOOH (lepidocrocite); δ-FeOOH; α-Fe$_2$O$_3$

(hematite) together with amorphous phases. Two methods have been developed for producing this precipitate. In that developed by the Societé de la Vielle Montagne, in Belgium, the iron in the liquor produced on leaching calcine is reduced to Fe^{II} with zinc concentrate (ZnS) and is then oxidised at 80–90°C with simultaneous neutralisation to pH 2–3.5, oxygen or air being used for oxidation. The rate of oxidation of Fe^{II} in acidic solutions by oxygen is slow and the presence of copper in the solution, derived from the ore, catalyses the reaction. The reason for reducing and then reoxidizing the iron is to keep the concentration of Fe^{III} in the solution low, certainly below 1 gl^{-1}. An alternative way of achieving this, the Electrolytic Zinc Goethite Process has been tested extensively. The hot solution containing the dissolved Fe^{III} is added together with the neutralising agent needed to control the pH in the range 2–3.5, to a continuous iron precipitation stage in which the Fe^{III} concentration is at, and is maintained at, a low value (< 1 gl^{-1}) and the solution temperature between 70°C and 95°C. The two hydrolysis reactions in which calcine is used to neutralise the acid produced are

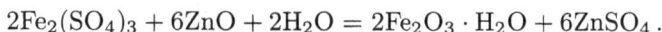

$$4FeSO_4 + 4ZnO + O_2 + 2H_2O = 2Fe_2O_3 \cdot H_2O + 4ZnSO_4$$

$$2Fe_2(SO_4)_3 + 6ZnO + 2H_2O = 2Fe_2O_3 \cdot H_2O + 6ZnSO_4 \,.$$

Thus neutralising the dissolved Fe^{III} requires 50% more calcine than is used in the Vielle Montagne method and an increased amount of residue from this extra calcine is lost with the precipitated iron. This residual zinc cannot be recovered by washing with acid because the goethite would dissolve.

When dealing with the behaviour of systems previously in this chapter thermodynamic data, including stability constants of complex species, have been used to predict behaviour at equilibrium. In the case of precipitation of goethite this is not possible. It is known that the mixture of hydrated ferric oxides described as "goethite" is precipitated when Fe^{II} is oxidised to Fe^{III} in a pH region where a solid oxide or hydroxide of Fe^{III} is predicted to be formed (Figs. 3.1 and 3.8). The mechanism by which the solid is formed and the species present in the solution which form it are not fully understood.

In very dilute solutions containing Fe^{III}, $< 10^{-3}$ M, a series of complex species involving OH^- as ligand is formed, $Fe(OH)^{2+}$, $Fe(OH)_2^+$ etc. In more concentrated solutions of Fe^{III} a dimerised species predominates

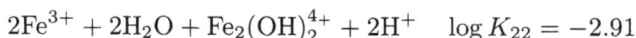

$$2Fe^{3+} + 2H_2O + Fe_2(OH)_2^{4+} + 2H^+ \quad \log K_{22} = -2.91$$

in which the two Fe^{3+} ions are assumed to be linked by two hydroxyl bridges with the remaining four coordination positions on the two iron atoms being occupied by H_2O molecules. The extent of this hydrolysis increases with increasing temperature[70]

$T°C$	15	25	35	45	51
$10^3 K_{22}$	4.9	7.3	10.2	16.3	25.1

In acidic sulphate solutions, sulphate acts as a ligand forming $FeSO_4^+$, with $\log K_1 = 2.03$, and $Fe(SO_4)_2^-$ with $\log K_2 = 0.97$ and these are probably the species present in solutions of relevance in zinc processing, 0.2–2.0 M SO_4^{2-}. In strongly acidic conditions complexes with HSO_4^- are also formed by Fe^{III}. These hydroxyl, sulphate and bisulphate species are in true solution equilibrium.

Mixed hydroxyl–sulphate species can also form and it has been shown that the hydrolysis behaviour of Fe^{III} sulphate solutions is approximately the same as that in a solution in which the anion does not form complex species with the iron[71] and it was concluded that the dimer, $Fe_2 (OH)_2(SO_4)_2$ was the prevalent species. Some such species seems necessary to explain the precipitation of basic iron sulphates and the, at least partial, control by the anion of the form of FeOOH produced, α-, β- or γ-.

The equilibrium constants mentioned above were measured by the method outlined in Sec. 2.4. As was stated there, additional evidence is required before it may be positively accepted that any of these species does actually exist in a solution. Some support for the polynuclear complexes has been provided by spectrophotometric and magnetic susceptibility measurements.[72] In its simple complexes, oxides and oxyhydroxides the Fe^{3+} ion is in a high spin state with 5 unpaired electron spins per iron ion. In bridged species two iron atoms are bound to an OH^- or O^{2-} ion at angles of 90° or 140° to 180°. When the bond angle is between 140° and 180° the spins can align themselves to each other by means of a superexchange interaction, the strength of which depends on the distance and angle between the Fe^{3+} ions.[72] This interaction, the antiferromagnetic coupling, causes the magnetic susceptibility of the iron to decrease. This decrease was measured using solutions of Fe^{III} sulphate, chloride and nitrate which had been acidified and then hydrolysed at a temperature below 70°C. The results, considered in conjunction with the absorption spectra of the solutions, led to the suggestion that the empirical formulae of the polynuclear species present were $Fe_2(OH)_2$ Cl_2 O(I); Fe_2 $(OH)_3$ $(NO_3)_3$ (II); Fe_2 $(OH)_3$ $(SO_4)_{1.5}$ (III) and Fe_3 $(OH)_2$ $(SO_4)_{3.5}$ (IV). It was proposed that in all of the species except (IV) edge-shared octahedral dimer units are antiferromagnetically linked with one another by hydroxyl- or oxo-bridging. When the temperature at which the hydrolysis was carried out was raised to above 70°C the polynuclear complexes hydrolysed further and formed precipitates. (I) gave β-FeOOH; (III) gave α-FeOOH; (IV) produced MFe_3 $(OH)_6$ $(SO_4)_2$ where M = Na, K or NH_4; and (II) gave α-FeOOH or α-Fe_2O_3 depending on the temperature at which the hydrolysis was carried out.

Other, earlier work led to the suggestion[69] that the polymers produced from the labile polynuclear species, which eventually form solid hydrolysis products which settle out, are ill-defined. If this is so chemical thermodynamics is not able to indicate conditions which might lead to the formation of an oxyhydroxide, oxide or basic sulphate except in very general terms based on predominance area diagrams and phase diagrams. A number of these are presented in Ref. 69.

The use of statistical methods of experimental design and evaluation of results under such circumstances was examined[73] by determining the optimum conditions for precipitating iron as "goethite" from solutions which could be produced by reducing a Greek lateritic nickel ore and leaching the product with sulphuric acid, a process used commercially on one plant at Moa Bay, Cuba. The variables selected for evaluation in a preliminary set of experiments were (i) pH of the solution at the temperature of the precipitation; (ii) temperature of precipitation; (iii) initial total iron concentration; (iv) oxygen flowrate; (v) initial Cu^{2+} concentration, the presence of Cu^{2+} being assumed to be necessary to catalyse oxidation of Fe^{II} by O_2 in the acidic solution. Variables which might have an effect on iron precipitation but which were assigned fixed values in this experimental design were, reaction time; stirring speed; initial Ni^{2+} concentration and initial Co^{2+} concentration. Clearly the classical method of experimentation, changing one variable at a time would require a large number of experiments. In the first, preliminary, factorial design of experiments 8 sets of conditions were used and it was found that under laboratory conditions oxygen flowrate could not be used as a variable since only a large stoichiometric excess could be used if the gas was to be dispersed in the liquid. In the second, 16 sets of conditions were used and initial total concentration of iron was found not to be significant within the limits set. In the third, 8 sets of conditions were used. Optimum conditions for precipitation of "goethite" from the nickel–cobalt solution were found to be identical with those used in electrolytic zinc plants. The goethite particles obtained were aggregates of irregular shape and of varying size, between about 1.4 and 20 μm along the longest dimension. This is the reason why the slurries had excellent filtering properties. Each particle was composed of crystallites which were needles or flat plates with length between about 0.08 and 1 μm. The higher the pH of the solution from which the goethite was precipitated the larger were the crystallites and the sharper were the X-ray diffraction patterns they gave.

Precipitation of Jarosite

In the jarosite process, as used for removing iron from process liquors at present the pH of the solution is adjusted to about 1.5 and maintained at that value while a salt of a monovalent cation causes jarosite to precipitate at a temperature of about 95°C. The jarosites are a group of basic Fe^{III} sulphates of composition $MFe_3(SO_4)_2(OH)_6$ where M is a monovalent cation from the group H_3O^+, Na^+, K^+, NH_4^+, Ag^+, Rb^+ and $1/2\ Pb^{2+}$. Published free energy data for the jarosites have been assembled[69] and it is clear that the hydronium jarosite $H_3OFe_3(SO_4)_2(OH)_6$ is not stable in the presence of alkali metal ions.

The equilibrium constant for the dissociation of a jarosite into its constituent ions can be used to calculate the solubility of iron under given conditions but the most significant factors controlling iron removal from a solution are pH and temperature. Again, uncertainty over the species present in the solution used in the electrolytic zinc process make it impossible to estimate activity coefficients for the ions in the equation for K_{so}.

In one of the papers on which the commercial jarosite process was based[74] it was shown that over the temperature range 160–200°C the amount of iron remaining in a sulphate solution when precipitation ceased depended on whether $Na_2 SO_4$, $Na_2 CO_3$, NH_4OH or K_2SO_4 was added as the precipitant for jarosite (source of the 1-valent cation), the residual concentration decreasing in that order. Above 180°C, however the difference between the additives was small. Ammonia was regarded as being probably the preferred material in an industrial process and at 180°C addition of more than the stoichiometric quantity required for complete precipitation of the iron present as ammonium jarosite had little effect on the final iron concentration in solution. Use of less than that quantity resulted in much higher final concentrations. When using the stoichiometric quantity of ammonia an increase of temperature above 180°C had little or no effect on the final iron concentration which was much higher, however, if a temperature of 170°C or lower was used. In a series of experiments carried out to determine the effect of the acidity of a solution before hydrolysis on the final iron concentration a solution containing initially 12.4 g l^{-1} H_2SO_4 was used and either more sulphuric acid or zinc calcine (ZnO) was added to adjust this. The same amount of NH_4^+ and of iron was present in each experiment. The final acidity of a solution leaving the autoclave was determined by both the initial acidity and the degree of hydrolysis and precipitation which had occurred. The extent of hydrolysis and in consequence of equilibrium iron concentration are influenced greatly by acidity levels, indicating the necessity for leach solutions to be low in free acid before passing to the hydrolysis stage of the process. The results obtained using additions of sodium and potassium compounds were similar to those obtained with ammonia.

In the absence of added 1-valent cations precipitation of ferric oxide occurs from zinc sulphate leach solutions at temperatures around 200°C, but the extent of hydrolysis is not sufficient to lower the final iron content of the solution to a level which is of practical importance. The extent of the precipitation can be increased in two ways. The first is to use a temperature higher than 200°C. Increased pressure and corrosion problems are disadvantages of this, but the lowering of the solubility of zinc sulphate above 200°C makes this alternative impractical. The second method is to lower the free acidity in the solution which can be done by adding zinc calcine which will give a final iron concentration of the desired level, less than about 3 gl^{-1}, if the final acid concentration is below about 40 gl^{-1}. Below 180°C, an excessive amount of alkali (ZnO) is required to reach 3 gl^{-1} Fe^{III} in solution but provided the final acidity is below about 20 gl^{-1} ferric oxide is precipitated. At higher acidities hydroxonium jarosite, or a mixture of this with ferric oxide is precipitated and under these conditions the final dissolved iron concentration tends to be too high.

In a process developed recently, although not yet used commercially, for recovery of copper from concentrates using nitric acid or NO_2 as oxidant, the jarosite process is used to remove iron from the leach liquor. The temperature employed is around 180°C and ammonia is added to the reactor.

A process such as the jarosite or goethite process is merely one step in a sequence of steps in a process which accepts zinc sulphide concentrate as a feed material and eventually produces zinc metal cathodes. The ability to remove dissolved iron from the zinc sulphate solution greatly increased the profitability of the electrolytic zinc process, but this was because the jarosite and goethite processes were introduced into the overall circuits and used in such a way as to make the best possible use of their capabilities. This required an appreciation of the factors influencing the chemical behaviour of the solutions at elevated temperatures.

Precipitation of Hematite

The hematite process has been used at an electrolyte zinc plant in Japan since 1972. The acid leach solution is reduced with SO_2 at 95–100°C to give Fe^{II} and after precipitating any copper present with H_2S and raising the pH to 4.5 with limestone the clarified solution of $FeSO_4$ is heated at 180–200°C for 3 hours at 1.8 MPa (470 lb in^{-2} g) oxygen pressure. The predominance area diagram for the Fe–H_2O system, Fig. 3.1, shows that at 25°C hematite is the stable oxide of iron under such oxidising conditions, of those considered. Goethite, α-FeOOH, is not well defined thermochemically so that calculations concerning it are unreliable, but above 130°C it is not a stable phase in the Fe_2O_3–H_2O–SO_3 phase diagram.[75]

3.9. pH at Elevated Temperatures

The developments in nuclear technology and geochemistry which led to the work on the properties of aqueous solutions of electrolytes at high temperatures considered in Sec. 3.6, and the application of processes operating at high (by aqueous solution standards) temperatures in hydrometallurgy have resulted in a need to be able to measure pH under these conditions. The glass electrode has been used to measure pH at temperatures up to about 100°C ever since it became commercially available in the 1940's and has made pH a readily measurable parameter in many studies of aqueous solutions within this temperature range.

To avoid difficulties concerning the precise definition of pH[76] all practical measurements of pH are normally made with respect to a series of seven standard buffer solutions which have been calibrated on a conventional scale between 0°C and 95°C. The published results have been collected and summarised.[77] The suitability of the buffers for use at higher temperatures has now been examined and those found to be stable at these temperatures, and suitable for studying using the experimental method employed, have been calibrated.[78]

The method of measurement and assignment of pH to the solution was similar to that used by Bates. The hydrogen electrode provided a measurement of hydrogen ion activity and the silver–silver chloride electrode acted as a reference electrode in KCl solution. Three different concentrations of this in each buffer solution were used in measurements. Thus the cell was

$$\text{Pt}(H_2)|\text{buffer, KCl(m)}|\text{Ag–Ag Cl, Pt}$$

where Pt was replaced by Pd or Ir when its catalytic activity made this desirable. Sufficiently accurate values of $E^o_{Ag,AgCl}$ have been published for the temperature range 25–275°C so that the emf of the hydrogen electrode could be obtained. The cell has no liquid junction and therefore its half cell reactions are

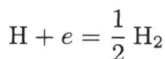

$$H + e = \frac{1}{2} H_2$$

$$\text{AgCl} + e = \text{Ag} + \text{Cl}^-$$

with half-cell potentials

$$E_{H^+,H_2} = E^o_{H^+,H_2} - \frac{RT}{F} \ln(\{H_2\}^{1/2}/\{H^+\})$$

$$E_{Ag,AgCl} = E^o_{Ag,AgCl} - \frac{RT}{F} \ln(\{Ag\}\{Cl^-\}/\{AgCl\})$$

so that the full cell potential

$$E = E_{Ag,AgCl} - E_{H^+,H_2}$$

$$= E^o_{Ag,AgCl} - \frac{2.303RT}{F} \log(\{Cl^-\}\{H^+\}/\{H_2\}^{1/2})$$

$\{Cl^-\}$ in the buffer solution is determined by the molality of the KCl added to it and $\{Cl^-\} = m_{KCl}\gamma_{Cl^-}$ so that

$$p(\{H\}\gamma_{Cl^-}) = -\log(\{H^+\}\gamma_{Cl^-})$$

$$= \frac{E - E^o_{Ag,AgCl}}{2.303RT/F} - \log m_{KCl} + \frac{1}{2} \log\{H_2\}.$$

Thus what was measured was $p(\{H\}\gamma_{Cl^-})$ for a series of chloride concentrations in each buffer and the values at a particular temperature were extrapolated to zero chloride concentration to give $p(\{H\}\gamma_{Cl^-})^\circ$ which is a totally thermodynamically defined function. The pH of the buffer solution in the absence of chloride is

$$p\{H\} = p(\{H\}\gamma_{Cl^-})^\circ + \log\gamma_{Cl^-}^\circ$$

and $\gamma_{Cl^-}^\circ$ is assumed to be given by

$$\log\gamma_{Cl^-}^\circ = -AI^{1/2}/(1+1.5I^{1/2})$$

where A is the Debye–Hückel limiting slope and is temperature dependent. This assumption[76] is the only nonthermodynamic part of the definition of $p\{H\}$ and makes γ_{Cl^-} effectively equal to the mean ionic activity coefficient of NaCl in solutions up to 0.1 m. The value of I, the ionic strength of the buffer solution in the absence of KCl was calculated from the $p(\{H\}\gamma_{Cl^-})^\circ$ data and any relevant equilibrium constants for the buffering salt.

The seven standard buffer solutions are given in Table 3.12.

Table 3.12. Standard buffer solutions and their pH values at $25°C$ on the practical scale.

Substance	Molality	pH at $25°C$
$KH_3Oxalate_2 \cdot 2H_2O$	0.05	1.679
KH Tartrate	Sat. at $25°C$, 0.0341	3.557
KH Phthalate	0.05	4.008
KH_2PO_4 Na_2HPO_4	0.025 0.025	6.865
KH_2PO_4 Na_2HPO_4	0.008695 0.03043	7.413
$Na_2B_4O_7.10H_2O$	0.01	9.180
$Ca(OH)_2$	Sat. at $25°C$, 0.0203	12.454

Of these the tetroxalate and tartrate solutions were calibrated but they both decompose at relatively low temperatures. The phthalate solution is thermally very stable and an ideal buffer but was subject to reduction by hydrogen during its calibration. This would not affect its use under normal conditions. The phosphate and borate buffers were found to be probably satisfactory for use at high temperatures but their pH values were too high for silver–silver chloride electrodes to be used as reference electrodes during their calibration. The solubility of lime, $Ca(OH)_2$, decreases at elevated temperatures.

Table 3.13.   pH values on the practical scale at elevated temperatures of some buffer solutions.[78]

Temp°C	Oxalate		Tartrate		Phthalate	
	pH	$-\log \gamma^o_{Cl}$	pH	$-\log \gamma^o_{Cl}$	pH	$-\log \gamma^o_{Cl-}$
95	1.779	0.114				
100	1.792	0.115	3.676	0.093	4.254	0.103
105	1.804	0.116	3.690	0.094	4.284	0.104
110	1.817	0.117	3.705	0.095	4.313	0.105
115	1.830	0.118	3.721	0.096	4.340	0.107
120	1.843	0.120	3.738	0.097	4.366	0.108
125			3.753	0.099	4.391	0.110
130			3.768	0.100	4.416	0.111
135			3.781	0.101	4.441	0.113
140			3.791	0.103	4.465	0.115
145			3.800	0.104	4.490	0.116
150			3.804	0.106	4.514	0.118
155			3.806	0.107	4.537	0.120
160			3.802	0.109	4.560	0.122
165					4.582	0.123
170					4.602	0.125
175					4.619	0.127
180					4.633	0.130
185					4.645	0.131
190					4.651	0.134
195					4.654	0.136
200					4.651	0.138

The experimental results published[78] relating to the tetroxalate, tartrate and phthalate buffers gave values of $p(\{H\})\gamma_{Cl-}$. The values of pH on the practical scale are given in Table 3.13.

3.10.   Precipitation of Solids at Elevated Temperatures

3.10.1.   *Introduction*

The temperature at which a reaction is carried out is usually determined by kinetic considerations, although when a precipitate is formed improved crystallinity and consequent ease of solid–liquid separation may lead to use of an elevated temperature. In some cases thermodynamic factors may make it essential to use a high temperature. Some examples are contained in Sec. 3.8.6, precipitation of iron. There is a general tendency for less hydrated species to be present in an aqueous system as the temperature is raised and this often leads to precipitation of a solid. In cases where a basic salt is formed this behaviour is commonly described as "high temperature hydrolysis".

The critical temperature of water is 374.1°C. At 25°C, the dielectric constant of water is 78.6. At 300°C, it is about 20. Thus water is a less polar solvent at high temperatures than at 25°C and it is not possible to predict that a successful theoretical treatment of solutions of salts which behave as strong electrolyes at 25°C will also be successful for solutions of them at high temperatures. The appropriate test is to apply the Debye–Hückel equation to determine whether it provides a measure of the long range interactions between ions at elevated temperatures. The limiting law is expressed by

$$- \log \gamma_i = A''(\varepsilon_r T)^{-3/2} z_i^2 I^{1/2} \tag{2.22}$$

and

$$- \log \gamma_\pm = A z_+ z_- I^{1/2} . \tag{2.25}$$

The Debye–Hückel coefficient A is used in Pitzer's equations (Sec. 2.2.3) in the form appropriate for the calculation of osmotic coefficients, in order to take account of the long range interactions between ions. The short range interactions are dealt with using a virial coefficient form of equation. If there were strong ion pairing at high temperatures the second virial coefficient parameters $\beta^{(0)}$ and $\beta^{(1)}$ would become large negative numbers. If ion pairing became very strong the virial coefficient form of equation would fail completely.

The properties of sodium chloride solutions at temperatures up to 300°C have been collected from the literature and examined in terms of the values of $\beta^{(0)}$ and $\beta^{(1)}$.[25] The properties at 300°C are qualitatively very similar to those at 25°C. The greatest change is in the Debye–Hückel parameter A_ϕ itself. The value of this doubles between 25°C and 300°C and although changes in T and the density of water are substantial the decrease in the value of the dielectric constant is by far the greatest cause of this. All of the effects related to the limiting law are intensified as the temperature increases. The value of $\beta^{(1)}$ increases steadily with temperature and at 300°C $\beta^{(0)}$ is still positive with a value close to that at 25°C. However, while $\beta^{(0)}$ increases between 25°C and 100°C it decreases between 150°C and 300°C. The term for triple ion interaction, C^ϕ is very small throughout the whole range of temperature studied and there was no indication of extreme behaviour up to 300°C. In view of this evidence the behaviour of strong electrolytes at temperatures up to 300°C can be regarded as being qualitatively similar to that at 25°C.

3.10.2. *True Solubility and Temperature*

Two different effects must be distinguished, the effect of temperature on true solubility, and hydrolysis, which may occur together. The effect of temperature on the solubility of salts depends mainly on the anion. The solubilities of potassium chloride, bromide and iodide and of sodium chloride increase steadily from 25°C through 370°C and this is true also for the halides of most other metals for which data are available. There are exceptions however. The solubility of barium chloride begins to decrease below 370°C whereas those of barium bromide, calcium chloride and strontium chloride continue to increase. The solubilities of nitrates also increase continuously with increasing temperature.

Among sulphates, ammonium sulphate is exceptional in that its solubility increases continuously with increasing temperature, through the critical temperature of water. The solubilities of most metal sulphates tend towards zero at the critical temperature, in many cases passing through a maximum at a temperature between 150°C and 300°C. Metals which do not hydrolyse show this very clearly, for example sodium, potassium and lithium sulphates. The lanthanides, which also do not hydrolyse, show decreasing solubilities above 20°C and low solubility at 100°C. The difference in the effects of temperature on the solubility of salts of different anions has been attributed to the "ionic character" of the salts.[79] Those which melt above 830°C are said to have a high proportion of ionic bonding and these show decreasing solubility at higher temperatures. Compounds which melt below 630°C are said to have an appreciable fraction of covalent bonding and have solubilites which increase steadily with temperature. This is a useful guide to the probable behaviour of a salt for which data are not available. It predicts the behaviour of calcium carbonate, sodium carbonate, calcium hydroxide, barium hydroxide, sodium hydroxide, potassium hydroxide, of selenates and selenites, and of fluorides and other halides. Many anions of course decompose at a temperature below 830°C.

3.10.3. *Hydrolysis and Temperature*

The behaviour of sulphates which hydrolyse under some conditions depends on the experimental method used to study the dependence of solubility on temperature. If a concentrated solution is heated and the solubility measured then the dependence on temperature is a curve similar to that found for the alkali metal sulphates. The solids which deposit from the solutions when the solubility falls are not basic sulphates but simple sulphates, either anhydrous or monohydrates. This behaviour is found with copper, nickel, magnesium, iron (II), manganese (II), cobalt (II), zinc and cadmium.

Solutions obtained by leaching usually have metal concentrations up to about 0.5 M and do not behave in this way. On heating hydrolysis occurs and basic sulphates form. Cobalt (II), nickel and copper each form at least two such compounds, the composition depending on the initial pH of the solution before it is heated to between 100°C and 200°C. The difference in behaviour when using the two experimental procedures can be attributed to the different changes in pH which occur on hydrolysis of concentrated and dilute solutions. The temperature at which precipitation occurs can also control the degree of hydration of the solid formed. Thus on heating uranyl nitrate solutions the compounds formed are:[80] at 330°C, $UO_3 \cdot 0.5H_2O$; 310°C, $UO_3 \cdot 0.8$–$1.0H_2O$; 270°C, $UO_3 \cdot 0.8$–$1.0H_2O$; 200°C, $UO_3 \cdot H_2O$; 100°C, $UO_3 \cdot H_2O$; 40°C, β-$UO_3 \cdot 2H_2O$. Anhydrous UO_3 was not formed at or below 330°C. Except for $UO_3 \cdot 0.8$–$1.0H_2O$ these solids are well defined thermochemically although as nonstoichiometric oxygen-deficient compounds.[50]

Control of the composition of a compound precipitated from a solution by selection of the composition of this and its temperature can be used for large scale processes. In the early 1960's, it was feared that a cartel might be set up among producers of bauxite to force up its price. Work was undertaken therefore to have a process available to supply an alternative feed to aluminium smelters if that occured.[81,82] The solution would preferably be produced from low grade ores such as kaolin by leaching with sulphuric acid and so would contain iron and other impurities. It was proposed that the solution would be treated with sulphur dioxide to reduce the iron to Fe^{II} and basic aluminium sulphate of composition $3Al_2O_3 \cdot 4SO_3 \cdot 9H_2O$ precipitated by heating the solution at 200°C. The ratio of sulphate to aluminium must be controlled in order to obtain this particular salt. Ferrous sulphate does not hydrolyse in the solution at 200°C. Alumina would be obtained by roasting the basic sulphate. The cost of the plant to operate the leaching, precipitation and calcining steps would be about twice that required for the Bayer process universally used at present. However the cost of the ore used would be much less than that of bauxite.

3.11.   Speciation Diagrams

3.11.1.   *Introduction*

Predominance area diagrams such as those constructed in Secs. 3.3 to 3.6 are not satisfactory for indicating how the constituents present in a solution change as the molar ratio of a ligand to a metal ion is altered. Speciation diagrams are used to show this, the simplest kind being that in Fig. 3.10. The equilibrium constants used in constructing this were measured using 2 M ammonium nitrate as the supporting electrolyte.[59] The curves show the fractions of the nickel present in the solution as the simple hydrated ion and each of the six ammines as a function of p{NH$_3$}.

In some cases published data for the stability constants of species in a system are contradictory. This was the case for the chlorocomplexes of Cu^I which are of importance in hydrometallurgy. The formation of these in solutions having sufficiently high concentrations of chloride makes the Cu^I state thermodynamically stable at ambient temperatures, whereas the ion Cu$^+$ is not (Sec. 3.5.3). In such solutions the Cu^{II}–Cu^I couple can be used as an oxidising system (Sec. 5.3.5). The position was clarified[83] by reexamination of published experimental data on the stabilities of the Cu^I complexes using an adaptation

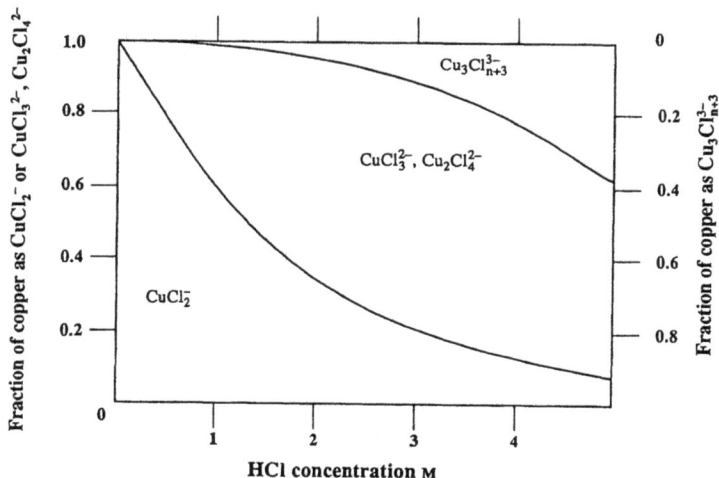

Fig. 3.11.   Species containing Cu^I present in solutions of CuCl in HCl of different concentrations. Cu$_3$Cl$_{n+3}^{3-}$ was used to represent the triply charged ions, having $n = 3$.

of the methods of Pitzer and his coworkers, Secs. 2.2.2 to 2.2.4. After deciding which of the proposed species were likely to actually exist, equilibrium constants were calculated which are consistent with tabulated[50] values of ΔG° of formation of the species, Fig. 3.11.

A characteristic feature of hydrometallurgical processes is the control of chemical conditions in order to achieve a particular result. Often it is to cause a metal to form an anionic species in the solution while leaving those metals from which it is to be separated as cations. This is the case when uranium is extracted from sulphate leach liquors as the species $UO_2(SO_4)_2^{2-}$ by solvent extraction with an amine from acidic solutions. In this case the conditions necessary can readily be seen on an E-pH diagram.

Solvent extraction from chloride solutions is an excellent method of separating many metals from one another. Its basis is the fact that for a metal M^{z+} the species MCl_z is uncharged and if the metal coordinates with a larger

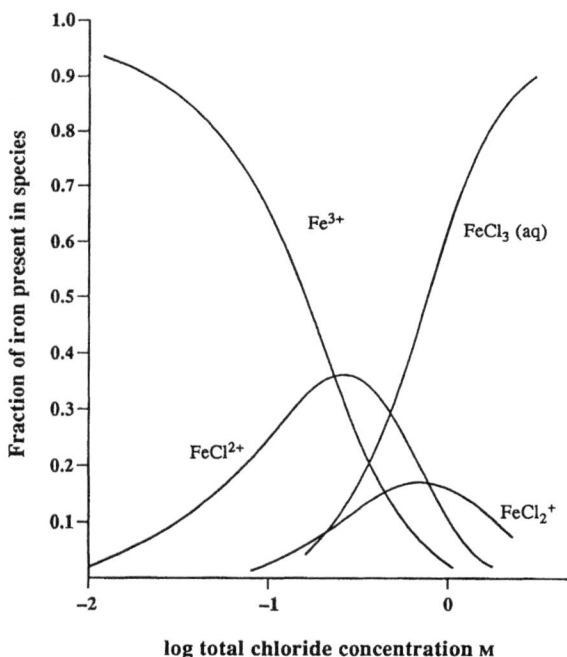

Fig. 3.12. Speciation diagram for Fe^{III} in chloride solutions. $[Fe^{III}] = 0.086$ M in 1 M $HClO_4$ with ionic strength adjusted to 2.6 with $NaClO_4$.[85]

number of chloride ions than z it becomes anionic. Since the values of the equilibrium constants for the formation of the complex species are unique to the individual elements these form uncharged and anionic species at different concentrations of chloride ion. A review of this and other aspects of chloride hydrometallurgy has been published.[84] In order to select conditions for experiments designed to lead to a process in which two or more metals are to be separated from one another making use of these differences, it is convenient to plot speciation diagrams showing the fraction of a metal present in each of its chlorocomplexes over the appropriate range of chloride concentrations. This may extend to 10 M. Examples are given in Figs. 3.12; 3.13 and 3.14 for Fe^{III}, Cu^{II} and Zn^{II} respectively. The data used in the diagram for Fe^{III} were obtained using $[Fe^{3+}] = 0.086$ M in 1 M $HClO_4$ with ionic strength adjusted to 2.6 with $NaClO_4$. The data for Cu^{II} and Zn^{II} relate to solutions containing the metals at concentrations below 0.01 M in LiCl solutions up to 10 M, with no adjustment of the ionic strength. For process development this

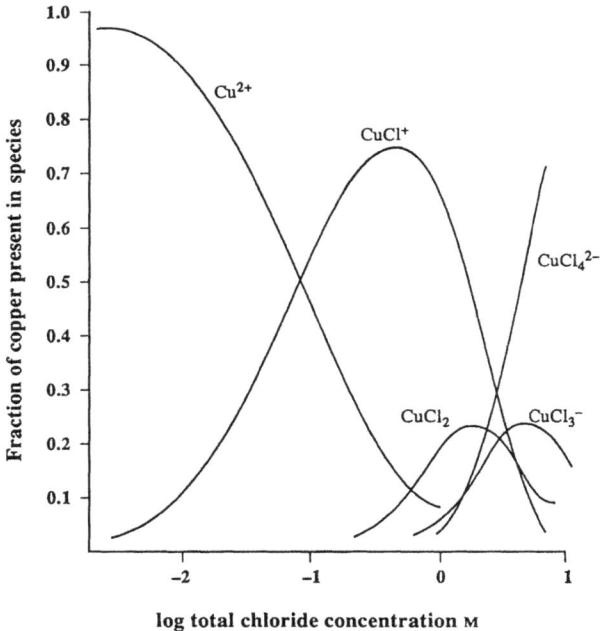

Fig. 3.13.   Speciation diagram for Cu^{II} in chloride solutions. $[Cu^{II}]$ less than 0.01 M in LiCl solutions up to 10 M with no adjustment of ionic strength.[86]

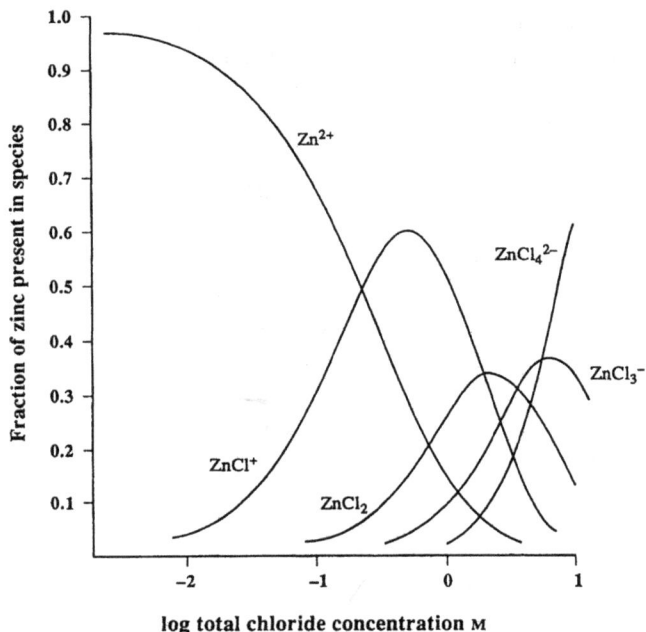

Fig. 3.14. Speciation diagram for Zn^{II} in chloride solutions. $q[Zn^{II}]$ less that 0.01 M in LiCl solutions with no adjustment of ionic strength.[87]

is usually the best approach but a much higher metal concentration would be used. When using thermodynamic equilibrium constants, where values have been extrapolated to infinite dilution, or $\Delta G°$ values for the species, the effects of changes in activity coefficients of species with concentration must be taken into consideration.

3.11.2.   *Speciation in Multicomponent Solutions*

The speciation diagrams considered in the previous section were drawn for a single metal ion and ligand, Cl^-. A much more powerful type of diagram would show the speciation of all of the metals and anions in a solution and how the speciation changes as the activity of one metal, anion or uncharged ligand is changed. Such diagrams can be calculated and plotted by computer.

The chemical models for calculating equilibria and the computer programs which were available in 1979 for carrying out the calculations have been compared.[88] Almost all of the computerised models are based on association between ions, or ions and molecules, resulting in the formation of distinct chemical species to an extent described by equilibrium constants. In a solution containing a number of components many such equilibria may have to be considered and each is represented by a mass action expression based on the chemical equation representing it, and its equilibrium constant. When these equations are used to calculate speciation in a solution two conditions must be met:

(i) The mass balance condition requires that the calculated sum of the simple uncomplexed and the associated species should be equal to the number of moles of a metal or ligand initially stated to be present in a given quantity of water. These numbers must of course result in electrical charge neutrality.

(ii) The chemical equilibrium condition requires that the most stable arrangement for the system being considered should be found.

This most stable condition is defined either by the equilibrium constants for all of the mass action expressions of the system, or through the use of the Gibbs free energies for all of the components and the associated species derived from them, present in the system at equilibrium. In thermodynamic terms these alternatives are equivalent but they permit different numerical approaches to be used.

In the equilibrium constant approach the mass action expressions and equilibrium constants are substituted into the mass balance conditions, giving a set of nonlinear equations which must be solved simultaneously. The Gibbs free energy approach is a transformation of variables through the thermodynamic relationship for a reaction.[88,89]

$$\Delta G = \Delta G_r^\circ + RT \ln K = 0.$$

The total free energy of the system is then minimised for the set of species and their mole numbers, subject to the mass balance requirement. Whichever thermodynamic approach is used the problem is to find a solution to a set of nonlinear equations. The thermodynamic formalism and mathematical methods which can be employed have been reviewed.[90,91]

Because thermodynamic relationships are employed in these methods of calculating speciation, activities of species must be used, not concentrations. Combination of Pitzer's ion interaction model and equations involving the concept of complex formation has been used[92] to produce a large set of data for the calculation of activity coefficients of major cations, trace level cations and a wide variety of anions, applicable to solutions of ionic strength up to 1 m. The set was chosen for use in calculations relating to natural waters, an area in which speciation calculations are widely used.

In calculations relating to E-pH predominance area diagrams in Secs. 3.5 to 3.6 molal activites were used for species in solution and fugacities for gases. The same units must be employed in calculations of speciation. Conversion between concentration and activity is an entirely different calculation, to be carried out separately. This is necessary because of the mass balance condition (i) referred to above.

Database Systems

There are now a number of fully Integrated Thermochemical Database (ITD) systems in the fields of chemical and physical metallurgy which are accessible to the public. They offer extensive thermochemical databanks and powerful software. With these, thermodynamic calculations can be carried out relating to chemical equilibria and phase diagrams for multicomponent multiphase systems. A very large amount of thermochemical data for substances at high temperatures has been measured because of the application of chemical thermodynamics to pyrometallurgy. Consequently ITD facilities have been used extensively for calculating and plotting predominance area and phase diagrams for systems at high temperatures[94] and in the application of thermodynamics to the development and use of materials, including alloys, ceramics and coatings.[95] In the field of hydrometallurgy E-pH diagrams are regularly calculated and plotted.

ITD facilities can be applied to very complex inorganic chemical problems of interest to hydrometallurgists. Although process liquors contain numerous elements in some cases, the main difficulties in applying thermodynamics to processes for treating them have been the facts that they are usually moderately concentrated in at least one component, and that the theory for the calculation of activity coefficients for solutions of mixed electrolytes was unsatisfactory. It is probable that use of Pitzer's methods will overcome both of these difficulties, particularly if the computer is used to carry out the large number of simple calculations involved.

A different kind of complexity arises when the behaviour of elements in ground water or effluents from hydrometallurgical or other industrial plants is to be studied. Usually there is particular interest in the presence of certain elements at very low concentrations because of the legal limits imposed on maximum levels which may be discharged. In addition in many cases a large number of elements may be present and their concentrations in ground water will vary throughout the year according to the rainfall and other factors. When river water reaches the sea it mixes with sea water in the estuary and both pH and chloride concentration increase. The behaviour and fate of toxic and other harmful substances then becomes important. Speciation diagrams were employed in an analysis of the behaviour of some metals and anions based on a survey of the Carnon River system in South West England.

The Carnon River drains a mineralised area of Devonian slates and shales intruded by mineral lodes containing sulphides. The principal substances are ZnS, $CuFeS_2$, FeS_2, $FeAsS$ and PbS. The ore bodies have been mined for a very long period and one mine water drainage system collects underground waters from old mine workings over an area of more than 18 km^2.[96] The waters from this drainage system flow directly into the Carnon River. The only active mine in the area in the early 1980's was responsible for the two major mine water inputs into the Carnon River system. Bacterial oxidation of the sulphide minerals results in highly acidic mine waters with high concentrations of trace elements compared with most other river waters. The Carnon River flows at a rate of approximately 0.5 to 2.0 m^3 s^{-1} into Restronguet Creek and approximately 3.3 tons per day of Fe were transported into the estuary waters during 1982, when the survey was carried out.

A model was required which would indicate the speciation of the following elements which were present mainly as labile ions in solution (water filtered to remove particles larger than 0.45 μm); Fe, Mn, Cu, Pb, Zn, Cd, Co, Ni and As. It had to include other cations which might compete for ligands in solution or affect the behaviour of the system in some other way. The following cations, anions, acids and the anions derived from them were selected for consideration:

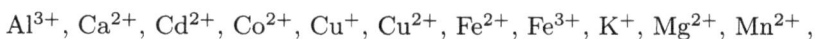

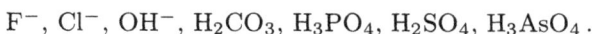

Al^{3+}, Ca^{2+}, Cd^{2+}, Co^{2+}, Cu^+, Cu^{2+}, Fe^{2+}, Fe^{3+}, K^+, Mg^{2+}, Mn^{2+},

Ni^{2+}, Pb^{2+}, Zn^{2+}

F^-, Cl^-, OH^-, H_2CO_3, H_3PO_4, H_2SO_4, H_3AsO_4.

The concentrations of these components in river and estuarine water used in speciation calculations are given in Table 3.14 and were based on analyses of samples taken monthly for one year, together with the pH and *Eh* values.

Table 3.14.   Component concentrations of river and estuarine water used in speciation calculations, gl^{-1}.

Component	Al	Ca	Cd	Co	Cu
concentration in fresh water	1.03(−4)	1.46(−3)	8.90(−8)	6.45(−7)	9.36(−6)
estuarine water	4.95(−5)	1.15(−2)	8.90(−10)	1.70(−9)	3.15(−8)
	Fe	K	Mg	Mn	Na
	1.34(−4)	1.48(−4)	3.16(−4)	1.88(−5)	2.47(−6)
	9.13(−7)	8.96(−3)	5.28(−2)	9.10(−8)	4.22(−1)
Component	Ni	Pb	Zn	F	Cl
concentrations in fresh water	9.71(−7)	4.83(−8)	1.11(−4)	7.08(−5)	7.69(−3)
estuarine water	1.70(−8)	1.48(−8)	5.66(−7)	7.40(−5)	5.04(−1)
	SO_4^{2-}	P	As		
	1.20(−3)	4.33(−6)	1.60(−6)		
	2.82(−2)	1.42(−5)	1.33(−8)		

pH Eh(mV)
4.8   +500
8.12   +370
The partial pressure of CO_2 gas in the earth's atmosphere, $10^{-3.52}$ atm was entered in the database.
Numbers in brackets indicate powers of 10.

The symbol Eh is used to emphasise that it is the redox potential on the hydrogen scale.

All of the solids and solute species for which $\Delta G°$ values could be found were listed using Refs. 48 and 50, calculated from equilibrium constants in the former case. In all 17 anionic ligands, 14 simple hydrated metal ions and 211 associated species were assembled. This was too large a number for the program to handle and preliminary calculations were carried out and plotted in order that those species which were not significant within the range of conditions of interest could be rejected. It was assumed that the regions of interest would be those where the activities of species shown on the diagrams plotted were similar to the measured concentrations of the metal contained in each of the water samples. To ensure a safe margin of error in selecting species for inclusion in subsequent calculations those containing 0.01% of the total quantity of that metal in the solution were included.

Some of the plotted results of these preliminary calculations are shown in Figs. 3.15 to 3.24, redrawn from the computer plots. These speciation diagrams were produced by C.A. Johnson in cooperation with the National Physical Laboratory using software developed there. Equivalent software is now incorporated in MTDATA, the NPL Databank for Metallurgical Thermochemistry. The present author is grateful to Dr T.I. Barry for his assistance and cooperation in this and other work.

Discussions of Figs. 3.15 to 3.24

Copper

Three diagrams are included for copper. Figure 3.15 was calculated and plotted for the river water containing [Cu total] $= 10^{-5}$ moles, indicated by $\log n$ Cu $= -5.00$, over the pH range 3 to 8. This range was selected to cover water emerging from a zone containing metal sulphides leaching *in situ* by bacterial oxidation, producing a solution having pH about 3.0, to estuarine water with pH about 8. All of the species containing copper and the anions selected

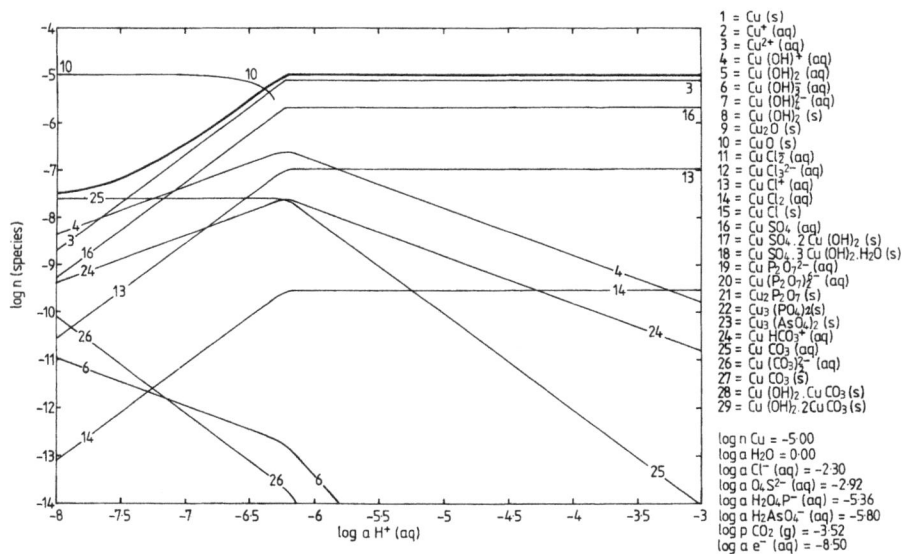

Fig. 3.15. Speciation diagram for Cu under conditions representing river water changing its pH over the range 3 to 8. Total quantity of copper present $= 10^{-5}$ moles written as $\log n\text{Cu} = -5.00$. $\log\{H_2O\} = 0.00$; $\log\{Cl^-(aq)\} = -2.30$; $\log\{SO_4^{2-}(aq)\} = -2.92$; $\log\{H_2PO_4^-(aq)\} = -5.36$; $\log\{H_2AsO_4^-(aq)\} = -5.80$; $\log p_{CO_2}(g) = -3.52$; $\log\{e^-(aq)\} = -8.50$.

for consideration, for which ΔG^o values were found are listed and numbered in the key for the copper set. The activities of the anions used in calculating Fig. 3.15 are shown, these derived from dissolved CO_2 being indicated by p_{CO_2}. The oxidation potential for which the diagram was calculated is given on the pe scale (Sec. 3.3.4), as $\log ae^-(aq)$. *Eh* for river water is $+0.500$ V so that $0.500 = 0.05917$ pe giving pe $= 8.45$ or $\log$ pe $= -8.45$, approximated as -8.5. Since no species containing both Cu and F^- is listed, F^- was not considered in the calculations.

Between pH 3 and 6 the topmost line, drawn heavier than the others and not numbered, indicates the total quantity n of copper in solution, 10^{-5} since all species occuring in this range of pH are in solution. Between pH 6 and 6.5 species 10, $CuO(s)$, precipitates and at pH 8 appears to contain all of the copper present. The heavy line, however, indicates that at this pH the solution still contains about $10^{-7.5}$ moles of copper, present as a number of species among which $Cu(CO_3)(aq)$, $Cu(OH)^+(aq)$ and $Cu^{2+}(aq)$ are predominant.

Figure 3.16 was calculated for river water having the same values of nCu and of anions other than Cl^- as were used in Fig. 3.15, but with $\log a$H$^+$ set

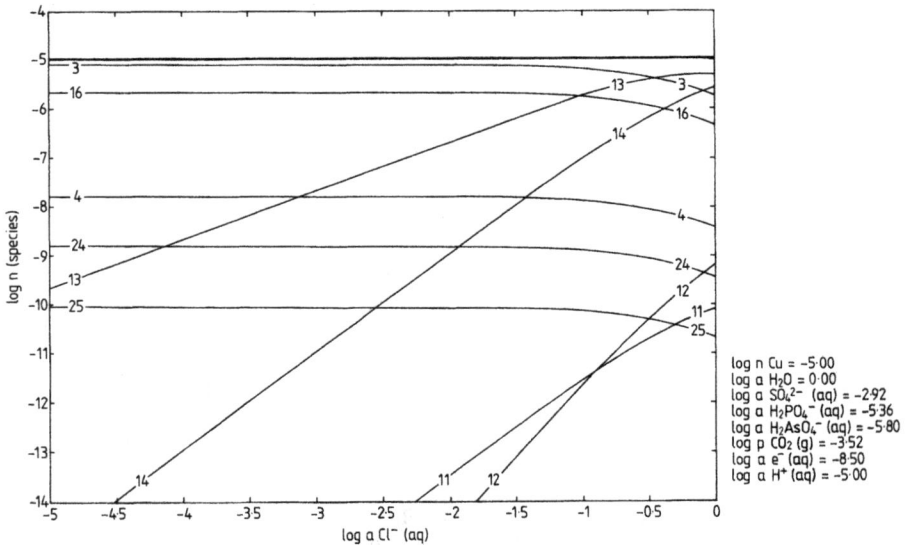

key:
log n Cu = -5.00
log a H₂O = 0.00
log a SO₄²⁻ (aq) = -2.92
log a H₂PO₄⁻ (aq) = -5.36
log a H₂AsO₄⁻ (aq) = -5.80
log p CO₂ (g) = -3.52
log a e⁻ (aq) = -8.50
log a H⁺ (aq) = -5.00

Fig. 3.16. Speciation diagram for Cu as a function of $\{Cl^-\}$. Conditions as for Fig. 3.15 except that $\log\{H^+(aq)\}$ is constant at -5.00 and $\{Cl^-\}$ is varied.

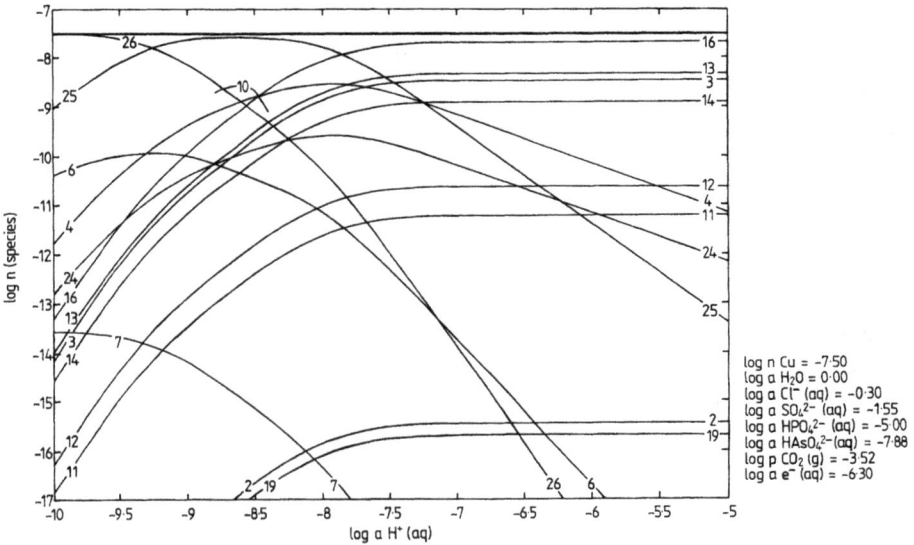

Fig. 3.17.   Speciation diagram for Cu as a function of pH. $\log n$ Cu $= -7.50$; $\log \{H_2O\} = 0.00$; $\log\{Cl^-(aq)\} = -0.30$; $\log\{SO_4^{2-}(aq)\} = -1.55$; $\log\{HPO_4^{2-}(aq)\} = -5.00$; $\log\{HAsO_4^{2-}(aq)\} = -7.88$; $\log p_{CO_2}(g) = -3.52$; $\log\{e^-(aq)\} = -6.30$.

Key to species considered in Copper systems, Figs. 3.15–3.17.

Number	Species	Number	Species
1	$Cu(s)$	16	$CuSO_4(aq)$
2	$Cu^+(aq)$	17	$CuSO_4 \cdot 2Cu(OH)_2(s)$
3	$Cu^{2+}(aq)$	18	$CuSO_4 \cdot 3Cu(OH)_2 \cdot H_2O(s)$
4	$Cu(OH)^+(aq)$	19	$CuP_2O_7^{2-}(aq)$
5	$Cu(OH)_2(aq)$	20	$Cu(P_2O_7)_2^{6-}(aq)$
6	$Cu(OH)_3^-(aq)$	21	$Cu_2P_2O_7(s)$
7	$Cu(OH)_4^{2-}(aq)$	22	$Cu_3(PO_4)_2(s)$
8	$Cu(OH)_2(s)$	23	$Cu_3(AsO_4)_2(s)$
9	$Cu_2O(s)$	24	$CuHCO_3^+(aq)$
10	$CuO(s)$	25	$CuCO_3(aq)$
11	$CuCl_2^-(aq)$	26	$Cu(CO_3)_2^{2-}(aq)$
12	$CuCl_3^{2-}(aq)$	27	$CuCO_3(s)$
13	$CuCl^+(aq)$	28	$Cu(OH)_2 \cdot CuCO_3(s)$
14	$CuCl_2(aq)$	29	$Cu(OH)_2 \cdot 2CuCO_3(s)$
15	$CuCl(s)$		

at -5.00 and $\log a\mathrm{Cl}^-$ varying over the range -5 to 0. Thus the river water could mix with cleaner water or sea water and a set of diagrams could be drawn for varying $\{\mathrm{Cl}^-\}$ at different values of pH if required. The steady rise in the amounts of $\mathrm{CuCl}^+(\mathrm{aq})$ and $\mathrm{CuCl}_2(\mathrm{aq})$ as $\{\mathrm{Cl}^-\}$ increases leads to these species being predominant over $\mathrm{Cu}^{2+}(\mathrm{aq})$ at the higher values of $\{\mathrm{Cl}^-\}$ plotted. It is of interest to note that even though the value of E is approximately $+500$ mV significant quantities of the $\mathrm{Cu^I}$ chloro species $\mathrm{CuCl}_2^-(\mathrm{aq})$ and $\mathrm{CuCl}_3^{2-}(\mathrm{aq})$ form at the higher values of $\{\mathrm{Cl}^-\}$.

Figure 3.17 was calculated for conditions other than pH approximating to estuarine water. Points of interest are that CuO precipitates only in small quantity at a pH of approximately 8.6 ± 0.3, redissolving on both sides of that range. In more alkaline solutions $\mathrm{Cu(CO_3)}(\mathrm{aq})$ and $\mathrm{Cu(CO_3)}_2^{2-}(\mathrm{aq})$ predominate, with much smaller quantities of hydroxy species present. On the acid side $\mathrm{Cu(SO_4)}(\mathrm{aq})$ and $\mathrm{CuCl}^+(\mathrm{aq})$ are predominant with smaller amounts of $\mathrm{Cu}^{2+}(\mathrm{aq})$ and $\mathrm{CuCl}_2(\mathrm{aq})$.

Zinc and Cadmium

Three diagrams are given for zinc; Fig. 3.18 for river water over the pH range 3 to 8, Fig. 3.19 for the same water at constant pH $= 5.00$ and $\log\{\mathrm{Cl}^-\}$ 0 to -5, and Fig. 3.20 for estuarine water in the pH range 5 to 10. The diagrams show the chemistry of the system with no new principles being involved. In Fig. 3.18, a single line is labelled as representing $\mathrm{Zn(OH)}^+(\mathrm{aq})$ (species 3) and $\mathrm{Zn(HCO_3)}^+(\mathrm{aq})$, (species 24), because the data are superimposed. Species 5 is species $8+\mathrm{H_2O}$. The behaviour of the chlorospecies in Fig. 3.19 is of interest and may be compared with the much more stable chlorospecies of cadmium, Figs. 3.21 and 3.22.

Aluminium

Figures 3.22 and 3.23 show how the chemistry of aluminium in the river water is dominated at pH values lower than about 5.5 by its fluorocomplexes and by precipitation of $\mathrm{Al(OH)}_3(\mathrm{s})$ at pH values between 5.5 and 8.

3.11.3. *Precipitation of Trace Metals in Estuarine Water*

The speciation diagrams for Cu, Zn and Cd show that each of these metals remains soluble over the ranges of conditions covered except at pH values above

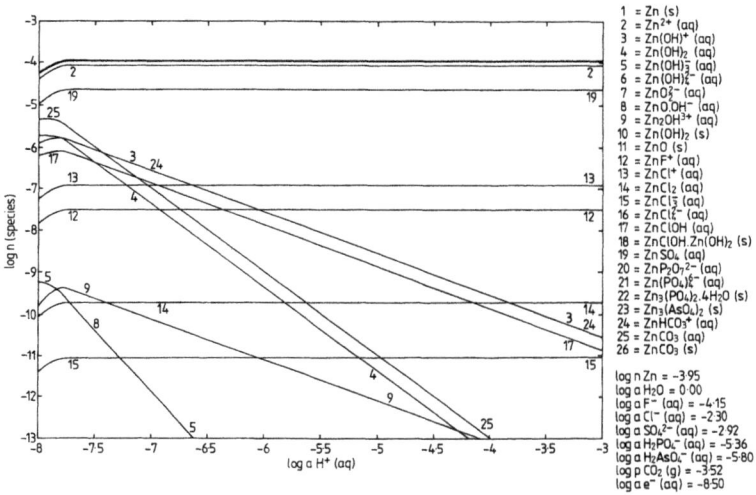

Fig. 3.18. Speciation diagram for Zn as a function of pH. $\log n\,\mathrm{Zn} = -3.95$; $\log\{\mathrm{H_2O}\} = 0.00$; $\log\{\mathrm{F^-(aq)}\} = -4.15$; $\log\{\mathrm{Cl^-(aq)}\} = -2.30$; $\log\{\mathrm{SO_4^{2-}(aq)}\} = -2.92$; $\log\{\mathrm{H_2PO_4^-(aq)}\} = -5.36$; $\log\{\mathrm{H_2AsO_4^-(aq)}\} = -5.80$; $\log p_{\mathrm{CO_2}}(\mathrm{g}) = -3.52$; $\log\{\mathrm{e^-(aq)}\} = -8.50$.

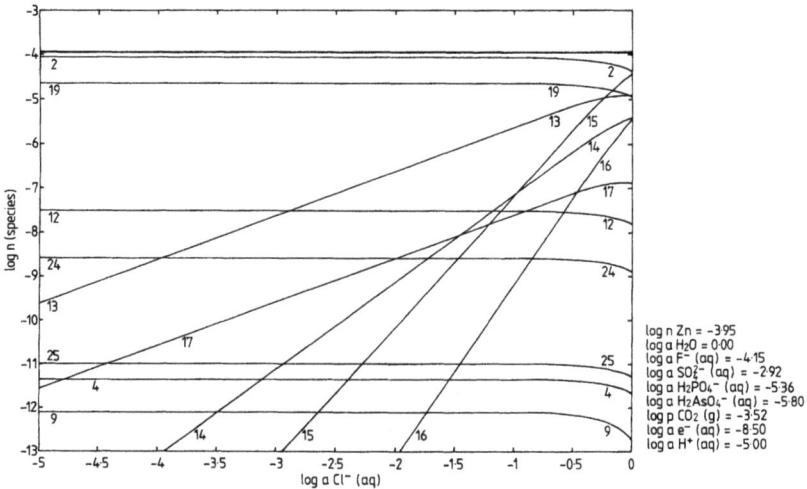

Fig. 3.19. Speciation diagram for Zn as a function of $\{\mathrm{Cl^-}\}$. Conditions as for Fig. 3.18 except that $\log\{\mathrm{H^+(aq)}\}$ is constant at -5.00 and $\{\mathrm{Cl^-(aq)}\}$ is varied.

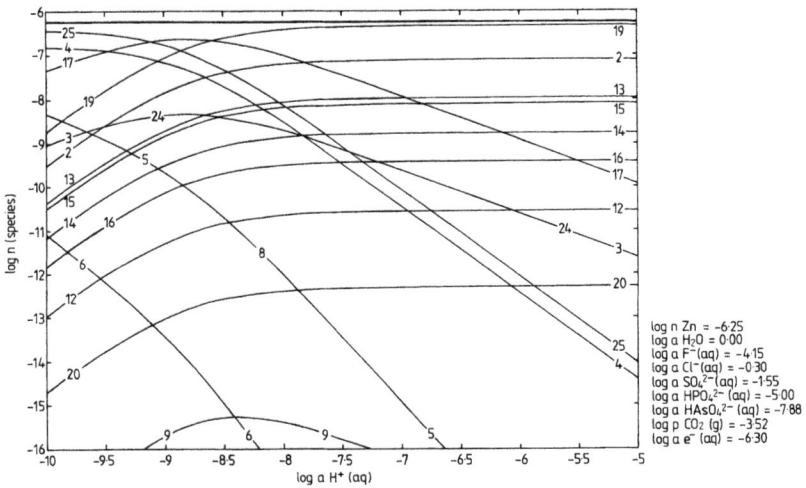

Fig. 3.20.   Speciation diagram for Zn as a function of pH. $\log n\,\text{Zn} = -6.25$; $\log\{H_2O\} = 0.00$; $\log\{F^-(aq)\} = -4.15$; $\log\{Cl^-(aq)\} = -0.30$; $\log\{SO_4^{2-}(aq)\} = -1.55$; $\log\{HPO_4^{2-}(aq)\} = -5.00$; $\log\{HAsO_4^{2-}(aq)\} = -7.88$; $\log p_{CO_2}(g) = -3.52$; $\log\{e^-(aq)\} = -6.30$.

Key to species considered in Zinc systems, Figs. 3.18–3.20.

Number	Species	Number	Species
1	$Zn(s)$	14	$ZnCl_2(aq)$
2	$Zn^{2+}(aq)$	15	$ZnCl_3^-(aq)$
3	$Zn(OH)^+(aq)$	16	$ZnCl_4^{2-}(aq)$
4	$Zn(OH)_2(aq)$	17	$ZnCl(OH)(aq)$
5	$Zn(OH)_3^-(aq)$	18	$ZnCl(OH)\cdot Zn(OH)_2(s)$
6	$Zn(OH)_4^{2-}(aq)$	19	$ZnSO_4(aq)$
7	$ZnO_2^{2-}(aq)$	20	$ZnP_2O_7^{2-}(aq)$
8	$ZnO\cdot OH^-(aq)$	21	$Zn(PO_4)_4^{6-}(aq)$
9	$Zn_2(OH)^{3+}(aq)$	22	$Zn_3(PO_4)_2\cdot 4H_2O(s)$
10	$Zn(OH)_2(s)$	23	$Zn_3(AsO_4)_2(s)$
11	$ZnO(s)$	24	$ZnHCO_3^+(aq)$
12	$ZnF^+(aq)$	25	$ZnCO_3(aq)$
13	$ZnCl^+(aq)$	26	$ZnCO_3(s)$

6 in Fig. 3.15 and over a small range of pH in estuarine water where some CuO is indicated as being deposited. In practice CuO does not precipitate when the pH of a solution containing Cu^{II} is raised, the product being Cu(OH)$_2$ which has a higher solubility product than CuO. The precipitated hydroxide

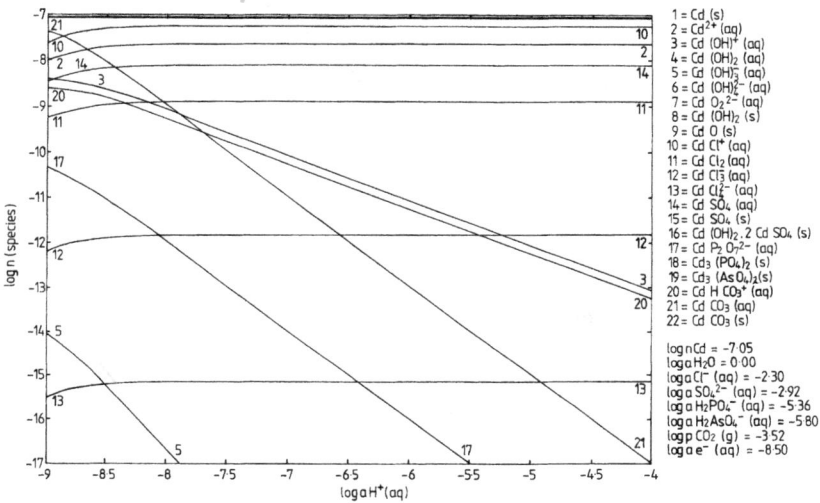

Fig. 3.21. Speciation diagram for Cd as a function of pH. $\log n\,\mathrm{Cd} = -7.05$; $\log\{H_2O\} = 0.00$; $\log\{Cl^-(aq)\} = -2.30$; $\log\{SO_4^{2-}(aq)\} = -2.92$; $\log\{H_2PO_4^-(aq)\} = -5.36$; $\log\{H_2AsO_4^{2-}(aq)\} = -5.80$; $\log p_{CO_2} = -3.52$; $\log\{e^-(aq)\} = -8.50$.

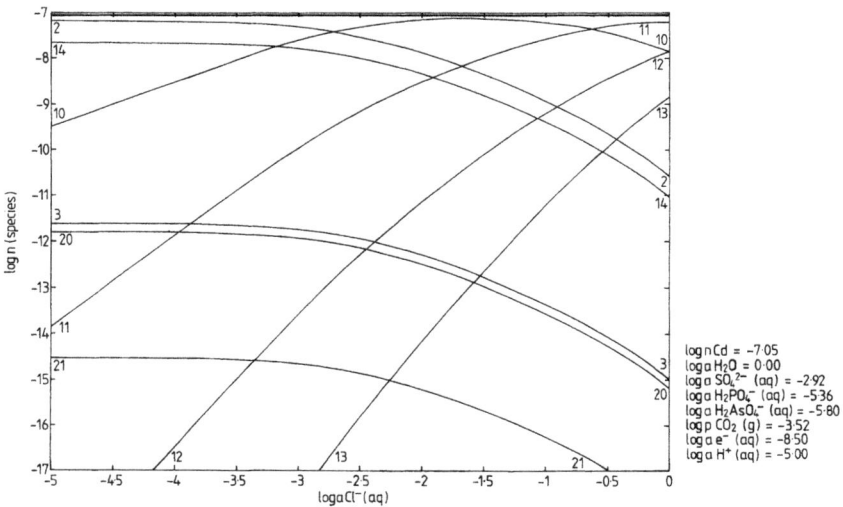

Fig. 3.22. Speciation diagram for Cd as a function of $\{Cl^-\}$. Conditions as for Fig. 3.21 except that $\log\{H^+(aq)\}$ is constant at -5.00 and $\{Cl^-\}$ is varied.

Key to species considered in Cadmium systems, Figs. 3.21–3.22.

Number	Species	Number	Species
1	$Cd(s)$	12	$CdCl_3^-$ (aq)
2	Cd^{2+} (aq)	13	$CdCl_4^{2-}$ (aq)
3	$Cd(OH)^+$ (aq)	14	$CdSO_4$ (aq)
4	$Cd(OH)_2$ (aq)	15	$CdSO_4$ (s)
5	$Cd(OH)_3^-$ (aq)	16	$Cd(OH)_2 \cdot 2CdSO_4$ (s)
6	$Cd(OH)_4^{2-}$ (aq)	17	$CdP_2O_7^{2-}$ (aq)
7	CdO_2^{2-} (aq)	18	$Cd_3(PO_4)_2$ (s)
8	$Cd(OH)_2$ (s)	19	$Cd_3(AsO_4)_2$ (s)
9	$CdO(s)$	20	$CdHCO_3^+$ (aq)
10	$CdCl^+$ (aq)	21	$CdCO_3$ (aq)
11	$CdCl_2$ (aq)	22	$CdCO_3$ (s)

is thermochemically ill-defined and no value of ΔG° of formation is listed for it.[50] However it is likely that copper does not precipitate from Carnon River water as its hydroxide when mixing with sea water takes place.

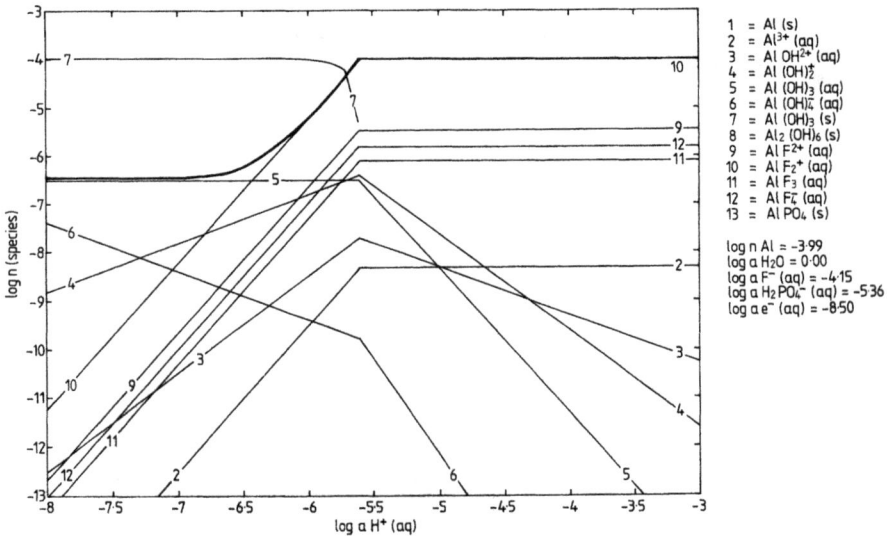

Fig. 3.23. Speciation diagram for Al as a function of pH. $\log n\, Al = -3.99$; $\log \{H_2O\} = 0.00$; $\log\{F^-(aq)\} = -4.15$; $\log\{H_2PO_4^-(aq)\} = -5.36$; $\log\{e^-(aq)\} = -8.50$.

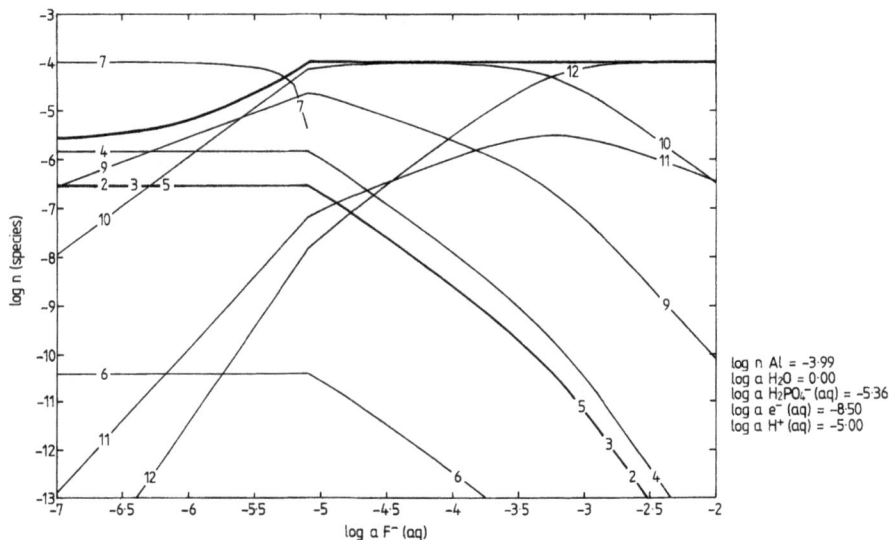

$\log n\ Al = -3.99$
$\log a\ H_2O = 0.00$
$\log a\ H_2PO_4^-\ (aq) = -5.36$
$\log a\ e^-\ (aq) = -8.50$
$\log a\ H^+\ (aq) = -5.00$

Fig. 3.24. Speciation diagram for Al as a function of $\{F^-\}$. Conditions as for Fig. 3.23 except that $\log\{H^+(aq)\}$ is constant at -5.00 and $\{F^-\}$ is varied.

Key to species considered in Aluminum systems, Figs. 3.23–3.24.

Number	Species	Number	Species
1	$Al(s)$	8	$Al_2O_3 \cdot H_2O(s)$
2	$Al^{3+}(aq)$	9	$AlF^{2+}(aq)$
3	$Al(OH)^{2+}(aq)$	10	$AlF_2^+(aq)$
4	$Al(OH)_2^+(aq)$	11	$AlF_3(aq)$
5	$Al(OH)_3(aq)$	12	$AlF_4^-(aq)$
6	$Al(OH)_4^-(aq)$	13	$AlPO_4(s)$
7	$Al(OH)_3(s)$		

Note: $Al(OH)_4^-(aq)$ is equivalent to $AlO_2^-(aq) + 2H_2O$.

In the intertidal region of the estuary an orange sediment has been deposited which is rich in iron and the trace elements present in the river water. It contains up to Fe, 15; Cu, 0.4; Zn, 0.2 and As 0.3%. $FeAsO_4$ is not indicated by the thermodynamic calculations as being a product under the conditions considered and it can be concluded that the arsenic, which is present in solution as arsenate, is coprecipitated with iron (III) hydroxide. Such coprecipitation is widely used in hydrometallurgical processes as a means of scavenging very low concentrations of impurity metals from process liquors.

The deposition of copper and zinc from the river water in the sediment has been attributed to adsorption on amorphous hydrated hydroxides of Fe^{III}.[96] Extensive studies on liquids emerging from regions of broken rock containing copper and iron sulphide minerals indicate that the pH value of the solution is maintained at about 3.0 by the buffering action of a solution of $Fe_2(SO_4)_3$ formed by oxidation, a rise in pH causing deposition of basic sulphates, which redissolve when the pH falls. These basic sulphates hydrolyse completely in the presence of large quantities of water which raises the pH. Thus fresh Fe^{III} hydroxide will be precipitating throughout a large volume as mine water is diluted. Also any Fe^{II} present in solution in ground water which has been in a reducing environment, one rich in organic material for example, will oxidise and also precipitate as Fe^{III} hydroxides. Such freshly precipitated hydrous solid has a much greater adsorptive power than aged material.

The surface adsorption reactions were treated[96] in terms of complex formation reactions between surface sites and the cations M^{2+} in solution. Rewriting the chemical equilibrium reaction slightly and writing the extensive surface as S

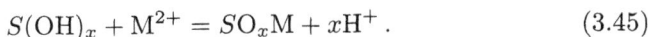

$$S(OH)_x + M^{2+} = SO_xM + xH^+ . \tag{3.45}$$

The equilibrium constant is

$$K = \{SO_xM\}[H^+]^x / \{S(OH)_x\}[M^{2+}]$$

where $\{SO_xM\}$ and $\{S(OH)_x\}$ represent the surface density of chemisorbed metal ions and of surface sites respectively in mole m^{-2}. Thus on the basis of this model the amount of metal ion, Cu^{2+} and Zn^{2+}, adsorbed on the hydrated amorphous iron (III) hydroxide depends on the pH of the solution.

3.11.4. *Adsorption and Coprecipitation of Metals from Solutions*

In physicochemical texts adsorption is usually discussed in terms of a gas adsorbing on a solid. The gas is considered as being attracted to the solid by the same kind of forces as those which produce liquefaction of gases and cohesion in liquids. This is described as physical adsorption and is easily reversible when the gas pressure is reduced. In other cases the adsorbed gas is not readily removed and is regarded as being chemically bound to the solid, or chemisorbed. Langmuir assumed that a uniform layer of gas only one molecule thick can form on the surface of the solid and by use of a model in which gas molecules collide with the surface and a constant proportion of them adheres for an appreciable time he derived the Langmuir adsorption isotherm.

Next, adsorption of a solute at the surface of a solution is considered. As a result of this adsorption the free energy of the surface is reduced. If a solute lowers the surface tension it will tend to concentrate at the gas–liquid interface and is described as surface active. If the solute causes an increase in surface tension its concentration at the interface will be lower than its bulk concentration and negative adsorption occurs. The exact thermodynamic relationship between adsorption and surface tension is given by the Gibbs adsorption isotherm.

When the relationship between the amount of a solute adsorbed on a unit mass of solid suspended in a solution, and the concentration of the solute is considered the models used to derive the Langmuir and Gibbs adsorption isotherms have generally been found to be of little or no use. Instead, the Freundlich adsorption isotherm is applied

$$\frac{x}{m} = kc^n$$

where x is the amount of solute absorbed by mass m of solid, c is the equilibrium concentration of the solute in the solution, and k and n are arbitrary constants for a particular system, n being generally less than one. When applied to adsorption of a gas on a solid c is replaced by pressure. In many texts n is replaced by $1/n$, n then being greater than one. The Freundlich adsorption isotherm is usually described as being empirical and "does not always fit the experimental facts, but is at least as successful as any other single equation".[97] An extensive account of early work on adsorption of solutes on solids in suspension in solutions was given by Freundlich.[98]

His isotherm can readily be derived by considering an equilibrium reaction in which a solute T replaces a species R adsorped or bound to the extensive solid surface S

$$SR + nT = STn + R$$

$$K = \frac{[ST_n][R]}{[SR][T]^n}$$

$$\text{Then } \frac{[ST_n]}{[SR]} = K\frac{[T]^n}{[R]}$$

$[ST_n]/[SR]$ is the fraction of the available surface sites occupied by T and $[T]^n/[R]$ is the stoichiometric equivalence between T and R for adsorption sites. Under most practical conditions R is H^+ or OH^-, the ions being derived from H_2O molecules bound to the solid by the chemical forces providing the heat of wetting of the "clean" solid. A "clean" solid has no atoms, molecules or ions interacting with the unsatisfied binding forces of the atoms at the surface of the solid crystal lattice. The ionisation of the bound water molecules on the surface need not correspond to the bulk values of $pK_w = 14$ at $25°C$. When $R = H^+$ this model corresponds to the equilibrium proposed for the adsorption of Cu^{2+} and Zn^{2+} on precipitated hydrous Fe^{III} hydroxide.

Speciation calculations related to natural waters frequently include an allowance for adsorption of metals on solids, as well as interaction with organic materials in simple or polymeric form, usually regarded as carboxylic acids and referred to as "humic acids". Neither of these interactions can be considered as having any thermodynamic basis since the "equilibrium constants" are not related to values of free energies of formation of defined substances. Thus their inclusion can invalidate conclusions based on thermodynamic principles.

The theoretical basis of adsorption of solutes on solid surfaces has been developed very much further as a result of the great importance of the froth flotation process in the beneficiation of minerals for the metallurgical and chemical industries. This is not of direct importance here but it must be accepted that the process of chemisorption is, in general, reversible. Thus, it cannot be concluded that because the copper and zinc present in the Carnon River is precipitated together with iron hydroxides in the intertidal zone of Restronguet Creek the metals are permanently immobilised. The model proposed for the adsorption reaction, expressed in Eq. (3.45), suggests that the extent of replacement of H^+ on the solid by M^{2+} is pH dependent. Thus although scavenging trace metals from a liquor in a hydrometallurgical process by precipitating a substance which adsorbs them may be an efficient method of solution purification it cannot necessarily be relied upon to immobilise toxic metals in a form suitable for permanent safe disposal.

A considerable improvement for this purpose would be to incorporate the toxic metal in the lattice of a substance which is crystallised from the solution. It can then be liberated from the solid only if the lattice is destroyed. Crystallisation at high temperatures usually gives better crystallinity than is obtained at lower temperature and since the use of pressure hydrometallurgy is now becoming more generally acceptable, this procedure may become more common. A considerable body of information exists on the incorporation of impurities in the structure of jarosites during the growth of the crystals[99] but it is pointed out that care has to be taken to distinguish between incorporation of an element in the jarosite lattice and precipitation of it as a separate phase. Incorporation of a metal in the lattice as an impurity can depend on speciation in the solution and this can also be the case with a metal which is a legitimate constituent. For example in a solution continuing silver as the species $AgCl_2^-$, silver jarosite is not precipitated.

Chapter 4

Kinetics of Heterogeneous Reactions in Hydrometallurgy

4.1. Rate Controlling Processes

4.1.1. *Introduction*

The chemical conditions selected for carrying out a step in a hydrometallurgical process provide an adequate driving force for the reaction or change required, and kinetic considerations then become of particular importance since thermodynamic requirements have been satisfied. Many of these reactions are heterogeneous, for example leaching when a solid reacts with reagents in a solution and part or all of it dissolves. In a second kind of reaction a gas reacts with a solute and the rate of reaction may be controlled by the rate at which the gas dissolves in the liquid in which the reaction occurs. This is the case when hydrogen sulphide precipitates nickel sulphide from a nickel sulphate leach liquor or hydrogen precipitates copper metal from a solution of its salts. In some cases such a reaction is self-nucleating, reduction to form copper for example, but in others the reaction between the dissolved hydrogen and a metal species only occurs on a catalytically active surface, as in the case of reduction of nickel ammines. The kinetics of solvent extraction reactions are of great practical importance, a metal moving across the boundary separating two liquid phases, one aqueous and the other organic. This reaction is frequently an exchange of ions, metal and H^+. In the case of ion exchange resins the rate of anion or cation exchange depends on the diffusion of the ions within the resin beads.

Such reactions differ greatly from those generally considered in texts on physical chemistry but, of course, the treatment of the kinetics of heterogeneous reactions has very much in common with that of homogeneous reactions. The rate of the chemical change depends on the concentrations of the reacting substances. If one of them is a solid its effective concentration is related to its surface area, taking into account surface roughness and any internal surface. The rate, speed or velocity (the terms are synonymous in chemical kinetics) of the reaction falls off as reactants are consumed and for this reason the velocity is considered at a particular instant in time. If it is possible to measure the extent to which the reaction has occurred over a period of time during which the concentrations of the reactants are effectively constant then these measurements will give the reaction rate under the conditions used. Otherwise such measurements will show progressive slowing of the reaction and the rates must be obtained by taking tangents to the curve at selected times.

For the purpose of treating reaction rates mathematically it is most convenient to measure the order of the reaction with respect to each of the substances taking part. This is dealt with in Sec. 4.2.1.

4.1.2. *Transport and Chemical Control*

The theory of chemical kinetics was developed from experimental studies of homogeneous reactions in the liquid or gaseous state. In a heterogeneous reaction between one or more solute species in a solution and a solid surface some processes have to be considered which are not relevant to homogeneous kinetics. For example the solid surface may be regarded as stationary in which case the solute reactant has to reach it. This requirement is dealt with in terms of simple diffusion or of forced diffusion. If an oxidation reaction occurs at the surface and an ion transfers from the solid to the solution phase, Cu^{2+} formed from Cu_2S for example, an insoluble reaction product may be formed, elemental sulphur in this example, or sulphate ions under more strongly oxidising conditions. Whereas the latter product diffuses away from the surface the sulphur tends to remain on it, making it more difficult for the reactant to reach the Cu_2S surface. In some cases such as the nickel sulphides the sulphur forms a coherent layer and the leaching reaction then stops. In some leaching reactions oxidation of the solid does not produce a soluble species unless a suitable ligand is present also, an example being the oxidation of metallic gold in the presence of cyanide ions, forming $Au(CN)_2^-$, or thiourea. If the solid is an electronic conductor, the oxidising species and ligand do not have to interact with the same metal atom in order to form a soluble ion. The electron is removed from a conduction band in the solid and a ligand ion or molecule which is conveniently sited close to a metal atom forms the charged complex ion and diffuses away, restoring electrical neutrality in the solid.

The overall leaching reaction process may be broken down into the following steps: (i) transport of reactants in solution to the solid–liquid interface; (ii) adsorption of reactants to the surface; (iii) reaction at the surface; (iv) desorption of soluble products of the reaction; (v) transport of soluble products away from the solid–liquid interface. Stages (i) and (v) are controlled by the rates of diffusion of the solute species and, in a stirred system, by the hydrodynamics of the system. Stages (ii), (iii) and (iv) may be considered as chemically controlled processes. An electron transfer process is often the rate determining step in a chemically controlled reaction.

4.1.3. *The Nernst Theory for Transport Control*

Nernst considered that chemical processes occurring at the interface are always much faster than at least one of the transport processes, so that the reaction rate is transport controlled. In the simplest case reaction between the single reactant ion and the solid is so fast that equilibrium is reached almost instantaneously, so that the concentration c_1, of the reactant at the interface remains vanishingly small. The rate at which the reaction can occur is therefore controlled by the rate at which the reactant reaches the interface. If the solution is well stirred the concentration c in the bulk of the solution is uniform, and the concentration gradient between the values c and c_1 occurs across a thin, stationary, layer of liquid of thickness δ. If the surface area of the solid is A, by Fick's law the number, dN, of molecules of solute which cross from the bulk of the solution to the solid surface in time dt is proportional to the concentration gradient normal to the surface, dc/dy:

$$dN = -DA(dc/dy)\,dt.$$

The proportionality factor, D, is the diffusion coefficient, with dimensions (length)^2/time. If the solid is in contact with a volume V of solution, then

$$-dc/dt = (DA/V)(dc/dy).$$

Nernst assumed the concentration gradient to be linear and expressed by $(c - c_1)/\delta$, so that

$$-dc/dt = DA(c - c_1)/V\delta \qquad (4.1)$$

and the first-order rate constant k is given by

$$k = DA/V\delta. \qquad (4.2)$$

From experiment the velocity constant per unit area at unit volume, k_T, may be found; the subscript T indicating that the velocity constant refers to a transport controlled reaction. If the diffusion coefficient for the reactant is known, then the thickness, δ, of the layer across which diffusion is supposed to occur, can be found, since

$$k_T = kV/A = D/\delta. \qquad (4.3)$$

For a number of reactions in water at 20°C, δ is about 3×10^{-3} cm. The fact that the value is approximately the same for a variety of reactions which are chemically quite different is consistent with the supposition that the rates are controlled by a diffusion process. The magnitude of the value is physically improbable, however, since it implies a diffusion layer about 50 000 molecules thick. It seems unlikely that the solid could exert any significant effect on a water molecule separated from it by so many other molecules, yet this layer is supposed to be identical with the film of liquid which is held back by the solid surface when the liquid flows past it.

The effect of temperature on the value of k_T can be used to support Nernst's view of heterogeneous reactions. If the value of δ is independent of temperature, then dk_T/dT should equal the rate of change of the diffusion coefficient with temperature. The energy of activation for diffusion at 25°C is usually between about 12 and 27 kJ per mol, depending on both solute and solvent, so that the observed critical increment of energy, E_A, for a transport controlled process, should be within the same range, about 17 kJ per mol. This is frequently found to be the case. The critical increment of energy is the equivalent in a heterogeneous reaction of the activation energy in a homogeneous reaction, being determined by using the Arrhenius equation

$$E_A = RT^2(\mathrm{d}\ln k/\mathrm{d}T) \,.$$

When absolute reaction rate theory is used in the study of a homogeneous reaction the experimentally measured activation energy can be analysed in terms of the free energy and entropy of formation of the transition state. In the case of a heterogeneous reaction however heats of adsorption are also involved. The two situations are different therefore and strictly speaking should be distinguished. In the hydrometallurgical literature, however, the distinction has in the past often been ignored. See also Sec. 4.2.1.

Increasing the rate of stirring will decrease the thickness of the layer of water adhering to the solid surface, so decreasing the value of δ and increasing the reaction rate. For a particular rate of stirring the value of δ depends on the dimensions and geometry of the system. If the reacting solid is suspended powder it is necessary to consider relative motion between particles and the liquid, and this is strongly influenced by turbulence such as is caused by the introduction of baffles.

For massive solids the reaction rate increases with stirring rate (revolutions per minute) raised to a power a, where a is less than or equal to unity. For a simple system, benzoic acid dissolving in sodium hydroxide solutions, the values of δ have been found to be inversely proportional to the stirring rate (rpm), that is $a = 1$, and about 3×10^{-3} cm.

Considering a number of solids, each being insoluble in the pure solvent but reacting to give soluble products in the presence of a solute, then under the same experimental conditions they should dissolve at identical rates if the rates are controlled only by the rate of transport of the solute to the solid surface. This has been shown to be the case for mercury, cadmium, zinc, copper, silver, iron, nickel and cobalt metals dissolving in aqueous iodine solutions.[100] In the case of zinc, $a = 0.56$. The iodine acts as oxidant and provides the anion for the metal.

4.1.4. *Hydrodynamic Theory for Transport Control*

There are many examples known where reactions between solids and solutes are chemically controlled, at least in part. Even considering transport-controlled reactions, however, the assumptions on which Nernst's theory is based are not strictly correct. The main objection is to the assumption that the diffusion layer is stationary with respect to the solid surface, and the result that the thickness of the layer is about 3×10^{-3} cm in many cases of different kinds of reaction.

Very strong evidence has been presented[101] suggesting that fluid motion persists very close indeed to the surface; it has been observed[102] at distances of 10^{-5} cm from the solid wall. This means that the assumption that the concentration of the solute is a linear function of the distance, y, from the surface, up to the bulk concentration c when $y = \delta$, is at best only an approximation. However, there is ample evidence for the existence of a concentration gradient between the surface of the solid and points in the liquid at a distance from it which may again be called δ, in systems where transport-controlled reactions are taking place at the solid–liquid interface. The properties and thickness of this region are determined by the value of the diffusion coefficient of the solute, the viscosity of the solution, and by the way in which the liquid flows relative to the solid surface. Thus, for any particular reacting system the most important variable is the extent of agitation, which is dealt with in terms of hydrodynamics.

The theory of hydrodynamics relevant to the analysis of transport controlled reactions of solids with solute ions and molecules has been reviewed,[103] and a very complete treatment is also available.[104] Two main results are required: (i) the way in which the concentration of solute changes with distance from the solid surface, and the magnitude of the region over which the change occurs; (ii) the rate at which solute is transported by forced diffusion from the bulk of the solution to the solid surface. It is desirable that the theory should indicate what factors are involved and relate them quantitatively to give this information.

Unfortunately the equations for convective diffusion are complicated, and usually semiempirical methods must be used to obtain a solution. Because of the high degree of symmetry in the case of a rotating disc, however, a complete solution is possible, and this gives the velocity distribution throughout the whole mass of the fluid.[104] The disc rotates about an axis perpendicular to its plane surface and at the centre; it is large enough to make edge effects insignificant, and the volume of the solution in which it stands is also large so as to avoid wall effects.

Far from the rotating disc the liquid moves towards the disc in a direction normal to the surface, with constant velocity, v_y, and

$$v_y = -0.886\sqrt{(v\omega)} \tag{4.4}$$

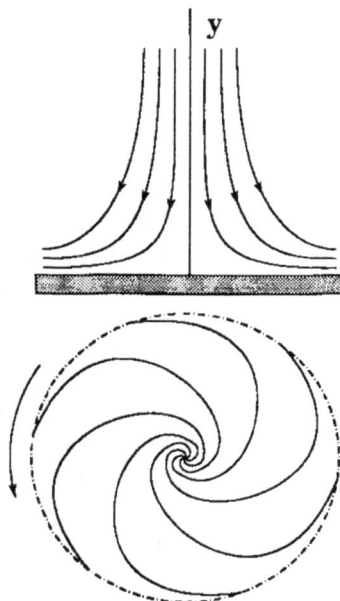

Fig. 4.1. (Top) Flow of liquid towards a disc rotating about axis y. Close to the surface of the disc liquid flows parallel to it. (Bottom) The broken circle indicates the outer edge of the disc. The swirling solid lines show the flow pattern of liquid very close to the rotating surface. Such lines are etched on a metal disc after appreciable reaction has occurred.

where v is the kinematic viscosity of the fluid and ω is the constant angular velocity of the disc (radians s^{-1}). The liquid acquires a rotating motion only when it has arrived very close to the disc, but then the angular velocity increases with closer approach until it reaches that of the disc itself. The centrifugal force which results from the angular momentum gives the liquid a radial velocity also which drives it outwards, across the surface and then away from the disc. Thus, liquid is continuously drawn towards the disc and driven away when it arrives very close to the surface (Fig. 4.1). The liquid is therefore subject to two kinds of flow, one normal to the surface, with constant velocity v_y, and another parallel to the surface. The transition from one to the other indicates that a viscous boundary layer is present. It is found that at a distance from the surface equal to about $2.8\sqrt{(v/\omega)}$ the rate of flow normal to the surface is 80% of its maximum value, while the rate parallel to the surface is 10% of that of liquid at the surface. This distance may be taken as the approximate thickness of the viscous boundary layer. For water at 25°C, the layer is about 5×10^{-2} cm if the disc has an angular velocity of 25 radians per second.

Considering now convective transfer and steady state conditions, the concentration of the solute which reacts with the surface of the rotating disc must depend only on the distance from the surface and the distance from the axis of rotation, since the system has axial symmetry. Consideration of the way in which the liquid flows makes it reasonable to assume that the concentration is not dependent on the distance from the axis, and it is found that

$$D(\mathrm{d}^2c/\mathrm{d}y^2) = v_y(\mathrm{d}c/\mathrm{d}y). \tag{4.5}$$

When y is large, v_y is constant, and provided the angular velocity, ω, of the disc is great enough to give a reasonably high value of v_y, then the slow rate of the other process which can change the concentration, nonconvective diffusion, plays no part in altering solute concentrations. Thus, the concentration of solute in the bulk of the solution is constant if the liquid far from the disc flows towards it sufficiently fast. Then diffusion to the solute-deficient layer at the disc surface is too slow to have any effect far from the surface.

Very close the the disc v_y is much smaller than its maximum value, and the rate of diffusion becomes similar in magnitude, so that the rate of transfer of the solute becomes increasingly controlled by diffusion. The solution[105] of (4.5) is shown in Fig. 4.2 which gives the ratio of the concentration at distance y from the surface, c_y, to the bulk concentration c.[103,105] This ratio is shown as a function of distance expressed as a fraction of the thickness of the diffusion layer δ defined by Nernst's Eq. (4.1). The broken line refers to this equation.

The thickness of the diffusion boundary layer depends on the thickness of the hydrodynamic or viscous boundary layer, which was assumed to be constant over the whole surface of the disc. This is the case except for a zone at the edge of the disc, the extent of which is of the order of the thickness of the hydrodynamic boundary layer. Thus, the theory outlined above will apply only if the diameter of the disc is large compared with this thickness. The walls of the containing vessel must also be so far away from the edge of the disc that they do not interfere with the pattern of liquid flow, which must be nonturbulent. Under these conditions the solution of Eq. (4.5) may be used for a system of finite size giving

$$-\mathrm{d}c/\mathrm{d}t = DA(c - c_1)/V(0.893\delta'). \tag{4.6}$$

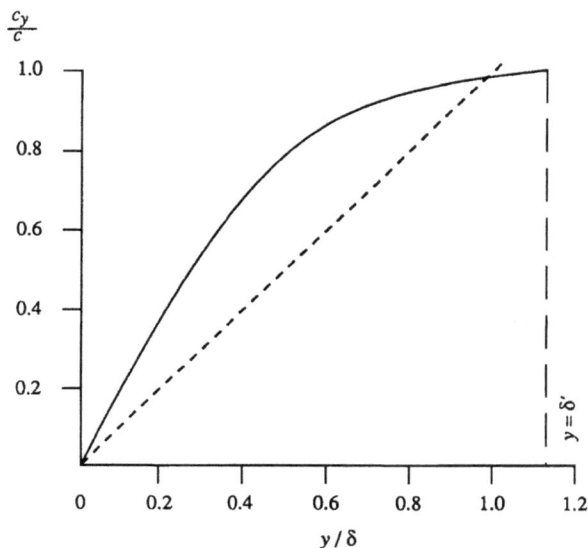

Fig. 4.2. (Solid line) Ratio of solute concentration at distance y from surface of rotating disc, c_y, to bulk concentration c, as a function of distance y, expressed as a fraction of the thickness of the diffusion layer. (Broken line) Plot of Nernst's Eq. (4.1).

equivalent to (4.1) where δ' is the thickness of the layer across which the concentration change occurs near the disc surface. Comparison of (4.1) and (4.6) shows that $\delta = 0.893\delta'$. Integration of (4.5) gives

$$\delta' = 1.805(D/v)^{1/3}(v/\omega)^{1/2}$$

so that

$$\delta = 0.893\delta' = 1.612 D^{1/3} v^{1/6} \omega^{-1/2} . \tag{4.7}$$

The rate of flow of solute from the bulk solution to a solid surface is given by the Nernst theory as

$$I = DA(c - c_1)/\delta \tag{4.8}$$

so that in the simplest case when the concentration at the surface is zero

$$I = D\pi r^2 c / 1.612 D^{1/3} v^{1/6} \omega^{-1/2} \tag{4.9}$$

$$\approx 1.9 \, r^2 c D^{2/3} v^{-1/6} \omega^{1/2} \tag{4.10}$$

where r is the radius of the disc. Thus, if a rotating disc of a solid is used, it is possible to calculate the maximum rate at which it can react with a solute, if the rate is entirely transport controlled. For other kinds of geometry of the system the way in which δ and δ' depend on D, v, and the characteristic velocity of the system will be similar to (4.7) as long as flow is nonturbulent.

4.2.   Experimental Methods for Studying Leaching Kinetics

4.2.1.   *Introduction*

The methods used to study the kinetics of heterogeneous reactions are similar to those employed for homogeneous reactions. The changes in the rates of reaction are studied as the concentrations of the various reactants, and the temperature, are altered. Since the reaction may be wholly or partially transport controlled, either the conditions of agitation must be kept constant throughout the experiments or the stirring rate must be so large that further increase has no effect on the reaction rate. Alternatively, the solid may be fashioned into a disc and hydrodynamic theory used in the form discussed above.

When the reaction is between the solid and a solute in the liquid the activity of the solid as a reactant is a function of its surface area. Whereas in a homogeneous reaction it is supposed that reactant molecules collide and either separate or form some kind of complex in which the chemical change occurs, in a heterogeneous reaction the collision is between solute and the solid surface. In a very large number of cases such collisions lead to adsorption of the solute, even when this does not react so as to decompose the surface. If chemisorption occurs there is some kind of chemical interaction between the adsorbed substance and the solid, so that it is possible that atoms and radicals which may be formed as a result of the chemisorption can take part in the reaction. This can lead to the provision of alternative reaction paths.

When two or more solutes react at a solid surface either with the solid or with one another the same considerations apply. The provision of the new reaction path may, however, make possible a reaction between two solute species which does not occur when they collide in the solution, often, probably, by making available intermediate stages of the reaction with low activation energy. Another way in which adsorption facilitates reactions between two solutes is by making it possible for the ions or molecules to remain on adjacent adsorption sites for a relatively long time, giving increased opportunities for them to reach a condition favourable for reaction.

The rate of a reaction occuring at a solid surface when two solute species react either with one another or with the solid depends on the amount of each at the solid–liquid interface, c_1. The value of c_1 depends on the rate at which the substance is removed by reaction and by the rate at which a fresh supply arrives. Under constant conditions of stirring this is controlled by the concentration in the bulk of the solution, c. If the two reactants are A and B, when the concentration of A is increased from a small value while that of B remains constant, the reaction rate will rise as more A becomes available at the surface, but the extent of the rise for a given increase in concentration will eventually drop as the rate becomes dependent also on the rate at which B can reach the surface. Finally, the amount of A available will be so large that the reaction rate depends only on the rate at which B reaches the surface.

The order of such a reaction with respect to a reactant will change therefore, without any alteration in the mechanism of the chemical reaction, the order being defined as the power to which the concentration of the reactant is raised in the rate equation, and being determined by any of the usual methods. The concentration at which the order changes for the reactant A depends on the concentration of B and the rate of transport to the surface, that is the conditions of agitation. The concentration of B at which the order with respect to B changes depends similarly on the concentration of A and the agitation.

The change in the rate of a chemically controlled heterogeneous reaction with temperature depends on at least three quantities.[106] These are the real activation energy in the adsorbed layer, which determines the rate at which the adsorbed molecules or ions react, the heat of adsorption of the reactants, and the heat of adsorption of the products. If the reactants have a large heat of adsorption, as the temperature rises the quantity present at the surface will decrease rapidly, which tends to slow the reaction. If the products are soluble but tend to block a large proportion of the solid surface by adsorption, increasing the temperature will expose more surface if the heat of adsorption is large, thus tending to speed up the reaction.

The apparent activation energy obtained by applying the Arrhenius equation to experimental data concerning the effect of temperature on the reaction rate is clearly not equivalent to the activation energy determined in a similar manner for a homogeneous reaction. It should be referred to as the critical increment of energy to make the distinction clear. Yet another complication may arise if one of the reactants is a dissolved gas, such as oxygen, hydrogen, or carbon monoxide, in equilibrium with the gaseous material at constant partial pressure. At the different temperatures employed to measure the critical increment of energy the solubility of the gas may change somewhat, altering the concentration in the bulk of the solution. This may restrict the temperature range which can be used for the measurement.

The two extreme cases of heterogeneous reaction are: (i) when the reaction is entirely transport controlled and the critical increment of energy is equivalent to the energy of activation for diffusion, about 17 kJ per mol and (ii) when the reaction is entirely chemically controlled and the value will normally be considerably larger. It is possible also to have an intermediate situation in which the reaction rate is controlled in part by the rate of transport of a reactant to the interface and in part by chemical factors. This will give a critical increment of energy determined by the extent to which each process controls the rate.

It is clear that the values of the critical increments of energy determined experimentally cannot be analysed in terms of the theory of absolute reaction rates unless a great deal of additional information is available. This is very seldom the case.

When excess solid is present thermodynamic equilibrium may be attained close to the surface after some time of reaction with a solute. If the concentration of solute at the solid surface, c_1, is not equal to this equilibrium concentration and is not the same as in the bulk solution, c, then the rate of the reaction depends on the rate of transport of solute to the surface, on the rate of the chemical reaction at the interface, and the rate at which soluble products are transported away from the surface; see Sec. 4.3.2.

The rate constant k_c of the chemical process may be written

$$-dc_1/dt = k_c A c_1^n / V \tag{4.11}$$

and the rate constant k_T of the transport process (compare (4.1))

$$-dc/dt = k_T A (c - c_1) / V . \tag{4.12}$$

Thus in the steady state condition

$$k_c c_1^n = k_T (c - c_1) \tag{4.13}$$

and c_1 may be expressed in terms of c and the two rates.

For a first-order chemical reaction, with $n = 1$,

$$c_1 = c k_T / (k_c + k_T)$$

and substituting into (4.12) gives

$$-\frac{dc}{dt} = \left(\frac{k_c k_T}{k_c + k_T}\right) \frac{A}{V} \cdot c. \tag{4.14}$$

When k_T is much larger than k_c the observed rate is controlled by the chemical process at the surface; when k_c is much larger than k_T the reaction is transport controlled.

When the chemical reaction is of higher order than unity the expressions are more complex. For a second-order reaction, with $n = 2$,

$$-\frac{dc}{dt} = \frac{k_T A}{V} \cdot c \left[1 + \frac{x}{2} - \sqrt{\left(x + \frac{x^2}{4}\right)}\right] \tag{4.15}$$

where $x = (k_T/ck_c)$. In this case, using Eq. (4.13) with $n = 2$,

$$c_1 = c\sqrt{\left(\frac{k_T}{ck_c} + \frac{k_T^2}{4c^2 k_c^2}\right)} - \frac{k_T}{2k_c}. \tag{4.16}$$

For such a second-order reaction, let the ratio $k_T/k_c = y$. Then

$$-\frac{dc}{dt}\frac{V}{A} = k_T c \left[1 + \frac{y}{2c} - \sqrt{\left(\frac{y}{c} + \frac{y^2}{4c^2}\right)}\right]. \tag{4.17}$$

This shows that when a reaction is controlled by both a transport and a chemical process and the order of the chemical reaction is not unity the reaction rate depends on the concentration of the reactant in the bulk of the solution in a complex manner.

If the order of a chemical reaction is not unity, therefore, the order must be determined under experimental conditions such that the rate is controlled only by the chemical reaction. This is the case when an increase in the stirring rate does not increase the rate of the reaction. The alternative is to determine the value of k_T by calculation or measurement and to obtain k_c as a function of c.

4.2.2.   Use of an Exposed Face of Mounted Solid

Several studies were carried out of the kinetics of the reaction in which gold or silver metal was oxidised by oxygen and dissolved in a solution of an alkali metal cyanide.[107-109] The reaction is, in the case of silver

$$4Ag + 8CN^- + O_2 + 2H_2O = 4Ag(CN)_2^- + 4OH^-$$

but in order to avoid the presence of HCN, which does not directly take part in the reaction and may be lost from the solution as gas, free alkali is added to raise the pH to between about 11 and 12.9. Below pH 11, hydrolysis of cyanide reduces the rate and above pH 12.9, complications arise due to the kinetics of reduction of oxygen.

The experimental procedure was to use a flat surface of the metal and a stirred solution of the reagents. For example in one case,[109] the silver was cold-worked, annealed and mounted in plastic so as to avoid the possibility of electrical leakage affecting the reaction rate. The sample was mounted on the base of the reaction vessel with the exposed flat silver face upwards. A stirrer was positioned above it so as to direct a flow of solution downwards to the metal surface. The oxygen pressure above the solution in the reactor was maintained at the desired value during the experiment and samples of the solution could be taken for analysis. Because the surface area of the metal reacting was constant, as was p_{O_2}, and the quantity of CN^- which reacted did not significantly change its concentration, the rate curves obtained by plotting $[Ag^+]$ in solution against time for an experiment were linear, the slope giving the reaction rate. The lines seldom passed through the origin. In some experiments a small amount of silver dissolved very rapidly before the rate settled down, a result attributed to traces of the metal left on the surface after polishing it in readiness for the next experiment being in a much more reactive state than the massive crystalline solid. In others there was an induction period before silver appeared in the solution, attributed to some kind of passive coating on the surface. Analysis of the experimental results to determine what controlled the rate of dissolution of the silver required the rate of dissolution of silver in convenient units, which proved to be mg of Ag per cm^2 per hour.

The results of the three investigations can be summarised as follows:

(i) The rate of dissolution depends on the surface area of metal in contact with the cyanide solution. Thus a heterogeneous reaction is involved.

(ii) The rate of dissolution depends on the speed of stirring, so that a transport process is at least partially rate controlling under the conditions studied.

(iii) The critical increment of energy is about 8 to 20 kJ g atom^{-1} which is a characteristic value for a diffusion process.

(iv) Approximately 0.9 g moles of oxygen were consumed for every 2 g atoms of metal dissolved.[108]

 (v) Approximately 2 g ions of free cyanide were removed when 1 g atom of metal dissolved.[107]

(vi) Approximately 0.9 g moles of hydrogen peroxide were produced for each 2 g atoms of metal dissolved.[107,108]

(vii) The effects of changing cyanide concentration at constant oxygen partial pressure and oxygen pressure at constant cyanide concentration were as shown in Figs. 4.3 and 4.4.

Figure 4.3 shows that at low cyanide concentrations, up to about 0.02 M, the rate of dissolution of silver is independent of the oxygen pressure; that is of the concentration of oxygen in the solution at these two partial pressures. Thus the rate of reaction is controlled by the rate at which CN$^-$ is transported to the surface of the metal. This is determined by the rate of stirring and a figure similar to Fig. 4.3 would show different rates for different stirring speeds. The curves also show that with CN$^-$ concentrations below 0.02 M adequate

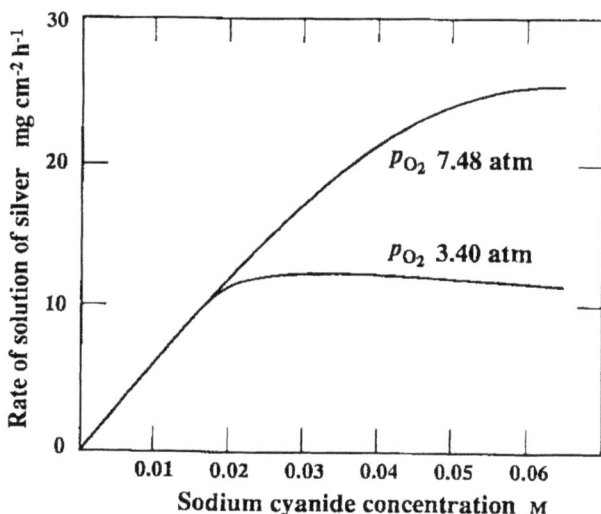

Fig. 4.3. Rate of solution of silver as a function of sodium cyanide concentration, with oxygen partial pressures 3.40 and 7.48 atm; 24°C.

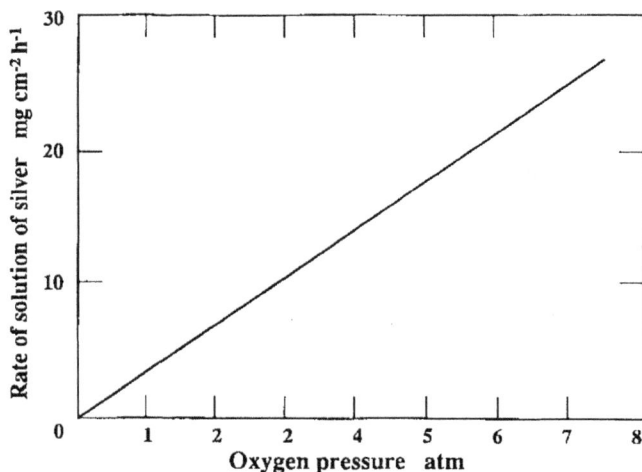

Fig. 4.4.   Effect of oxygen pressure on rate of solution of silver in 0.055 M sodium cyanide solution; 24°C.

oxygen is available at the silver surface to support the reaction at the velocity at which it is occuring. However if $[CN^-]$ is higher than 0.02 M the reaction rate levels off if p_{o_2} is 344 kPa (3.40 atm), reaches a constant value independent of $[CN^-]$ and falls slightly at the highest cyanide concentrations used. The plateau was caused by the rate of oxidation of silver being controlled by the rate at which dissolved oxygen was reaching the surface; the fall was attributed to the solubility of oxygen being lower at the higher cyanide concentration.

At the higher oxygen pressure used the reaction rate continued to increase with increasing $[CN^-]$ but eventually became independent of $[CN^-]$. As is shown in Fig. 4.4, with $[CN^-]$ at a value slightly lower than that at which the rate became independent of $[CN^-]$ with p_{o_2} 758 kPa (7.48 atm) the effect of oxygen pressure, and hence it is assumed $[O_2]$ in solution, is linear. Thus the rate of the reaction between silver, cyanide and oxygen is first-order with respect to $[CN^-]$ and to $[O_2]$ within the ranges of concentration where they are separately rate determining.

4.2.3.   *Use of an Exposed Face of a Rotating Disc*

The advantage of using a rotating disc of the reacting metal over the use of a stationary surface in a stirred vessel is that if the system is properly designed the rates at which reactants reach the surface can be accurately controlled and calculated. A number of leaching studies on metals and minerals using this technique were carried out in Russia and have been reviewed.[110] In order to avoid electrochemical reactions the metal discs, approximately 2 cm in diameter, were annealed and mounted in electrically insulated holders.

In the case of silver reacting with cyanide and oxygen it was found that at 25°C with 1 atm oxygen pressure above the solution, potassium hydroxide concentration 10^{-3} M, and the disc rotating at 1100 rpm, the reaction was first-order with respect to cyanide concentration within the range 3.85×10^{-3} to 9.24×10^{-3} M. The rate of dissolution of the silver did not increase if the concentration was raised above 9.3×10^{-3} M, which may be described as a critical value under the conditions employed.

Below the critical cyanide concentration the "reduced dissolution rate", (defined as v/c, where v is the rate of reaction, g atom cm^{-2} s^{-1}, and c the cyanide concentration, g mol L^{-1}) was found to be 2.91×10^{-6} litre cm^{-2} $s^{-1} \pm 2\%$ under the conditions defined above. The value was proportional to the square root of the number of revolutions per minute of the disc, which is in agreement with the theory for laminar-flow conditions. This was not true for very low rates of revolution because of disturbance of the flow by the oxygen blown into the apparatus.

When the cyanide concentration was below the critical value the rate of dissolution of the silver was shown to be independent of the oxygen pressure. Thus, within this region of cyanide concentration the dissolution reaction rate constant may be calculated from the reduced dissolution rate and the rate of revolution of the disc (18.3 r.s^{-1}). At 25°C, the reaction rate constant k was found to be 0.68×10^{-6} litre cm^{-2} $s^{-1/2}$ rev$^{-1/2}$.

The critical cyanide concentration was shown to depend on the oxygen pressure and to be directly proportional to it. The critical concentration was not affected by the rate of rotation of the disc, it decreased with rising temperature, and did depend on the cation present in the solution. The latter two facts may be ascribed to the dependence of the solubility of oxygen on the temperature and on the actual alkali metal cyanide present. The concentration of oxygen in the solution was the factor controlling the rate of oxidation of the silver, and although it may be assumed that under the conditions used experimentally the concentration was proportional to the oxygen partial pressure, this is only the case if no other factor affecting the solubility was altered.

When the cyanide concentration was above the critical value the rate of the reaction was found to be directly proportional to the oxygen pressure. Thus, the overall effect of increasing this is complex, the critical cyanide concentration is raised, and if it is not exceeded a small change in oxygen pressure will not affect the rate. If the cyanide concentration is above the critical value a small increase in oxygen pressure will increase the rate and also the critical value, which must still be exceeded.

The results outlined show that the rate of dissolution of silver from a rotating disc under laminar-flow conditions depends on the rate of transport of reactants to the surface of the metal. At a given concentration of oxygen raising the cyanide concentration from a low value will increase the amount getting to the silver metal per second, and so increase the rate of dissolution of the silver. Eventually the rate will become so fast that the availability of oxygen becomes the limiting factor and the rate becomes independent of cyanide and dependent on oxygen concentration. The value of the rate at which the change of control occurs depends, clearly, on the oxygen concentration.

The effect of temperature on the reaction rate was shown to be different in the two different regions of control. Below the critical cyanide concentration the critical increment of energy was about 15 kJ g atom^{-1} at temperatures around 30°C. At higher cyanide concentrations the rate was only very slightly dependent on temperature, since the increase in the diffusion coefficient with rising temperature was almost completely offset by the reduction in the solubility of the oxygen.

Copper behaves in a manner practically identical with silver. In particular, the rate of dissolution of copper is proportional to the square root of the rate of revolution of the disc, and the critical increment of energy at cyanide concentrations below the limiting value is 13.4 kJ g atom^{-1}.

With the rate of rotation of the disc below 150 rpm gold dissolved in a similar manner to silver and copper, the critical increment of energy being 14.2 kJ g atom^{-1}. The limiting reduced reaction rate at 25°C with 1 atm oxygen pressure was 1.35×10^{-6} litre cm^{-2} s^{-1} at 150 rpm.

When the rate of rotation was above 150 rpm, the limiting reduced rate under similar conditions of temperature and oxygen pressure fell sharply to about 0.5×10^{-6} litre cm^{-2} s^{-1}, and was then independent of the rate of rotation of the disc up to 1100 rpm. The critical increment of energy under these conditions was 58.5 kJ g atom^{-1}. This behaviour is consistent with the view that with 1 atm oxygen pressure and cyanide concentrations above the critical value, the rate of dissolution of gold was controlled by a chemical process instead of by the rate of transport of reactants to the solid surface.

The rates at which the reactants arrived at the silver, copper and gold discs can be calculated for the experimental conditions employed and compared with the rates at which they reacted with the metal. When the cyanide concentration was below the critical value and the rate of dissolution v of the metal was independent of the oxygen concentration

$$v = Q/t = kAcn^{1/2} \tag{4.18}$$

where Q is the quantity of metal dissolved, g atom; t the time of reaction, s; k the reaction rate constant, $1 \text{ cm}^{-2} \text{ s}^{-1/2} \text{ rev}^{-1/2}$; A the area of the disc surface reacting, cm^2; c the concentration of cyanide in the solution, g mol L^{-1}; n the number of revolutions per second of the disc.

The reaction rate constant can be calculated from Levich's theory (4.10)

$$k = \frac{6.18 \times 10^{-4} D^{2/3}}{m v^{1/6}} \tag{4.19}$$

where m, the stoichiometric coefficient of the reaction is the number of solute ions reacting with one metal atom, $= 2$. The diffusion coefficient D for potassium cyanide is unknown, but was taken as equal to the value for potassium nitrate, which at 25°C is $1.85 \times 10^{-5} \text{ cm}^2 \text{ s}^{-1}$ for $7–10 \times 10^{-3}$ M solutions. The kinematic viscosity of such a solution is $8.98 \times 10^{-3} \text{ cm}^2 \text{ s}^{-1}$. Then

$$k = 4.74 \times 10^{-7} \text{ litre cm}^{-2} \text{s}^{-1/2} \text{rev}^{-1/2}$$

when n is expressed in radians per second. Conversion to revolutions per second for comparison with experimental values requires that k be multiplied by $(2\pi)^{1/2}$ giving

$$k = 1.19 \times 10^{-6} \text{ litre cm}^{-2} \text{s}^{-1/2} \text{rev}^{-1/2}.$$

The experimentally determined values of k at 25°C, with the cyanide concentrations below the critical value and, in the case of gold, with a speed of rotation below 150 rpm were for copper, silver and gold respectively 1.035, 0.68, and 0.83×10^{-6} litre $\text{cm}^{-2} \text{ s}^{-1/2} \text{rev}^{-1/2}$. These values differ significantly from one another and from the theoretical figure. The low values of k found for the metals dissolving in cyanide solutions when the rate-controlling process was the rate of transport of cyanide ions to the metal surface were explained as being due to the formation of an insoluble film of a product on the surface. The rate of diffusion through this film was the factor controlling the rate of dissolution of the metal, and this rate depended on the concentration of cyanide ions at the interface between the film and the liquid. This in turn depended on the concentration within the diffusion boundary layer, so that the kinetics of the metal dissolution reaction were in general agreement with the theory giving the rate of transport of reactants to the reacting interface. The actual value of the reaction rate constant depended, however, on the thickness and permeability to reactants and products of the film produced on the metal.

The formation of an insoluble material during the reaction is easy to understand. The reaction of the metal can be represented as

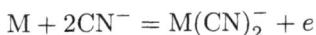

$$M + 2CN^- = M(CN)_2^- + e$$

the electron being accepted by the oxygen at the interface. It is unlikely that the metal would react in one step to give the soluble cyanide ion, since the intermediate reaction

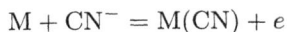

$$M + CN^- = M(CN) + e$$

could take place. The rate of appearance of the metal in solution depends on the rate of reaction between the insoluble cyanide and more cyanide ions, to give the soluble double cyanide. Alternatively a different product might be formed by the first stage of oxidation of the gold, by the oxidant, O_2. The rate of attack on the metal depends on the rate of transport of the reactant through the film to the metal surface. It was stated in Sec. 4.1.3 that a number of metals dissolve in aqueous iodine solutions at identical rates under identical experimental conditions.[100] No layer of product can form in these reactions.

The results obtained using the rotating disc technique to study the kinetics of dissolution of metals confirm those obtained by other methods and the evidence for diffusion control under some conditions is conclusive. The most important fact which emerges however is that even in simple cases the rate of reaction is significantly slower than the rate at which the solute which is controlling the rate of the chemical reaction reaches the solid–liquid interface, as calculated using the theory of Levich. When the kinetics of leaching minerals was investigated using the rotating disc method, particularly metal sulphides, the rates found were very much lower even when sulphur was not produced. This is due to the formation of insoluble materials during early steps in a series of consecutive reactions as suggested by Kakovsky[110] or by other physical or chemical processes controlling the rate of reaction in such a way as to make diffusion processes in the aqueous phase irrelevant to some degree.

4.2.4. *Use of Particulate Solid*

In hydrometallurgical processes leaching is carried out either by agitating particulate solid in a solution containing the reagents required or by causing the solution to flow through porous solid, which may be a porous rock or a heap of broken material. The most important practical requirement when designing a leaching process is to understand what chemical reactions take place and what factors control their rates. The most generally convenient way of determining this is to use samples of closely sized ground ore, or mineral if the behaviour of a single substance is being studied.

A procedure which has proved to be satisfactory in many cases has been described in an investigation of the leaching of chalcocite, Cu_2S, in acidic solutions of $FeCl_3$.[111] The mineral was synthesised by heating stoichiometric quantities of copper rods and sulphur in separate chambers of an evacuated silica tube in a horizontal tube furnace at 500°C until all of the sulphur had reacted, as could be seen by the absence of brown vapour in the tube after 24 hours. The temperature was raised to 900°C and after 2 days the furnace was allowed to cool and the chalcocite was removed. It was crystalline, nonporous and perfectly uniform in composition. It reacted during the leaching experiments in exactly the same way as a sample prepared by the same method except that it was melted in the silica tube at 1140°C before being cooled and removed.

Leaching was carried out in a wide-necked reaction vessel onto which a five-necked lid could be clipped. The central neck held a suitable seal and the shaft of a propeller mixer which could be glass, or steel coated with a thick (0.5 cm) layer of an epoxy resin. Thin coatings did not protect the steel adequately because of the presence of pinholes. Three of the other necks held a condenser to prevent evaporation, a baffle to provide a turbulent regime during leaching, and a thermometer. The fifth neck was used for sampling. The vessel was kept in a bath with the temperature held constant to within ±0.1 degree. The solution was allowed to reach thermal equilibrium with the bath before the chalcocite was added through the sampling neck. The solid was ground by hand using an agate pestle and mortar and sized on nylon sieves immediately before the required quantity was weighed out and added to the leach solution. For analysis, 1 ml samples of solution were sufficient when atomic absorption spectroscopy was used. In kinetic studies allowance was made for the volumes and copper content of the samples removed.

Since the rates of the chemical reactions were to be studied it was desirable that mass transport in the aqueous phase should play no part in controlling the rate. This is the case when the rate of leaching is independent of stirring speed. Using chalcocite within the particle size range 150–300 μm leaching experiments were carried out at stirring speeds of 250, 540, 900 and 1500 rev min^{-1}. At the slowest speed not all of the particles were held in suspension and reaction was slower than at the higher speeds. Results at the three higher speeds were identical within experimental error and for subsequent experiments 900 rev min^{-1} was used.

The leaching reaction between chalcocite and acidic ferric sulphate solutions was said by Sullivan[112] to proceed in two stages

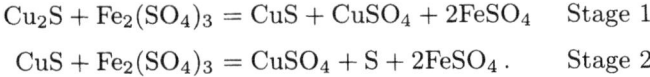

$$Cu_2S + Fe_2(SO_4)_3 = CuS + CuSO_4 + 2FeSO_4 \qquad \text{Stage 1}$$

$$CuS + Fe_2(SO_4)_3 = CuSO_4 + S + 2FeSO_4 . \qquad \text{Stage 2}$$

He reported that the particles of covellite, CuS, formed as an intermediate product during stage 1 retained the shape of the original chalcocite particles

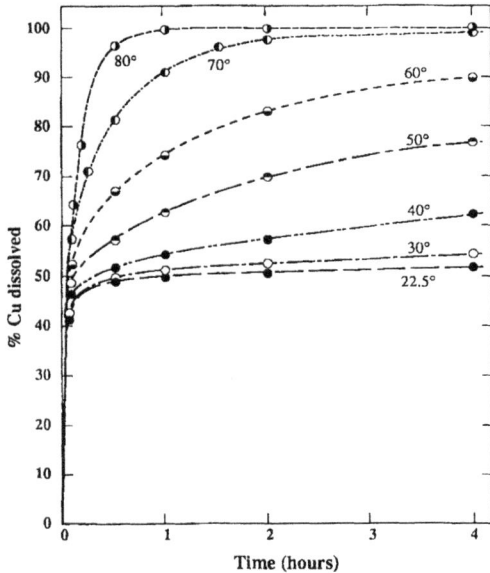

Fig. 4.5. Leaching of chalcocite; effect of temperature (FeCl$_3$, 0.5 M; HCl, 0.2 M; Fe^{3+}/Fe^{2+}, 10; 2.5 g 150–300 μm Cu$_2$S).

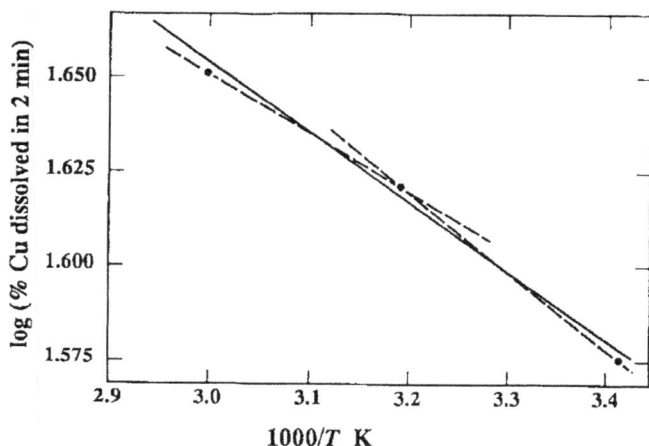

Fig. 4.6. Leaching of chalocite; Arrhenius plot for first stage of leaching reaction (data from Fig. 4.5).

and stage 1 proceeded much faster than stage 2. Results obtained[111] over a range of temperatures between 22.5°C and 80°C confirmed this, Fig. 4.5. The conditions used were chosen so that the redox potential and pH were fixed and remained effectively unchanged throughout the reaction. 0.5 M $FeCl_3$; 0.2 M HCl and 2.5 g of 150–300 μm Cu_2S were used and $FeCl_2$ was added to fix the Fe^{III}/Fe^{II} ratio at 10. The relevant predominance area diagrams are given in Figs 3.1, 3.2 and 3.3. The first stage of the leaching reaction was completed in 3 to 6 minutes at 40°C. The apparent activation energy of this stage was, therefore, determined from measurements of the amounts of copper dissolved from the chalcocite in 2 minutes at 20°C, 40°C and 60°C, Fig. 4.6. The value obtained was 3.43 ± 0.38 kJ mol^{-1}. The apparent activation energies of stage 2 were calculated from the rates, at different temperatures, at which various fractions of the total copper content of the original chalcocite were lost. Thus the time required for a sample of solid which had already lost 50% of its copper to lose another 10% of its copper gave the rate for 50–60% copper dissolved. The results obtained are given in Table 4.1.

Table 4.1.

% Cu dissolved	50–60	60–70	70–80	80–90
E_A, kJ mol^{-1}	101	103	112	122

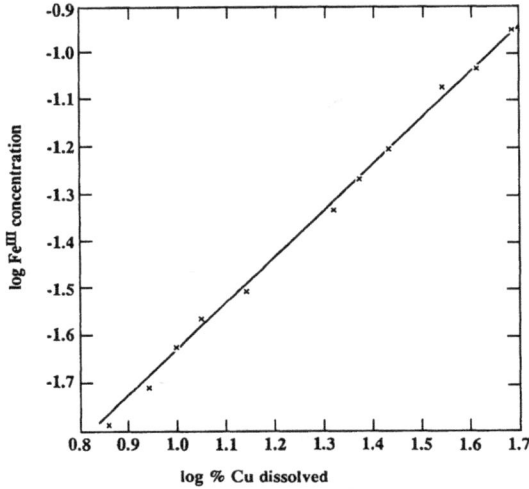

Fig. 4.7.   Leaching of chalcocite; order of reaction with respect to $[Fe^{3+}]$ in dilute solutions (HCl, 0.2 M; 2.5 g 150–300 μm Cu_2S; 40°C; 60 min; $FeCl_3$, 0.0156–0.125 M; line drawn with slope 1.00).

The increase in E_A as reaction proceeded was attributed to the progressive formation of sulphur during this stage of the reaction. It is unusual to be forced to obtain values of rates for use in the Arrhenius equation in such a fashion. Normally tangents can be drawn to the experimental rate curves.

The effect of Fe^{III} concentration on the rate of the first stage of leaching of chalcocite is shown in Fig. 4.7 over the range 1.25×10^{-4} and 1.56×10^{-2} M $FeCl_3$. Log % Cu in solution after 60 minutes at 40°C is plotted against log $[Fe^{3+}]$ and the line drawn has slope 1.00. Thus under the conditions used the first stage reaction is first-order with respect to Fe^{III} concentration. When the concentration of ferric iron was in the range 0.25–1.0 M the first stage of the reaction was complete in less than 5 minutes and no dependence of rate on concentration was observed. Over this concentration range the rate of the second stage was dependent on Fe^{III} concentration but the spread of the results indicated only that the order was about 0.2 or 0.3.

When 1 M $FeCl_2$ solution was used to leach 2.5 g of 150–300 μm Cu_2S at 40°C slow leaching occurred, Table 4.2, which can be explained by the amount of Fe^{III} present as an impurity in the salt and the free access of air to the reacting solution.

Table 4.2.

Time, min	5	30	60	120	240
% Cu dissolved	10.1	20.7	22.4	23.2	24.4

The effect of adding $FeCl_2$ to $FeCl_3$ solutions used to leach chalcocite was studied. The general conditions were 0.5 M $FeCl_3$; 0.2 M HCl; 2.5 g 150–300 μm Cu_2S; 40°C. The additions to give 0.05, 0.25, 0.5 and 1.0 M $FeCl_2$ caused almost exactly the same increase in the rate of dissolution of copper during stage 2 of the reaction as would have occurred if the same additional amount of $FeCl_3$ had been added. Other nonoxidising salts such as $MgCl_2$ increased the leaching rate in a similar fashion to $FeCl_2$ when added to the $FeCl_3$ solution. Thus the rate of leaching appears to be affected significantly when the total salt concentration is increased in a moderately concentrated solution of the oxidising agent. The rate of dissolution of copper was not affected by the concentration of HCl in 0.5 M $FeCl_3$, within the range 0.05–0.8 M HCl.

The results of the study of the kinetics of the reaction were used to prepare samples of leached chalcocite from which predetermined quantities of copper had been leached. These samples were examined in order to discover what happened to Cu_2S while copper was progressively removed from it. The conclusions drawn are outlined in Sec. 5.3.3.

4.2.5. *Electrochemical Methods of Following Leaching Reactions Involving Oxidation or Reduction*

When a leaching reaction involves electron transfer the solid loses electrons if it is oxidised and these are accepted by ions or molecules of the oxidising reagent in contact with the solid surface, which is itself reduced. Thus two half-cell reactions occur simultaneously by electron transfer across the interface between the solid and the solution. In metals and semiconductors the electrons produced by the oxidation of an atom are not restricted to any particular position; an equivalent number of electrons may be released to the oxidising agent anywhere on the surface. If the electron energy level is not identical over the whole surface, they will be released more readily from some areas than from others. A single piece of the solid then has anodic areas at which an element in the crystal lattice is oxidised and cathodic areas at which the electrons are transferred to the oxidising agent. This is the situation dealt with when considering the corrosion of metals and it was shown[113] that the theory of metal corrosion can be used in the case of gold dissolving in a cyanide solution saturated with air. The rate-controlling factors in that reaction were later considered in terms of the two half-cell reactions.[114]

In order to study the oxidative leaching of a metal or semiconductor it is not necessary to use a solute as the oxidant and analytically measure the rate at which a product appears in the solution. Instead, the reaction can be made to occur by electrolysis between the solid being studied, mounted so that it can be used as an anode, and a cathode. The extent to which oxidation has occurred at any time can be measured by recording the total current which has flowed. This method can be used conveniently if the anode material is mounted to form the reactive surface of a rotating disc. The speed of rotation can then be adjusted so as to move out of the region of rate control by a particular reagent as discussed in Sec. 4.2.3. Recently two other electrochemical techniques have been developed which can be effectively used to study leaching reactions involving electron transfer, chronopotentiometric and galvanostatic methods. The theoretical basis and practical application of these are not considered here but the ways in which they can be used are outlined. Several excellent texts are available.[115–117] These experimental methods can of course be used to study reduction reactions as well as oxidation, and to measure the diffusion coefficient of an electroactive species in a solid.[118]

It cannot be assumed that the results obtained by electrochemical methods necessarily correspond to these which are obtained by using an oxidising or reducing reagent in the solution to create the same electrochemical potential (E value) as that applied from the source outside the reacting system. It is necessary to determine experimentally whether this is the case if the distinction is important. If the rates are the same within experimental error then the reaction between the solute and the solid is electrochemical in nature and has an *electrochemical mechanism.* In this case the rate of the reaction, and therefore the current flowing, will not be proportional to the value of the overpotential applied, and the Butler–Volmer equation can be used to study the mechanism. The electrochemical model for the oxidative leaching of uranium dioxide in both acidic and alkaline solutions is discussed in Sec. 5.2.3.

Chronopotentiometry

The solid being studied is used in the form of a "working electrode". A constant electric current is passed between this and a "counter" or "auxiliary" electrode by use of a current source, the galvanostat. A reference electrode is also present and the potential between the reference and working electrodes is measured continuously and recorded or displayed on an oscilloscope. This is the main method, constant current chronopotentiometry, but there are variations on it; chronopotentiometry with current increasing linearly, current reversal- and cyclic-chronopotentiometry.

The layout of the electrodes in the electrochemical cell and a schematic diagram of the apparatus are shown in Figs. 4.8(a) and (b). The electrochemical cell has typically a volume of 75 ml and a removable PTFE cover with six ports for the working electrode, we, the auxillary, counter electrode, ce, the thermometer, inlet and outlet for nitrogen gas and one available for sampling when required but otherwise closed.

The working electrode usually can be prepared by polishing a flat surface of the material to be used as the electrode, joining multistrand copper wire to the back of the specimen with silver epoxy resin, mounting it in araldite, passing a piece of glass tubing over the copper wire to isolate it and to hold the electrode, and joining the tubing to the resin block with rapid setting epoxy resin (araldite).

The auxillary electrode is usually a platinum square with exposed area of surface about 1 cm^2. The reference electrode re is in a separate vessel containing a suitable electrolyte solution which makes electrical contact with the solution in the electrochemical cell via a Luggin probe, the tip of which can be moved to different positions within the cell but is usually kept as close to the working electrode as possible. Different reference electrodes and electrolytes can be used depending on what solution is present in the electrochemical cell. For example if the electrolyte used is 1 M sulphuric acid a mercurous sulphate reference electrode can be employed (0.6111 V versus Standard Hydrogen Electrode at 30°C) while if a hydrochloric acid electrolyte is used a saturated calomel electrode is appropriate (0.2378 V versus SHE at 30°C). The cell and the reference electrode assembly are held in a temperature-controlled bath and the solution in the electrochemical cell is purged with oxygen-free nitrogen for 10 minutes before experiments are started, to remove oxygen from it.

When a potential is applied across the working and counter electrodes so as to produce a constant flow of current, the electroactive species in the specimen is oxidised (or reduced) at a constant rate. The electrical potential between the working and reference electrodes assumes a value characteristic of

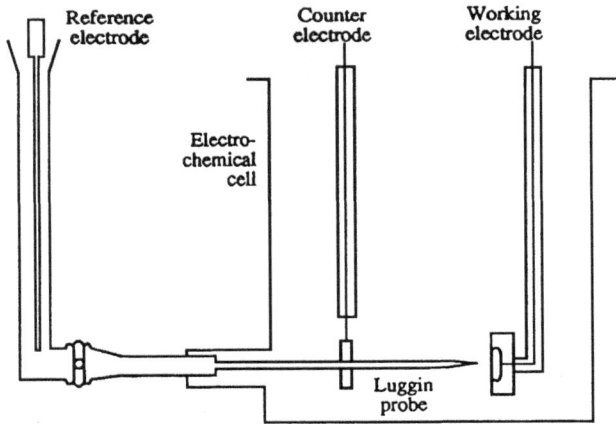

(a)

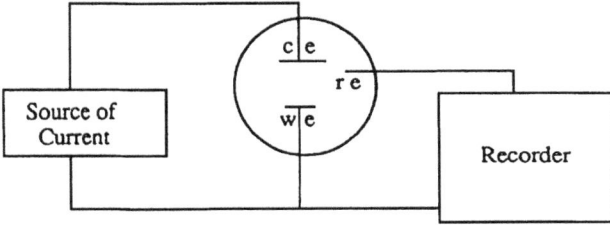

(b)

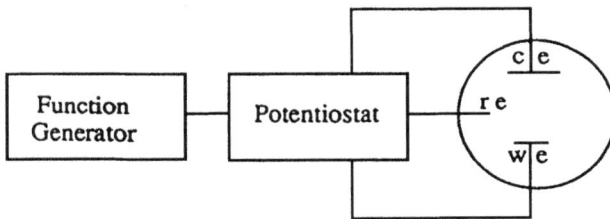

(c)

Fig. 4.8. (a) Layout of electrodes in the electrochemical cell for chronopotentiometry. (b) Schematic diagram of the apparatus for chronopotentiometry. (c) Schematic diagram of the apparatus for chronoamperometry and cyclic voltametry.

the electrochemical process taking place and varies with time as the concentration of the electroactive species at the solid–liquid interface changes. When the concentration decreases to zero at the surface of the working electrode it is assumed that the flux of the electroactive species within the solid to the surface is not sufficient to discharge (or accept) all of the electrons available at the interface. As soon as this is the case the potential drop between the working and the reference electrodes will move to a more positive (or negative) value until a new oxidation (or reduction) reaction can occur. The time elapsed between the application of the current and the shift of the potential is called the transition time, t_1.

The relationship between t_1 and the value of the current applied depends on the system being studied and the particular chronopotentiometric technique being used. In the case of constant current chronopotentiometry the Sand equation, see Ref. 117, has frequently been used for the analysis of experimental results.

$$j = \frac{n_i F \pi^{1/2} D_i^{1/2} c_i^*}{2t_1^{1/2}} \tag{4.20}$$

where

j = current density, A cm^{-2}
F = Faraday constant
D_i = diffusion coefficient of the electroactive species i, cm^2 s^{-1}
c_i^* = initial concentration of i, mol cm^{-3}
t_1 = transition time, s
n_i = number of electrons transferred per molecule of i
oxidised or reduced.

A typical chronopotentiogram of synthetic bornite (Cu_5FeS_4) in 1 M sulphuric acid is shown in Fig. 4.9. Three stages can be seen; the first is section AB of the curve. The potential increased rapidly as the experiment was started but after a few seconds it levelled off at a value which depended on the electrical resistance of the bornite electrode, the temperature and the applied current density. The oxidation reaction involving the electroactive species in bornite proceeded steadily during the period until the potential began to rise at the time indicated by B. This period of time, t_1, depended on the temperature and current density. The potential then stabilised at point C and the second stage of the reaction proceeded until the potential rose again at D. The

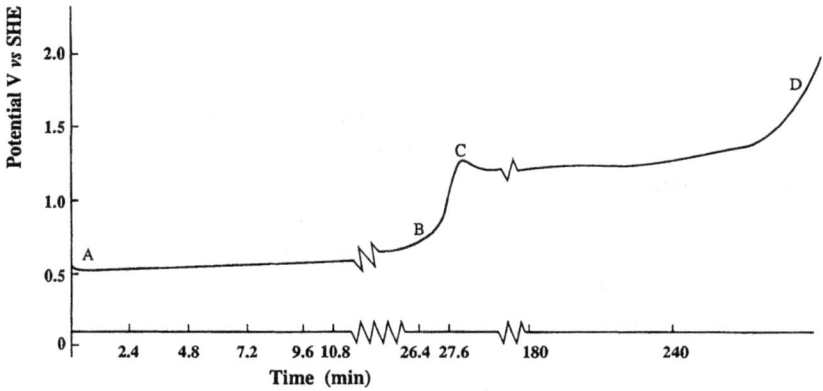

Fig. 4.9.   Chronopotentiogram of synthetic bornite in 1 M H_2SO_4; 30°C; current density 170 Am^{-2}. Note that breaks are shown on the time scale and chronopotentiogram. Courtesy Dr. S. Aguayo-Salinas, Universidad de Sonora, Mexico.

reaction after this was probably associated with decomposition of water at the anode surface.

The chronopotentiometric technique showed that oxidation of bornite in acidic solutions occurs in two stages at two distinct applied potentials. The chemical reactions which take place in those stages have to be inferred from other evidence. Chemical analysis of samples of solution taken at intervals during stage 1, AB, of an experiment using current density 140 Am^{-2} showed that after 38 minutes, 1.28×10^3 μg cm^{-2} of Cu had been dissolved from the bornite surface while no Fe was detected in the solution. After 43 minutes, the values were 2.52×10^3 μg Cu and 3.56 μg Fe cm^{-2}. Thus stage 1 involves initially dissolution of copper from the bornite but not iron. The electrode surface on reaching point C was pale orange with extensive cracking. During stage 2 of the leaching reaction, CD, the colour changed to dark greenish grey and both copper and iron appeared in the solution. The presence of elemental sulphur on the electrode surface was confirmed by X-ray powder diffraction and electron microprobe analysis.

Chronoamperometry and Linear and Cyclic Voltammetry

In chronoamperometry and sweep voltammetry the potential is controlled and the current which flows is measured as a function of time or potential respectively. The experimental arrangement is shown in Fig. 4.8(c).

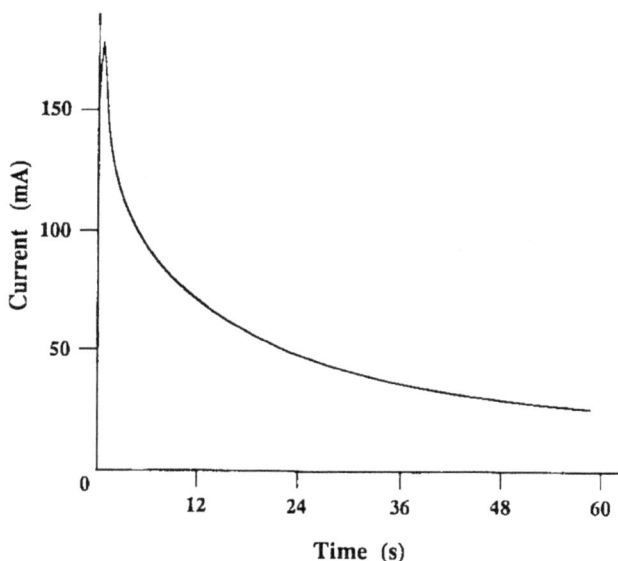

Fig. 4.10. Chronoamperogram of natural bornite in 1 M H_2SO_4; 30°C; potential 0.6111 V versus SHE; surface area 0.5973 cm^2. Courtesy Dr. S. Aguayo-Salinas.

Chronoamperometry. When no potential is being applied across the working and counter electrodes the system is described as being at the open circuit potential E_1. At zero time a potential E_2 is applied across the electrodes and in consequence oxidation (or reduction) of a species may occur at the working electrode if the potential is sufficiently large. If it does, a current flows as shown in Fig. 4.10. If it is assumed that the rate at which the reaction occurs depends on the concentration of the electroactive species close to, or at, the interface between the electrode and the electrolyte, then the current required to carry out the oxidation (or reduction) immediately after time zero is the maximum value for E_2. As the electroactive species near the interface is removed by the reaction a concentration gradient is produced and this causes more of the species within the solid to diffuse towards the surface. The flux of the species and hence the current which flows is proportional to the concentration gradient at the electrode surface. Since the rate at which the species reacts is usually faster than the rate at which it can diffuse to the surface the current usually decays exponentially. For this experimental situation where the applied potential is constant and the reaction is diffusion controlled the Cottrell equation applies (see Ref. 117).

$$j = \frac{n_i F c_i^* D_1^{1/2}}{\pi^{1/2} t^{1/2}} \tag{4.21}$$

where t is the time, s, and the other symbols have the same meaning as in (4.20).

The curve shown in Fig. 4.10 was obtained by polarizing a bornite electrode at a potential of zero mV versus a mercurous sulphate reference electrode (0.6111 V versus SHE) for 60 seconds. The exponential type of decay of the current could be linked to a diffusion controlled reaction and the values of current density (mA cm^{-2}) and time read off and plotted as j versus $t^{-1/2}$, using the Cottrell Eq. (4.21). The value of D_i can be calculated by using appropriate values of n_i and c_i. The number of electrons transferred per molecule of i oxidised requires knowledge of the chemical reaction occurring. Since oxidation of bornite was shown by the chronopotentiogram in Fig. 4.9 to occur in two stages it may reasonably be supposed that the curve in Fig. 4.10 applies to the first stage. It was shown by chemical analysis that initially in this stage copper is removed from the solid but not iron. During the second stage both copper and iron were dissolved from the residual solid. The electroactive species in stage 1 is Cu, therefore, but the number of electrons transferred per Cu oxidised cannot be deduced because the nature of the other product, the solid produced in stage 1, is not known. The value of c^*, the initial concentration of Cu can be calculated using the density ρ of bornite, 5.08 g cm^{-3}, the atomic weight of Cu, 63.546 g mol^{-1}, the factor corresponding to the mass fraction of copper in bornite, 0.6333, and the proportion of the total amount of copper in bornite available for reaction in stage 1

$$c_{Cu}^* = [(0.633 \times 5.08)/63.546] \times \text{proportion of Cu available} \, .$$

This proportion is not known from the evidence provided by chronopotentiometry or chronoamperometry.

The curves shown in Fig. 4.11 were obtained by polarizing at potentials greater than 2 V versus SHE. They show clearly two steps. The first corresponds to stage 1 of the dissolution of bornite and lasted for about 6 seconds. The second stage started after 12–18 seconds had elapsed. The response varied with the potential applied and from bornite sample to sample and was, therefore, not suitable for quantitative study.

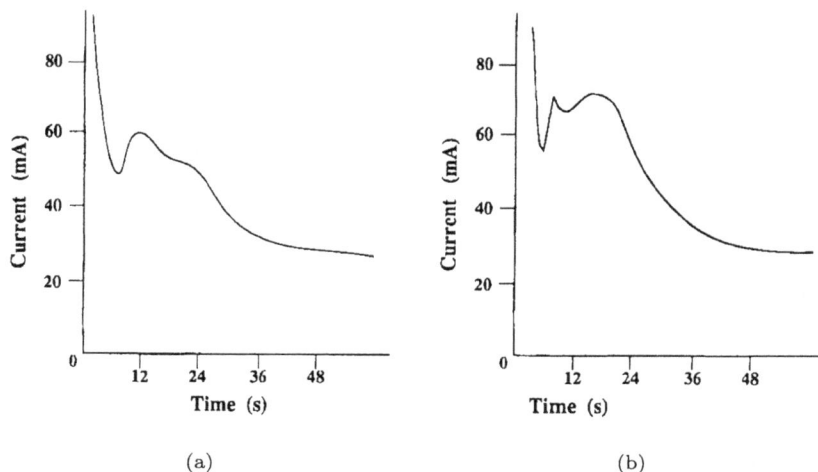

(a)

(b)

Fig. 4.11. Chronoamperograms of natural bornite in 1 M H$_2$SO$_4$; 30°C; (a) poten-tial 2.2111 V versus SHE; surface area 0.2185 cm^2; (b) potential 2.4111 V versus SHE; surface area 0.3329 cm^2. Courtesy Dr. S. Aguayo-Salinas.

Linear sweep voltammetry. The potential across the working and counter electrodes is varied linearly with time and the potential–current curve is recorded. As the potential E_1 is applied at time zero only non-Faradaic currents flow initially. Eventually the potential reaches a value at which some electroactive species begins to react and a current flows and increases in value until the applied potential reaches the value $E_{1/2}$, the half wave potential which is characteristic of the species reacting, after which the rate of increase of the current falls until the current reaches a maximum value I_p at potential E_p. The current then falls as the applied potential is increased further because diffusion of that species to the surface of the solid cannot maintain the rate of the reaction required to provide that current.

In cyclic linear sweep voltammetry the potential is raised linearly to a maxi-mum value and is then reversed. As the potential becomes smaller the products of the oxidation reaction become less stable as the potential approaches and passes $E_{1/2}$. This causes a cathodic current to flow as they are reduced. The shape of the current-potential curve as the potential becomes smaller is usually similar to that obtained while the potential is increased.

The solution of the diffusion equation for linear and sweep voltammetry depends on whether the reactions occurring are reversible, quasireversible or

irreversible. The solutions given[119] show that in most cases the current density
is proportional to the square root of the diffusion coefficient of the electroactive
species.   The curves obtained using linear potential sweep methods probably provide the most useful information for studying oxidation and reduction
processes involving solids in an aqueous environment by electrochemical techniques. It has been stated[117] that by proper treatment of linear potential sweep
data using convolutive or semiintegral mathematical techniques the voltammetric current–potential (or current–time) curves can be transformed into forms
closely resembling the steady state voltammetric curves. These are frequently
more convenient for further data-processing and interpretation of the results
in terms of chemical reactions.

A typical cyclic voltammogram for synthetic bornite in 1M sulphuric acid
is shown in Fig. 4.12. This is for a sweep speed of 5 mVs^{-1}; a change in
sweep speed gives changes in peak currents but the voltammograms remain

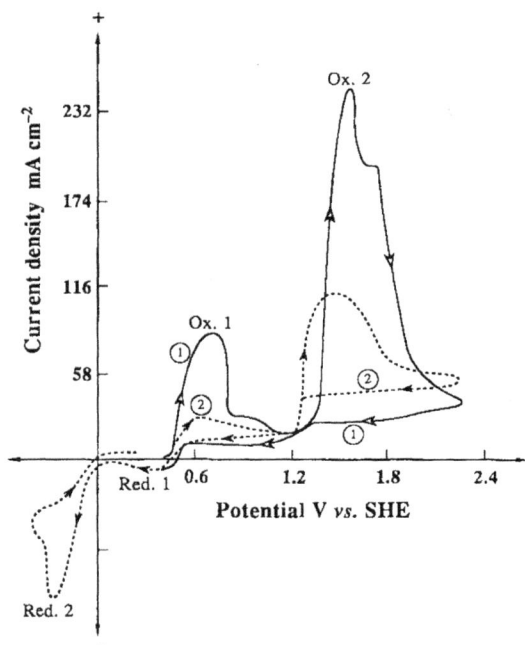

Fig. 4.12.   Cyclic voltammogram of synthetic bornite in 1 M H_2SO_4; 30°C, sweep speed
5 mVs^{-1}. Courtesy Dr. S. Aguayo-Salinas.

essentially similar. The figure shows two successive cycles of oxidation with one reduction cycle finally. The first oxidation peak (Ox 1) has a half wave potential, at which the current is half of its maximum value, of 0.511 V versus SHE. For steady state mass transfer

$$E_{1/2} = E^{\circ} - \frac{RT}{zF} \ln \frac{\{Ox\}}{\{Red\}}$$

so that ΔG° for the chemical reaction can be calculated from

$$\Delta G^{\circ} = -0.511 \, zF.$$

The second peak (Ox 2) corresponds to the decomposition of the product of stage 1 of the reaction to give Cu^{2+} and Fe^{3+} in solution and elemental sulphur. The second oxidation sweep shows a generally similar shape with considerably lower current densities because of depletion of the electroactive species near the surface of the solid during the first sweep. There is also a shift in the values of $E_{1/2}$ for the same reason. When the scan was reversed two reduction peaks (Red 1) and (Red 2) occurred within the range of potential studied. The oxidative leaching of bornite in acidic solutions is considered further in Sec. 5.3.4.

4.3. Kinetics of Leaching Ground Ores and Concentrates

4.3.1. The Problem of Non-uniform Particle Size

The feed to a hydrometallurgical leaching circuit has usually been ground in a ball- or rod-mill and subsequently passed through a classifier to remove particles larger than a certain size, these being returned for further grinding. The solid then has a range of particle sizes which is statistically determined by a number of factors, including the mechanical properties of the rock and the individual minerals contained in it. Using a given set of chemical conditions and temperature the rate of leaching of a desired metal from the feed depends on the surface area of the solid being treated. It is necessary therefore to use some method of describing the continuous range of particle sizes and the fraction of the total number of particles having each size.

This is done by passing a weighed sample of the dried ground material through a set of standard sieves and weighing the amount of solid remaining on each sieve. The smallest sieve size used is generally 200 or 325 mesh, which have 75 or 45 μm openings, and material which passes through these is not usually analysed further, although subsieve sizing methods are available. The weights of solid on the individual sieves are converted to give the cumulative mass fraction finer than each given mesh size. That is, the total weight of solid, expressed as a fraction (or percentage) of the weight sieved, which has particle diameters smaller than the holes in a particular sieve. The mass fractions for a set of sieves can usually be represented by the Schuhmann equation

$$y = (d/K)^m \qquad (4.22)$$

where y is the mass fraction having diameter smaller than d; m is the distribution constant and K another constant, the size modulus, both constants for the particular solid used. The values of m and K for the ore are obtained by plotting log y against log d and it is then possible to calculate an average particle radius r_i for selected weight fractions w_i.

4.3.2.   Equations for Modelling Leaching Kinetics

The Shrinking Particle Model

For purposes of process control it is convenient to fit a mathematical equation to the curve showing recovery of the metal values by leaching as a function of time. In Sec. 4.2, such curves were related to the rates of transport of reactants to the solid surface or the rate of a chemical reaction such as an electron transfer step. For process control under defined conditions the equation need not refer to such factors.

It has been pointed out[120] that in a system where the surface of a solid reactant advances or recedes during the course of the reaction, the change affects the kinetics. So also does the progressive formation of a solid reaction product around the reacting particles. Mathematical equations were obtained for a number of models based on such behaviour and used to describe the shapes of the rate curves obtained when chalcopyrite was leached under various conditions. Such equations are now widely employed. However when determining the chemical conditions to be used, and in particular when developing a new process, the chemical behaviour of the system is of paramount importance.

The equation for the shrinking particle model was derived as follows.[120] The particles were assumed to be spherical but the final equation is applicable to particles of any isometric shape. The total number of moles in an unreacted sphere, n, is given by

$$n = 4\pi r^3/3V \tag{4.23}$$

where V is the molar volume equal to M/ρ where M is the molar mass and ρ the density of the solid. The rate of reaction at the surface of the sphere of radius r can be written

$$-dn/dt = 4\pi r^2 ck' \tag{4.24}$$

k' being the rate constant for the first-order reaction with a reactant having concentration c in the solution. Wadsworth included a factor, k_o, to modify the simple geometric surface area of the sphere and indicate the concentration of potentially reactive surface sites. It includes for example a factor to allow for surface roughness. Differentiating (4.23) with respect to time and equating to (4.24) gives the linear rate

$$-dr/dt = Vck' \tag{4.25}$$

Vk' being the linear rate constant k_1. If c is the concentration of the reacting solute in moles cm^{-3} the linear rate of decrease in r is given in cm s^{-1}. If c remains constant, Eq. (4.25) represents the constant velocity of movement of the reaction interface, which is the definition of linear kinetics.

If the initial radius of the reacting particle is r_o and α is the fraction of it which has reacted then

$$\alpha = 1 - (r^3/r_o^3). \tag{4.26}$$

Differentiation with respect to time gives

$$d\alpha/dt = -3(r^2/r_o^3)(dr/dt) \tag{4.27}$$

and combining (4.25), (4.26) and (4.27) gives

$$d\alpha/dt = (3ck_1/r_o)(1-a)^{2/3}. \tag{4.28}$$

For the condition $\alpha = 0$ when $t = 0$, (4.28) may be integrated assuming c is constant giving

$$1 - (1-\alpha)^{1/3} = kt \tag{4.29}$$

where $k = ck_1/r_o$ (time^{-1}). Plotting the left hand side of (4.29) against t should give a straight line of slope k, k having units $1/t$.

All particles of a particular mineral in a slurry which have the same initial diameter will react in the same way according to this derivation. Thus when Eq. (4.29) is used for a slurry containing a distribution of particle sizes the weight fraction w_i of each selected average particle radius r_i (Sec. 4.3.1) is required. If the initial average radius of the particles in weight fraction w_i is r_{io} then (4.29) is replaced by

$$1 - (1 - \alpha_i)^{1/3} = (ck_l/r_{io})t \tag{4.30}$$

where α_i is the fraction reacted in the weight fraction w_i. The total quantity of solid reacted is given by $\alpha = \sum_i w_i \alpha_i$.

The Shrinking Core Model

There are many cases where a mineral particle contains several metals, only one of which is dissolved during a leaching process. This can lead to the formation of a porous solid reaction product surrounding each particle of unreacted mineral. The special case in which the diameter of the composite particle remains equal to the original diameter r_o of the particle before reaction commenced has been considered.[120] The shrinking core, of radius r, continues to react at a rate controlled by the rates of diffusion of reactants to the unreacted core, through the reaction products.

If the particle is spherical the rate of reaction can be written

$$-\frac{dn}{dt} = \frac{4\pi r^2}{\sigma} D \frac{dc}{dr} \tag{4.31}$$

where n is the number of moles of unreacted mineral in the core and σ is the stoichiometry factor, the number of moles of the diffusing species required to liberate one mole of the metal to be leached from the core. Integration of this equation between limits r and r_o, assuming steady state conditions, gives

$$-\frac{dn}{dt} = \frac{4\pi Dcrr_o}{\sigma(r_o - r)} \tag{4.32}$$

where the concentration of reactant at the interface is small compared with c. Combining (4.23) and (4.31) gives the equation

$$-\frac{dr}{dt} = \frac{VDcr_o}{\sigma r(r_o - r)} \tag{4.33}$$

for the rate of movement of the boundary between the core and reaction products in terms of the radius of the unreacted core. Combining (4.33) with (4.26) and (4.27) gives the rate of reaction in terms of the fraction already reacted, α,

$$\frac{d\alpha}{dt} = \frac{3VDc}{\sigma r_o^2} \frac{(1-\alpha)^{1/3}}{1-(1-\alpha)^{1/3}}. \tag{4.34}$$

Integration for the boundary condition $\alpha = 0$ when $t = 0$ gives

$$1 - \frac{2}{3}\alpha - (1-\alpha)^{2/3} = \frac{2VDct}{\sigma r_o^2}. \tag{4.35}$$

Equation (4.35) has been applied[120] to data obtained on leaching monosize 12 and 47 μm chalcopyrite in 0.5 M $Fe_2(SO_4)_3$, solutions at 90°C. Plotting the left hand side of the equation against time (hours) gave a straight line in each case, the slopes being proportional to r^{-2}. The elemental sulphur formed by the reaction

$$CuFeS_2 + 4Fe^{3+} = Cu^{2+} + 5Fe^{2+} + 2S$$

is said to form a strongly adherent coating on the surface of the mineral (Sec. 5.3.5). The shrinking core model is also said to fit the rate of dissolution of Al_2O_3 from bauxite in the Bayer process. The bauxite particles contain aluminium oxides together with some SiO_2, Fe_2O_3 and TiO_2. It is digested with NaOH solution when the alumina and silica dissolve, leaving the Fe_2O_3 and TiO_2 as a residue which disperses as very fine particles towards the end of the dissolution process.

4.3.3. *Calculation of the Rate of Dissolution of Zinc Calcine in Dilute Acid Solutions*

A successful attempt has been made to calculate the rate at which a solid dissolves in an aqueous solution, based on the ionic diffusivity and ionic mobility of each of the chemical species taking part in the dissolution process.[121] These calculations assume that the rate of dissolution is controlled by mass transport and in consequence the procedure can be used only in cases where no chemical reaction is involved. The example chosen was the dissolution of zinc oxide in dilute acid solutions. This reaction is used at two points in the electrolytic zinc process, first when the oxide produced by calcining zinc sulphide concentrates is dissolved in dilute sulphuric acid, and second after $ZnFe_2O_4$

formed during the roasting has been dissolved in stronger sulphuric acid and excess free acid is neutralised by zinc calcine to precipitate the iron from the hot solution (Sec. 3.8.6).

The data used were the ionic diffusivity and ionic mobility of each of the species H^+, HSO_4^-, SO_4^{2-} and Zn^{2+} together with the equilibrium constants and activity coefficients for the equilibria

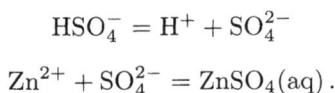

$$HSO_4^- = H^+ + SO_4^{2-}$$

$$Zn^{2+} + SO_4^{2-} = ZnSO_4(aq).$$

It was assumed that the rate of dissolution is controlled only by mass transport and that this can take place by three mechanisms, diffusion, convection and electrical migration. The latter mechanism was not considered in Sec. 4.1.3 because it can be taken as a modification of the value of D for an ionic species within the boundary layer, caused by the presence of the electric field in that layer, see also Sec. 5.2. This field is due to the concentrations of ions present at the solid–liquid interface and these concentrations were calculated on the assumption that the slowest step in the leaching process is the transfer of species across the boundary layer. Thus the equilibria between HSO_4^- and SO_4^{2-}, between Zn^{2+} and $ZnSO_4(aq)$ and between ZnO and Zn^{2+} by the solubility product reaction

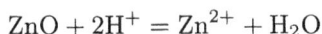

$$ZnO + 2H^+ = Zn^{2+} + H_2O$$

can all be assumed to have been achieved at the interface.

The flux of species due to each transport mechanism was considered separately and the net flux due to all three was obtained by combining the three equations. A mathematical model was set up on the assumption that the particles are homogeneous, nonporous and nearly spherical with radius r. A shape factor was introduced to take into account larger surface areas due to deviation from spherical symmetry. The particles were assumed to shrink as they dissolve, with no change in shape. No layer of insoluble reaction product formed on the surface. The mathematical model used the net flux based on the combined mass transport equations. Some assumptions were made in order to simplify the model equation and to obtain boundary conditions. This resulted in a set of 7 equations containing 7 unknowns; 3 surface concentrations, 3 fluxes and the electric field in the boundary layer. The equations were solved simultaneously to obtain the 7 unknowns. Once the fluxes of the species considered were known the rate of dissolution of ZnO could be calculated, using the radii of weight fractions for the calcine used in leaching experiments to test the model.

The boundary conditions used in the flux equations were (i) the concentrations of the chemical species at the surface of the particles. These were determined by assuming that the dissolution reaction is at chemical equilibrium. (ii) The concentration of each species in the bulk of the solution was obtained by assuming thermodynamic equilibrium had been achieved. (iii) The stoichiometric relationship between the fluxes. Since sulphate is neither generated nor consumed in the reaction the net flux of all species containing sulphate (SO_4^{2-}, HSO_4^-, $ZnSO_4(aq)$) must be zero at steady state.

The flux equations were used to calculate the particle radius, and so the fraction of ZnO reacted, as a function of time. Calculations were also carried out for the dissolution of ZnO in perchloric, nitric and hydrochloric acid solutions.

In order to test the model and calculations the rate of dissolution of calcine in each of the four acids was studied. The stirring conditions were selected so that all of the solid was in suspension and the rate of dissolution was then independent of stirring speed. Good agreement between the experimental curves and those calculated was obtained in nitrate and perchlorate solutions and reasonable agreement in sulphate solutions up to 80% dissolution in each case. Agreement was poor for the later stages of dissolution because the particles did not leach topochemically as was assumed in the model. Partially leached calcine particles were found to have large and deep pores. The simplified model could not be used to predict leaching rates in hydrochloric acid because it did not consider the formation of chloro complexes of zinc or adsorption of Cl^- ions on the surface of the solid. Failure to take into account electrical migration as a contribution to mass transport produced a strong deviation of the predicted leaching kinetics from the results obtained using 0.1 M $HClO_4$, which are in excellent agreement with those predicted by the complete model.

4.4. Kinetics of Reactions in Which Metals are Precipitated from Solutions by Gases

4.4.1. *Introduction*

The Sherritt Gordon process, produces nickel and cobalt metal powders by reaction between hydrogen gas and nickel ammine sulphates of composition nominally $Ni(NH_3)_{2.1}SO_4$, in moderately concentrated aqueous ammonium sulphate solution. Examination of the thermodynamics of the reaction, Sec. 3.7.3 and Fig. 3.9, shows that both Ni^{2+} and the ammine should react with H_2 at 1 atm pressure at 25°C. The reactions do not take place. At a temperature of about 130°C or above some metal is formed as a plating on metal surfaces in the solution. Thus an elevated temperature is required to cause reaction to occur on a catalytically active surface. The reaction is not self-nucleating.

In the commercial process for producing nickel powder the nickel concentration in the solution initially is about 45 gl^{-1} Ni and the reduction reaction is complete in about 30 minutes. The H_2 required to reduce the Ni^{II} must transfer from the gaseous into the solution phase and the rate at which this can be achieved depends on the engineering design of the agitation equipment. Thus the kinetics of reactions such as this may be controlled by mass transfer from the gaseous to the liquid phase or, alternatively, the solution may be kept saturated with the gaseous reactant by ensuring that the reducing reaction is carried out under conditions where it is slower than the rate at which fresh H_2 can replace that oxidised.

4.4.2.   *Reduction of Nickel Salts in Aqueous Solutions by Hydrogen*

Some of the laboratory work on the kinetics of the reduction of nickel in ammoniacal sulphate solutions by hydrogen which was used in the development of the process to produce nickel powder has been published.[122] The experimental procedure used is the basis for most work carried out in autoclaves in the field of hydrometallurgy since then. A 1 US gallon (3.79 l) stainless steel autoclave was used, fitted with a stirrer having 2 impellers, one an axial flow type with 3 blades bent to throw downwards and one a radial flow 4 straight blade impeller placed 7.5 cm above the first. Both impellers were 7.5 cm in diameter. The rotation speed was normally 730 rpm; usually this was "above the critical speed below which mass transfer would become the rate-controlling factor".

2.5 litre of solution and the catalyst used in the experiment were placed in the autoclave which was closed and purged with H_2. Rotation of the stirrer was started and the autoclave was heated externally, the temperature of the solution being measured by a thermocouple fitted into a thermowell through the head of the autoclave. When the reaction temperature was reached the pressure gauge indicated the vapour pressure of the solution (because it read zero at atmospheric pressure, hence the use of pounds per square inch gauge, lb in^{-2} g; expansion of the hydrogen at 1 atm is neglected). Hydrogen was then admitted through an opening on the head to obtain the desired partial pressure of the gas. During the reaction solution samples were drawn at convenient time intervals for analysis, using a porous metal disc at the bottom of the sample tube to avoid passing solid to the sampling valve.

The effect of the NH_3 to Ni^{2+} ratio $(R_A)Ni$ was investigated using 70 gl^{-1} Ni^{2+} and molar ratios 1, 2, 3 and 4. The temperature of reaction was maintained at 205°C and the total pressure at 4.42 MPa (640 lb in^{-2}) corresponding to a partial pressure of H_2 2.76 MPa (400 lb in^{-2}). The catalyst used was 100 gl^{-1} of nickel powder (72.3% smaller than 44 μm). Not all of the Ni^{II} was reduced when the NH_3:Ni ratio was 4 and reduction stopped sharply after about two thirds of the nickel had been reduced when the ratio was 1. When the ratio was 3 the final 20% of the nickel reacted more slowly than when the ratio was 2. It had been reported previously that only 10% of the nickel in sulphate solutions could be precipitated from acidic solutions by reduction with hydrogen.[123]

The effect of temperature was studied using $[Ni^{2+}]$ 80 gl^{-1}, NH_3:Ni molar ratio 2 and a partial pressure of H_2 2.42 MPa (350 lb in^{-2}); $FeSO_4$ at 1.2 gl^{-1} Fe was added to produce catalytic seed. The effect of the partial pressure of hydrogen was also investigated using nickel powder as catalyst and a number of experiments concerning the effects of changing conditions on the activity of seed from $FeSO_4$ were carried out.

Using an NH_3 to Ni^{2+} ratio approximately 2 with nickel powder as the seed catalyst between 150°C and 180°C, the kinetics of precipitation from ammoniacal sulphate solutions were represented by

$$-d[Ni^{2+}]/dt = kA_s p_{H_2} \tag{4.36}$$

where A_s is the initial surface area of the seed on which the nickel was deposited.

The same group of workers found a similar rate expression for the precipitation of nickel from aqueous ammoniacal carbonate solutions between 135°C and 180°C.[124] This system is used to produce nickel metal powder having low apparent densities, for special applications. In the latter case the system was described as self-nucleating because it was considered that the presence of small particles of basic nickel carbonate, $2NiCO_3 \cdot 3Ni(OH)_2$, provided nucleation centres for deposition of nickel.

Equation (4.36) shows that the rate of reduction is independent (zero order) of the concentration of nickel. In acidic solution however the order of reaction with respect to nickel was 1 and to p_{H_2} one half.[125] The effects of pH of the solution and of catalyst were not mentioned. Precipitation of nickel from acidic acetate buffered solutions over the pH range approximately 0.5–3.2, using carbonyl nickel metal spheres smaller than 44 μm as seed, in the temperature range 130° to 160°C led to the proposal of the rate equation[126]

$$-d[Ni^{2+}]/dt = k_1 A_S [Ni^{2+}] p_{H_2}^{1/2} - k_2 [H^+]^n \qquad (4.37)$$

where the value of n is less than 1. The rate of reduction of nickel was independent of the rate of stirring using a turbine type impeller in a baffled system so that it was assumed that the rate was not controlled by diffusion in the solution. Equation (4.37) shows the reaction to be first-order with respect to nickel concentration in solution whereas (4.36) indicates zero order in ammoniacal sulphate solution. However the initial concentration of nickel used in the acid solution was only 0.01 M so that rate control by availability of this at the catalytically active surface is not surprising.

Values of the apparent activation energy for precipitation of nickel from acid solutions have been given as in Ref. 125, 17.22 and in Ref. 126, 104.5 kJ mol^{-1}; and from ammoniacal media as 41.38 kJ mol^{-1} for carbonate;[124] and in Ref. 122, 20.9–62.7, later in Ref. 58 corrected to 42.64 kJ mol^{-1} for sulphate solutions.

The effect of agitation on the rate of precipitation of nickel from aqueous ammoniacal sulphate solutions by hydrogen has been considered in terms of "stirring intensities" expressed as Reynolds numbers, Re.[127] With a total nickel concentration 0.5 M the rate of reduction at 200°C was first-order with respect to [Ni^{2+}] when values of Re were greater than 18 000, and the order became fractional and, finally, zero as the value of Re was progressively decreased to zero. At the same time the apparent activation energy decreased from 74.0 to 16.7 kJ mol^{-1}. This indicates that a chemically controlled reaction changed to diffusion control as the "stirring intensity" was decreased. The dependence of the rate on p_{H_2} was 0.5 order, increasing to first-order when the value of Re was decreased from 18 000 to 0. This was interpreted[128] as indicating that as the stirring intensity decreased the concentration of hydrogen in the solution was lowered.

In any normal study of the kinetics of a reaction occurring in an aqueous solution the initial concentrations of the reactants are defined. The kinetics of the reduction of nickel in ammoniacal solutions by H$_2$ was studied using this procedure.[128] The total volume of the sealed autoclave and its ancillary equipment was measured with the autoclave body at a number of temperatures between 20°C and 250°C. The change in volume with temperature was linear. The measurements were carried out by introducing N_2 into the autoclave at 20°C to pressures between 1.7 and 53 cm Hg, measured using a manometer, and when the temperature and pressure had become constant the gas was released very slowly $(0.1 \ 1 \ min^{-1})$ through a gas flowmeter. The volume of the vessel at higher temperatures was measured by filling it with H_2 at the required initial pressure. Pressure and temperature were recorded continuously and when both were constant the autoclave body was heated by an electric "furnace" clamped around it using a current controlled to give a temperature rise of $0.5°C \ min^{-1}$. The pressure–temperature relationship was obtained between 20°C and 250°C on both heating and cooling cycles. A cooling rate of 0.5°C min^{-1} was obtained by controlling the flow of fluid through the cooling coil in the vessel. Measurements were made using a number of initial pressures of H_2.

The solubility of hydrogen in water and solutions of ammonium sulphate at temperatures between 182°C and 241°C at a range of values of p_{H_2} was measured in the autoclave. This was a nominally 2 L vessel constructed of type 316 stainless steel. The magnetically driven stirrer ("Magnedrive" unit, Autoclave Engineers) was designed to draw gas from the top region of the vessel down a tube and disperse it by means of the impeller blades below a hood. This facility was not used; the gas tube was sealed. Transfer of gas occurred across the surface of the liquid. The stirrer was a radial flow 4 straight blade impeller 32 mm in diameter and a baffle was fitted. The standard stirring speed used was 1100 rev min^{-1}, measured with a tachometer which was frequently calibrated with a stroboscope. A measured volume of water or the ammonium sulphate solution at 20°C was sealed in the autoclave which was flushed three times with N_2, time being allowed during the second period for displacement of dissolved O_2. Hydrogen was then introduced to the required initial pressure and stirring was begun. The temperature and pressure were recorded continuously and when both were constant the heaters on the autoclave were switched on to give a temperature rise of 0.5°C min^{-1}. The pressure–temperature relationship was obtained between 20°C and 250°C on both the heating and cooling cycles, a cooling rate of 0.5°C min^{-1} being maintained by controlling the flow of cold fluid through cooling coils in the vessel. The rate of dissolution of H_2 was recorded continuously. Once the solution was saturated with the gas, progressive lowering of the stirring speed (or stopping the motor) and then increasing it did not change the concentration in the solution. Use of a stirring speed of 900 instead of 1100 rev min^{-1} had no effect on the equilibrium solubility of hydrogen. The quantity of hydrogen present in the gas phase in the autoclave was calculated by subtracting the vapour pressure of the solution from the total pressure to give the partial pressure of H_2 and using the ideal gas law with a correction for the compressibility of hydrogen by means of virial coefficients.[129,130] Since no data on the partial molar volumes of metal salts at temperatures between 180°C and 250°C were available the specific volumes of pure saturated liquid water were taken.

The values of the equilibrium solubilities of hydrogen obtained were very much higher than any which had been reported previously and are shown in Fig. 4.13. Several previous workers who measured the solubility of the gas reported that increased solubility of more than 100% occurred if the liquid

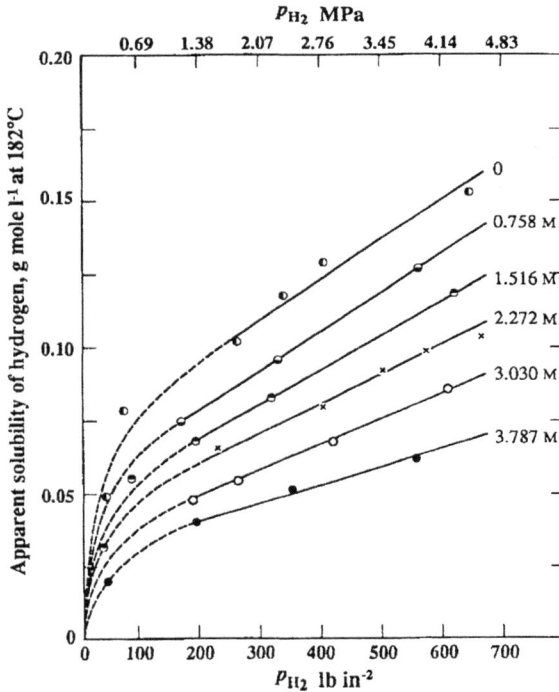

Fig. 4.13.   Apparent solubility of H_2 in aqueous ammonium sulphate solutions at 182°C. $(NH_4)_2SO_4$ concentrations shown on curves in g mol L^{-1} at 20°C.

was shaken vigorously. It was assumed that the values shown in Fig. 4.13 correspond to maximum supersaturation obtainable by the stirring system and equipment used. Since the amount of hydrogen lost from the gas phase to produce a solution exhibiting equilibrium solubility was not proportional to the volume of solution in the autoclave the measured apparent equilibrium solubilities could only be used in interpreting the kinetic data for reduction of metal salts when the volumes of solutions used were the same. In all experiments in which nickel or cobalt metal was produced the quantity of hydrogen lost from the gas phase was used to calculate the extent of reduction. In many experiments the metal content of the solution removed from the autoclave was determined by chemical analysis. In every case excellent mass balance was obtained, indicating that the apparent solubility values could be used to study the kinetics of the reactions.

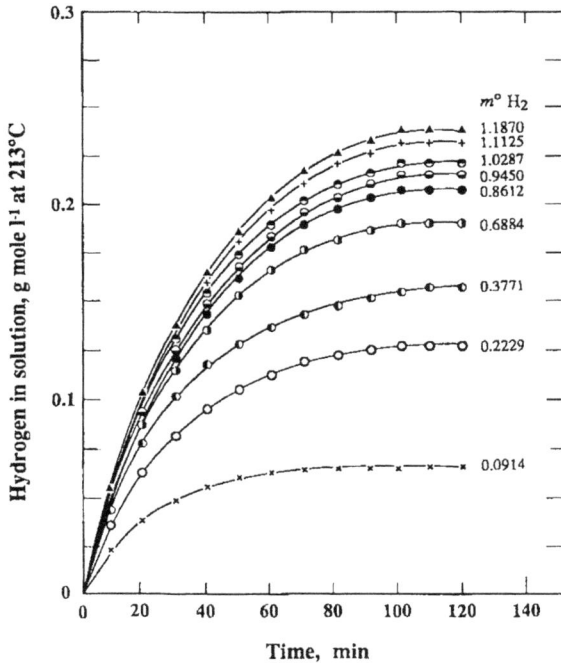

Fig. 4.14. Rates of dissolution of H_2 in solutions containing 0.5 moles NH_3 and 2.27 moles $(NH_4)_2SO_4$ per litre at 213°C. $m_{H_2}^\circ$ values are total quantities (g mol) of H_2 introduced into the autoclave at $t = 0$ (autoclave and contents at 213°C).

The rates at which hydrogen dissolved at 213°C in a solution which at 20°C contained 2.272 g mol L^{-1} of ammonium sulphate and 0.5 g mol L^{-1} of ammonia are shown in Fig. 4.14. The total quantities of H_2 introduced into the autoclave containing the solution at 213°C at zero time were in the range $m_{H_2}^o = 0.0914$ to 1.1870 g mol. The curves obtained at 204°C, 213°C and 222°C showed that a period between 60 and 100 minutes was necessary to saturate the solution but the equilibrium value was being approached 50 minutes after introduction of the H_2. Introduction of the gas caused a temperature rise of 5–15°C but this did not cause difficulties in studying the kinetics of metal deposition since little or none formed during the first 50 minutes after introduction of the hydrogen. The seed used to catalyse the reaction was selected to ensure this.

In experiments in which nickel ammine sulphates were reduced to nickel, ammonium polyacrylate, 0.1 g L^{-1}, was used to minimise plating of the metal on stainless steel surfaces in contact with the solution. 30 g of alumina powder

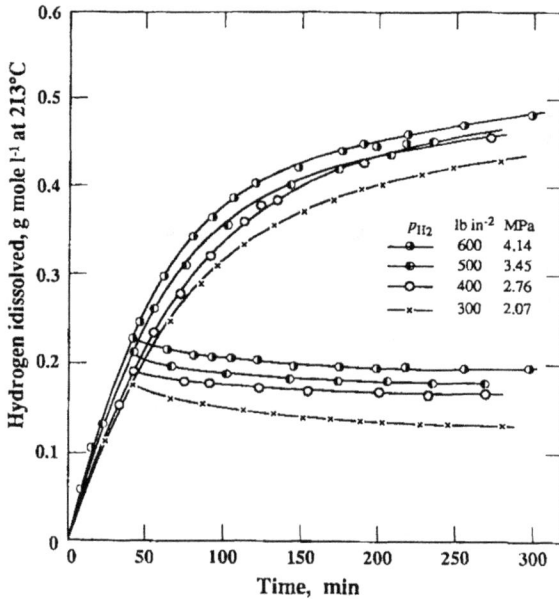

Fig. 4.15.   Reduction of nickel ammines by H_2 at 213°C. (Upper curves) Total moles of H_2 removed from gas space versus t. Initial values of p_{H_2} are shown. (Lower curves) Equilibrium solubilities of H_2 at these values.

was added as seed, and the stirring speed was 1100 rev min^{-1}. Results obtained using solutions containing at $20°C$ 0.32 M NiSO$_4$; 2.272 M (NH$_4$)$_2$SO$_4$ with [NH$_3$]/[Ni^{2+}] ratio (R_A), 2.1 are shown in Fig. 4.15, for several partial pressures of H$_2$. The upper group of curves shows the quantity of H$_2$, m$'$H$_2$ taken into the solution at each initial partial pressure of hydrogen. The lower group show the equilibrium solubility of H$_2$ at the corresponding values of p_{H_2}. The difference between the corresponding upper and lower curves for each initial p_{H_2} represents the quantity of H$_2$ consumed by the formation of Ni metal. The single curve shown in Fig. 4.16 was obtained by subtracting these quantities, converted to moles of Ni, from the initial concentration of nickel in solution. Data taken from that used to construct Fig. 4.16 and plotted as log [Ni^{2+}] versus time give a straight line showing that the reaction is first-order with respect to [Ni^{2+}] and zero order with respect to the concentration of H$_2$ dissolved in the solution.

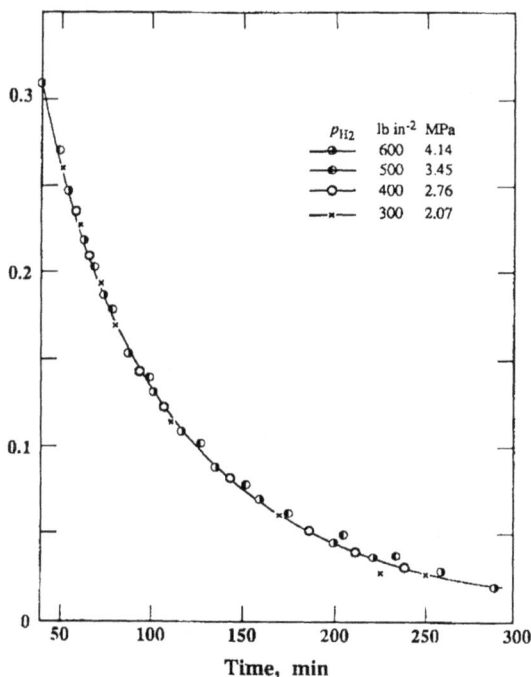

Fig. 4.16. Change in [Ni^{II}] with time. Data from Fig. 4.15.

Table 4.3.

$T°C$	$[Ni^{2+}](20°C)$	$[Ni^{2+}](T°C)$	$(R_A)_{Ni}$	p_{H_2}		$10^3 k$
	g mol L^{-1}			MPa	lb in^{-2}	min^{-1}
213	0.456	0.388	2.1	4.15	600	14.1
213	0.364	0.309	2.1	3.46	500	13.8
213	0.273	0.232	2.1	3.46	500	12.7
213	0.183	0.155	2.1	3.46	500	11.9
213	0.364	0.309	2.1	4.15	600	13.8
213	0.364	0.309	2.1	2.76	400	13.8
213	0.364	0.309	2.1	2.08	300	13.8
204	0.364	0.313	2.1	3.46	500	11.2
222	0.364	0.305	2.1	3.46	500	15.8
231	0.364	0.301	2.1	3.46	500	18.4
213	0.364	0.309	1.0	3.46	500	8.9
213	0.364	0.309	2.5	3.46	500	12.1
213	0.364	0.309	3.0	3.46	500	11.5
213	0.364	0.309	4.0	3.46	500	9.6

The apparent rate constants obtained under these and other conditions are given in Table 4.3. The value of k increases with increasing initial nickel concentration. This is attributed to the fact that the surface area of catalytically active solid increases with the amount of metal deposited. The values of k at the temperatures shown give a good straight line on plotting log k versus $1/T$, K. The apparent activation energy obtained $(31.3 \pm 0.8 \text{ kJ mol}^{-1})$ was larger than is usually accepted as indicating a diffusion controlled reaction in an aqueous solution.

The effect of changing the $[NH_3]$ to $[Ni^{2+}]$ ratio is shown in Fig. 4.17 together with a line showing how the fraction α_2 of the total nickel initially present as the species $Ni(NH_3)_2^{2+}$ changes with that ratio. This line was calculated by use of stability constant data[59] extrapolated to 213°C using the van't Hoff isochore in the absence of sufficient information to permit a more satisfactory procedure to be employed. Although a graph of k against $(R_A)_{Ni}$ is not symmetrically disposed about the value of $(R_A)_{Ni} = 2.0$, a graph of α_2 against k is approximately linear.

Under the electron microscope it could be seen that the nickel occurred as clusters of rods, needles, plates and other shaped particles attached to the alumina. Rods had average lengths about 5 μm and widths 0.5 μm. Thus during the initial period of the reaction while the hydrogen was dissolving and the pressure was changing rapidly, reduction occurred at some points on the surface

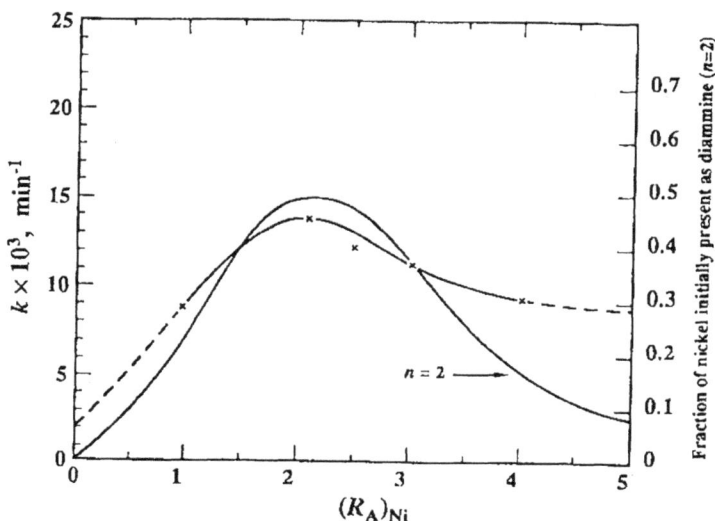

Fig. 4.17.   Effect of NH_3 to Ni^{II} ratio $(R_A)_{Ni}$ on the apparent first-order reaction rate constant k. The line indicated as $n = 2$ shows the fraction of the nickel initially present in solution as the species $Ni(NH_3)_2^{2+}$.

of the alumina particles so that nickel deposited on the particles. The reaction which was studied, however, was the growth of the strongly catalytically active nickel sites.

This makes it impossible to estimate the area of the active surface during the reduction reaction. Electron micrographs indicated that it would be of the same order as the surface area of the original alumina particles. These were approximately spherical with diameter 10^{-4} cm. Assuming a simple model in which $H_2(aq)$ diffuses through the solution under a diffusion gradient of 0.1 mol L^{-1} mm^{-1} to the surface of 30 g of alumina spheres the diffusion coefficient of hydrogen would have to be 2×10^{-7} cm^2 s^{-1} to maintain a steady rate of reduction of 6×10^{-3} g atom L^{-1} min^{-1} of metal ion. The diffusion coefficient of H_2 in nickel sulphate solution is not known but in 1.0 M $MgSO_4$ solution at 25°C with 1 atm pressure of H_2 a value of 2.86×10^{-6} cm^2 s^{-1} has been measured.[131] At 40 atm pressure of hydrogen and at 213°C, the value would be expected to be greater than that measured, probably by one or two orders of magnitude. Thus it is much larger than the value of 2×10^{-7} cm^2 s^{-1} required if this argument is valid.

This calculation indicates that under the experimental conditions used hydrogen was always available at the surface at which reaction was occurring in amounts greater than could be used by nickel ions arriving at the reactive sites. Hydrogen was probably present as a saturated solution in the nickel metal. The value of the apparent activation energy measured must, therefore, relate either to the diffusion of nickel ions to the surface or, more probably, to the charge transfer process between adsorbed nickel ammine ions and hydrogen dissolved in the metal.

4.4.3. *Reduction of Cobalt Salts in Aqueous Solutions by Hydrogen*

Because of the many similarities in the chemistry of Ni^{II} and Co^{II} it has been generally assumed that the kinetics of reduction of aqueous solutions of cobalt salts are similar to those found for nickel salts. The same rate equation was found for the two metals in ammoniacal sulphate solutions[132]

$$-d[Co^{2+}]/dt = kA_s p_{H_2} \cdot f(R_A)_{Co} \qquad (4.38)$$

which contains the additional term $f(R_A)_{Co}$ compared with 4.36. In that investigation[122] it was shown that the rate of reduction of nickel was affected by the ratio of free NH_3 to Ni^{2+} in the solution. The results plotted for the reduction of cobalt show 4 or 5 points covering the whole range of reduction up to 80 to 90% metal reduced. The points could be interpreted as showing a fractional order rate dependency on Co^{II} ion concentration since it is difficult to distinguish a straight line dependency from a curve with a small decreasing gradient.

Precipitation of cobalt from ammoniacal sulphate and from cobalt sulphate plus ammonium acetate solutions were compared using 0.4 M $CoSO_4$, p_{H_2} 3.09 MPa (448 lb in^{-2} g) and $(R_A)_{Co}$ 2.0. Precipitation rates of 41.4 and 33.6 g Co L^{-1} h^{-1} respectively were obtained at 200°C.[133] Using colloidal graphite as seed under comparable conditions in sulphate solution a rate of 11.0 g L^{-1} h^{-1} was reported.[134] No rate curves were published in Ref. 133 or 134. Courtney[135] maintained that colloidal graphite is not effective as a catalyst in this system but it was later[136] shown that colloidal graphite is as effective as carbonyl nickel powder for cobalt precipitation.

Optimum ratios of free ammonia to cobalt concentration $(R_A)_{Co}$ and apparent activation energies, E_A which have been reported are

$(R_A)_{Co}$	E_A kJ mol^{-1}	Temp. range °C	System	Ref.
2.4–2.6	37.6	120–175	Sulphate	132
2.0	65.6	175–245	Sulphate	134
2.0	74.0	170–200	Acetate	133
	16.3	200–232	Acetate	133
2.8	68.5	182–241	Sulphate	136

In a paper[137] dealing with the conditions required to produce cobalt powder in a commercial operation it is concluded that the optimum molar ratio of free NH$_3$ to Co is 2.3. A variety of organic reagents containing sulphur in their molecular structure and capable of forming cobalt sulphide in the solution can be used as the catalyst for nucleation. The physical properties of the cobalt powder produced were controlled by the densification technique and by the addition of surface active agents. Each additive modifies the physical properties of the powders in a specific way and so it is possible to produce cobalt powders with a wide range of physical characteristics.

The kinetics of precipitation of cobalt from ammoniacal ammonium sulphate solutions was investigated[136] using the apparatus and procedures described above for studies of nickel precipitation. A number of materials were investigated as seed for the deposition of the metal. It was pointed out that when colloidal substances were used care was necessary to ensure that flocculation or precipitation on steel surfaces did not occur. When this was achieved all of the wide range of substances used could act as nuclei for cobalt precipitation although only rarely was more than 95% of the cobalt initially present deposited.

Under similar experimental conditions to those employed for nickel precipitation the results obtained using cobalt were very similar. The slope of the line for cobalt indicating first-order dependence of the rate on cobalt concentration was 1.18, however. The dependence of the rate of reaction on p_{H_2} at 241°C was studied using a single initial cobalt concentration and was again independent of the concentration of molecular hydrogen in solution. The reaction was first-order with respect to cobalt concentration.

First-order apparent reaction rate constants using alumina powder seed are given in Table 4.4, and using cobalt powder seed in Table 4.5. The effect of changing the ratio of free ammonia to cobalt using cobalt powder seed

Table 4.4.

$T°C$	$[Co^{2+}](20°C)$	$[Co^{2+}](T°C)$	$(R_A)_{Co}$	p_{H_2}		10^3_k
	g mol L^{-1}			MPa	lb in^{-2}	min^{-1}
213	0.356	0.302	1.0	3.46	500	1.00
213	0.356	0.302	2.0	3.46	500	3.00
213	0.356	0.302	2.5	3.46	500	5.58
213	0.356	0.302	3.0	3.46	500	4.80
213	0.356	0.302	4.0	3.46	500	1.50
182	0.356	0.315	2.5	3.46	500	1.23
204	0.356	0.306	2.5	3.46	500	2.83
231	0.356	0.294	2.5	3.46	500	8.87

Table 4.5.

$T°C$	$[Co^{2+}](20°C)$	$[Co^{2+}](T°C)$	$(R_A)_{Co}$	p_{H_2}		Co seed	10^3_k
	g mol L^{-1}			MPa	lb in^{-2}	g	min^{-1}
213	0.416	0.353	2.5	4.15	600	30	6.55
213	0.356	0.302	2.5	4.84	700	30	6.90
213	0.356	0.302	2.5	4.15	600	30	6.90
213	0.356	0.302	2.5	3.46	500	30	6.90
213	0.356	0.302	2.5	2.76	400	30	6.90
213	0.296	0.251	2.5	3.46	500	30	7.18
213	0.296	0.251	2.5	2.76	400	30	7.18
213	0.236	0.200	2.5	4.84	700	30	6.30
213	0.236	0.200	2.5	3.46	500	30	6.30
241	0.356	0.289	2.5	4.15	600	30	14.70
241	0.356	0.289	2.5	3.46	500	30	14.70
241	0.356	0.289	2.5	2.76	400	30	14.70
213	0.356	0.302	2.5	3.46	500	100	15.90
213	0.356	0.302	2.5	3.46	500	50	8.50
213	0.356	0.302	2.5	3.46	500	10	4.30
213	0.356	0.302	2.5	3.46	500	0	0.85
231	0.356	0.294	1.0	3.46	500	30	4.60
231	0.356	0.294	1.5	3.46	500	30	7.40
231	0.356	0.294	3.5	3.46	500	30	9.45
231	0.356	0.294	4.5	3.46	500	30	6.70
182	0.356	0.315	2.5	3.46	500	30	2.21
193	0.356	0.311	2.5	3.46	500	30	3.45
204	0.356	0.308	2.5	3.46	500	30	4.84
222	0.356	0.298	2.5	3.46	500	30	8.05
231	0.356	0.294	2.5	3.46	500	30	11.80

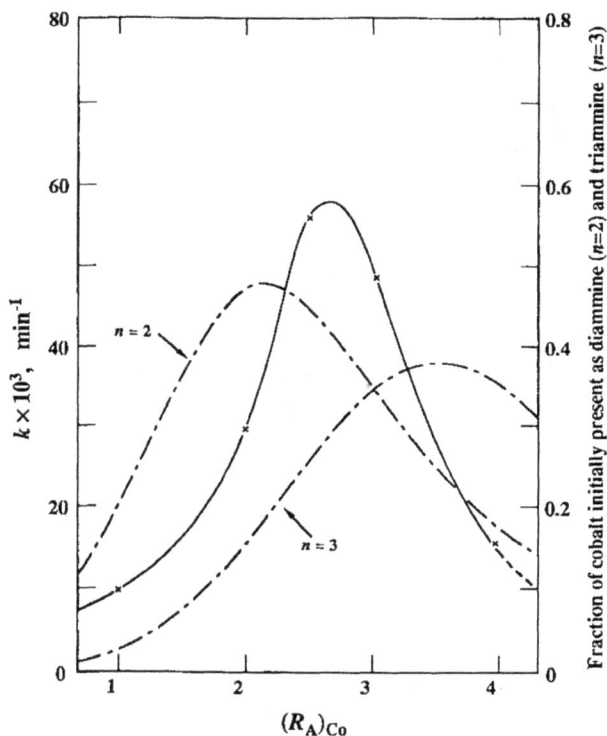

Fig. 4.18. Effect of NH_3 to Co^{II} ratio $(R_A)_{Co}$ on the apparent first-order rate constants. The lines indicated as $n = 2$ and $n = 3$ show the fractions of the cobalt initially present in solution as the species $Co(NH_3)_2^{2+}$ and $Co(NH_3)_3^{2+}$.

at 231°C is shown in Fig. 4.18 superimposed on lines showing the fractions of cobalt initially present as $Co(NH_3)_2^{2+}$ and $Co(NH_3)_3^{2+}$ at 231°C. Plotting log k against log $(R_A)_{Co}$ gives a slope of 1.0 on the left-hand side of the maximum and -1.4 on the right. The point of intersection corresponds to a value of 2.74 for the maximum value of k. However, this does not necessarily correspond with the optimum value for a process for the production of cobalt powder because of the tendency for the reaction to become slower, and stop as $(R_A)_{Co}$ progressively increases as 2.74 moles of NH_3 are released by reduction of Co^{2+} with formation of only $2H^+$. The simple approximate linear relationship between k and the fraction of the nickel present as the *diammine* found for that metal was found for neither α_2 or α_3 in the reduction of cobalt.

4.4.4.   *Reduction of Copper (II) Acetate in Aqueous Solutions by Hydrogen*

The discrepancy between the solubility of hydrogen in water at elevated temperatures and pressures previously published[138] and the results reported in Sec. 4.4.2. for solubilities in ammonium sulphate solutions, obtained using a pressure drop method in a vigorously stirred autoclave, suggests that the reduction in pressure should be used only as a measure of the amount of gas which has been transferred into the aqueous phase. Thus in order to establish unequivocally a model for a reaction in which the hydrogen gas after passing into solution causes reduction of a metal ion present in the solution, it is necessary to measure the amount of the metal ion which has reacted.

Visible and ultra violet spectrophotometry is a convenient method of measuring the concentrations of many transition metal ions. In a conventional instrument the presence of suspended solid, or gas bubbles, in the light path makes it impossible to measure the optical density accurately. A reaction for such an investigation should not form a solid product while measurements are being taken therefore. The acetates of both Cu^I and Cu^{II} are well defined compounds and their chemistry has been extensively studied, see Ref. 139, but reduction of the Cu^{II} to the Cu^I salt was not reported. Instead, reduction of solutions of $Cu(CH_3COO)_2$ in aqueous solutions by H_2 was said in Ref. 140 to occur by a homogeneous reaction and form Cu_2O. The mechanism proposed involved the formation of an intermediate hydride species which reacted with more copper (II) acetate

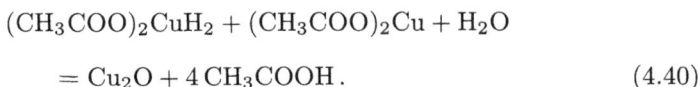

$$(CH_3COO)_2Cu + H_2 = (CH_3COO)_2CuH_2 \qquad (4.39)$$

$$(CH_3COO)_2CuH_2 + (CH_3COO)_2Cu + H_2O$$

$$= Cu_2O + 4\,CH_3COOH. \qquad (4.40)$$

The rate curves indicated that the reaction started with no induction period and did not accelerate as it progressed, thus there was no evidence for auto-catalysis and the rate equation given was

$$-d[Cu^{2+}]/dt = k[Cu^{2+}][H_2].$$

Reduction of aqueous copper (II) sulphate solutions by H_2 forms copper metal[123] but since acetic acid is a weak acid and copper acetate in solution is binuclear with carboxylate bridges the thermodynamics of the two alternative reactions requires investigation. The data for the free energies of formation of the relevant species are given in Ref. 50

Species	CH_3COOH	CH_3COO^-	$Cu(CH_3COO)_2$	$Cu(CH_3COO)^+$	Cu_2O	H_2O
$\Delta G°$ kJ	−396.46	−369.31	−693.84	−316.54	−146.0	−237.129

$$CH_3COOH = H^+ + CH_3COO^-$$

$$\log K_a = -4.756$$

$$Cu(CH_3COO)_2 + 2e = Cu + 2\,CH_3COO^-$$

$$E = +0.232 - 0.0592\log\{CH_3COO^-\} \qquad (4.41)$$
$$+ 0.0296\log\{(CH_3COO)_2Cu\}$$

$$Cu(CH_3COO)_2 + 2H^+ + 2e = Cu + 2CH_3COOH$$

$$E = +0.513 - 0.0592\log\{CH_3COOH\} \qquad (4.42)$$
$$+ 0.0296\log\{(CH_3COO)_2Cu\} - 0.0592\,\text{pH}$$

$$2Cu(CH_3COO)_2 + H_2O + 2e = Cu_2O + 2H^+ + 4CH_3COO^-$$

$$E = -0.0081 - 0.1184\log\{CH_3COO^-\} + 0.0592\log\{(CH_3COO)_2Cu\} \quad (4.43)$$
$$+ 0.0296\log\{H_2O\} + 0.0592\,\text{pH}$$

$$2Cu(CH_3COO)_2 + H_2O + 2e = Cu_2O + 2CH_3COOH + 2CH_3COO^-$$

$$E = +0.273 - 0.0592\log\{CH_3COOH\} - 0.0592\log\{CH_3COO^-\} \qquad (4.44)$$
$$+ 0.0592\log\{(CH_3COO)_2Cu\} + 0.0296\log\{H_2O\}$$

$$2Cu(CH_3COO)_2 + H_2O + 2H^+ + 2e = Cu_2O + 4CH_3COOH$$

$$E = +0.554 - 0.1184\log\{CH_3COOH\} + 0.0592\log\{(CH_3COO)_2Cu\} \qquad (4.45)$$
$$+ 0.0296\log\{H_2O\} - 0.0592\,\text{pH}\,.$$

The activity of water is included in the equations for reactions (4.43)–(4.45) since concentrated solutions of acetic acid in water might be used as the solvent in which to carry out the reduction. Use of the ion $Cu(CH_3COO)^+$ is thermodynamically less favourable than $Cu(CH_3COO)_2$ in equations. There is no value for $\Delta G°$ of CH_3COOCu so that only the overall reduction reactions can be considered. All of them can take place at 25°C with 1 atm H_2 pressure if suitable activities of reactants and products are selected. Values of $E°$ are much more positive for reactions (4.42) and (4.45) in which only acetic acid is formed than for the equivalent reactions in which acetate ion is a product. This is due merely to the difference between the free energies of formation of the acid and the ion and shows that reduction is thermodynamically more favourable at pH values lower than 4.7 than at higher values. The values of $E°$ for reactions (4.42) and (4.45), +0.513 V and +0.554 V, are so close that it seems probable that either copper metal or Cu_2O could be produced by reduction with hydrogen, depending on the total acetate concentration in the solution and the pH. These can be adjusted by adding ammonium acetate and acetic acid, making use of the buffering action of the weak acid — weak base system. It has been assumed above that the reaction which is more favoured thermodynamically will take place. However the reduction of copper (II) acetate by hydrogen does not occur at 25°C and 1 atm H_2; thus some kinetic factor prevents thermodynamic equilibrium being achieved under those conditions. If copper (I) acetate were formed as a first stage in the reduction process Cu_2O could be produced by hydrolysis of the salt.

Dakers and Halpern[140] studied the reduction reaction using a 1 gallon (US) stainless steel autoclave, samples of solution being withdrawn periodically. The solutions used generally contained about 0.5 M acetic acid and 0.5 M sodium or ammonium acetate to buffer the pH between 4 and 5. At higher pH values than 5, hydrolysis occurred and basic salts precipitated when the solutions were heated to the reaction temperature. In every experiment the only solid product was Cu_2O. It was not considered that copper (I) acetate, which might be formed as an intermediate, had a catalytic effect.

Copper (I) acetate is formed as a sublimate, together with CO_2, acetic acid and other decomposition products, when copper (II) acetate is heated *in vacuo* to a temperature above 220°C or by reducing Cu^{II} acetate with copper metal in pyridine or CH_3CN. It is a white crystalline compound which very easily volatilises and is immediately hydrolysed by water forming Cu_2O. Dakers and Halpern obtained Cu_2O as a bright red solid from solutions buffered with sodium acetate and a dark purple to black product when ammonium acetate was used.

Reduction of copper (II) acetate solutions by H_2 was also studied[141] using a spectrophotometer cell containing 2 chambers in the stainless steel block, one used as the reaction chamber and the second, giving the same length of light path as the first, for the reference beam of the instrument. A cylindrical stirring rod was held above the level of the quartz windows, driven by a magnetic stirrer, and created nonturbulent flow of liquid and no inclusion of gas bubbles. A pressure transducer was attached to the reaction chamber and the "apparent volume" of this was measured at 20°C, 50°C, 100°C and 150°C by allowing nitrogen to pass from the chamber when its pressure had become constant, to a water-filled gas volume meter. The volume of H_2, measured at ambient temperature and pressure, which caused a measured drop in the pressure at elevated temperatures was also measured similarly using the gas volume meter. The "apparent volume" of the reaction chamber was defined as the volume of the vessel if all the gas in the space were at the controlled temperature of the steel block. Using this volume, the volume of liquid in the reaction chamber and the change in pressure of H_2 gas during a reaction, as measured by the pressure transducer, it was possible to calculate the quantity of H_2 transferred from the gas space during the reaction. The output from the pressure transducer and the optical density of the solution at the selected wavelength were recorded continuously while the reaction continued. Optical density measurements were converted to Cu^{II} concentrations by using the known concentration at the start of the reaction, when H_2 was introduced into the cell.

Preliminary experiments carried out in the apparatus showed that:

(i) No solid reaction product formed when cupric acetate solutions were reduced by H_2 until reduction of Cu^{II} had been taking place for a considerable time.

(ii) If precipitation of solid did occur this happened when the concentration of Cu^{II} remaining in the solution was very small.

(iii) In solutions with low concentrations of acetate ions no solid reaction produced formed at the reaction temperature but Cu_2O precipitated as the solution was cooled.

(iv) X-ray powder diffraction measurements showed that the solid product formed by reduction of copper (II) acetate solutions by H_2 depended on the total concentration of acetate in the solution. With low concentrations, up to about 1 M, Cu_2O formed, the colour of which varied between red and dark purple, depending on the ultimate particle size as seen by electron microscopy. At high (5 or 6 M) total acetate concentrations metallic copper was always the only solid reaction product.

The general shape of the rate curves obtained from both the fall in the optical density of the solution due to reduction of Cu^{II} and the drop in H_2 gas pressure indicated that the rate of the reaction initially accelerated and then gradually became slower until the reaction was complete. The rate of loss of H_2 from the gas phase will be greatest at the beginning of an experiment when its concentration in the solution is zero and the pressure drop data fitted the equation

$$\ln(p_t - p_e) = k_g t \tag{4.46}$$

where p_t is the gas pressure in the reaction chamber after time t (min), p_e is the equilibrium pressure, when reaction had ceased, and k_g is the rate constant for dissolution from the gas phase. This is given by the slope of the line in Fig. 4.19 for the first stage of the reaction, between introduction of H_2 into the reactor at time 0 and about 3.5 minutes. The whole reaction took place with apparently three stages of exponential loss of H_2 from the gas phase. The change in Cu^{II} concentration under similar conditions of reaction is shown in Fig. 4.20, an accelerating reaction being followed an exponential lowering of $[Cu^{II}]$ and a final stage in which $[Cu^{II}]$ approaches zero.

Analysis of the kinetics of the reaction was based on two sets of experiments. In set 1, the rate of removal of H_2 from the gas phase was measured at the same time as the rate of lowering of the concentration of Cu^{II}. The first, accelerating stage was examined using these results. The second stage was analysed so as to test the supposition that the concentration of H_2 in the solution had reached such a value that the rate of the reaction depended only on the concentration of Cu^{II} remaining in the solution, that is

$$-d[Cu^{II}]/dt = k_h[Cu^{II}]. \tag{4.47}$$

In this set of experiments the pressure of H_2 in the reactor was maintained at a constant value.

Stage 1 — Accelerating Part of Reaction

Assuming that the volume of the solution remains constant during the reaction

$$\ln(m_e - m_t) = k_s t \tag{4.48}$$

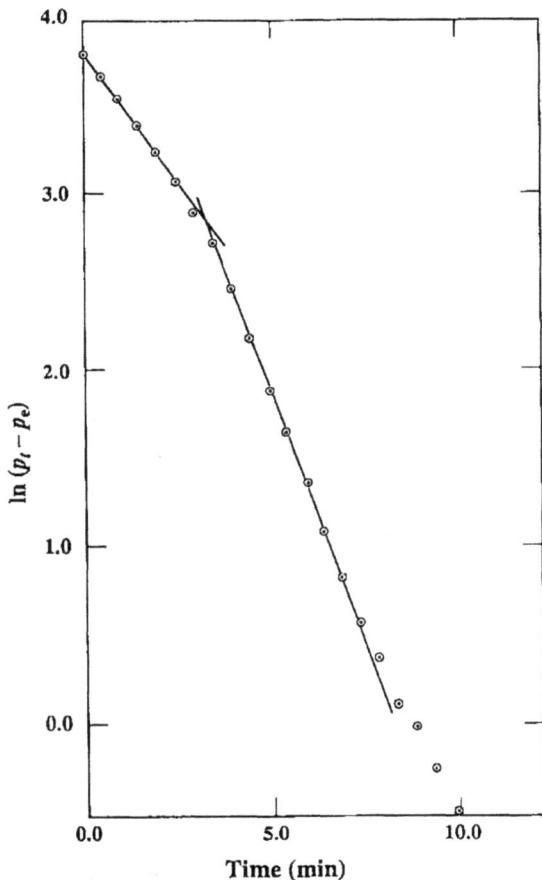

Fig. 4.19. Reduction of $Cu(CH_3COO)_2$ by H_2. Change of p_{H_2} with time at $150°C$. Conditions: Initial p_{H_2} 400 psi (2.76 MPa); initial $[Cu^{II}]$, 0.0491 M; total $[CH_3COO^-]$, 6.412 M; total $[NH_3]$, 1.713 M; initial pH, 4.0. Precipitation of $Cu°$ commenced 9.2 min after introduction of H_2.

from (4.46) where m is the total molal concentration of hydrogen in the solution, dissolved molecular H_2 plus H^+ equivalent to H_2 which has been oxidised by reaction with Cu^{II}. The subscripts have the same meaning as in (4.46). k_s is a rate constant for the increase in total hydrogen concentration in the solution. If β is the factor relating a unit change of pressure in the gas space to the change in molal concentration of hydrogen in solution, measured as described

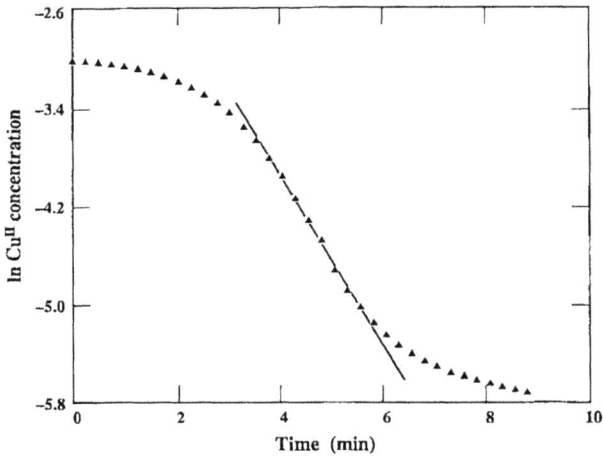

Fig. 4.20.   Reduction of $Cu(CH_3COO)_2$ by H_2. Change of $\ln[Cu^{II}]$ with time at $150°C$. Conditions: p_{H_2} constant at 400 psi (2.76 MPa); others as in legend to Fig. 4.19.

above, then if the H_2 pressure is p_i

$$m_t = \beta(p_i - p_t) \quad m_e = \beta(p_i - p_e)$$

and hence

$$\ln(p_t - p_e) = \ln[(m_e - m_t)/\beta].$$

Substitution of (4.46) into this gives

$$m_e - m_t = \beta e^{k_g t}. \tag{4.49}$$

The concentration of molecular hydrogen in the solution at time t is m_t minus the quantity which has reacted to form H^+, m_{rt}. Since no metallic copper is formed during the first part of the reaction one hydrogen molecule reduces two Cu^{II} to form Cu^I species. Thus

$$m_{rt} = m_t - 1/2\,\Delta[Cu^{II}]$$

where $\Delta[Cu^{II}]$ is the difference between the initial Cu^{II} concentration and $[Cu^{II}]$ after t min. Substitution from (4.49) gives

$$m_{rt} = m_e - \beta e^{k_g t} - 1/2\,\Delta[Cu^{II}]. \tag{4.50}$$

If the rate of reduction of Cu^{II} to Cu^{I} depends directly on the concentration of molecular hydrogen in solution and on the concentration of nothing else

$$-d[Cu^{II}]/dt = k_o[H_2] = k_o m_{rt} \tag{4.51}$$

so

$$-d[Cu^{II}]/dt = k_o(m_e - \beta e^{k_g t} - 1/2\,\Delta[Cu^{II}]). \tag{4.52}$$

Equation (4.52) is based on the assumptions that the rate of reduction of Cu^{II} at time t is directly proportional to the concentration of molecular hydrogen in solution and is independent of the concentrations of all species containing Cu^{I} or Cu^{II}. The value of the rate depends on the value of the reaction rate constant k_o which is defined in (4.51), on the extent to which the concentration of molecular hydrogen in the solution is lower than the equilibrium concentration at time t, and on the value of k_g, defined in (4.46). k_g depends on the design of the mechanism used to assist in the transfer of hydrogen from the gas phase into the solution, that is on the type and design of the stirring equipment.

The validity of this model was tested using simultaneous measurements of gas pressure and the optical density of the solution. The conditions used were: copper, 0.049 M; acetate, 6.411 M; ammonia, 1.713 M giving a solution buffered at pH 4.0 at 25°C; hydrogen partial pressure 400 lb in^{-2} (27.6 MPa), and temperatures 130°C, 140°C and 150°C. The rate of loss of hydrogen from the gas phase was exponential in 3 stages, as shown in Fig. 4.19. Values of k_g were taken from the slopes of the lines for the first stage of the reaction. The results of plotting $d[Cu^{II}]/dt$ against

$$\beta e^{k_g t} + 1/2\,\Delta[Cu^{II}]$$

are shown in Fig. 4.21. The data represented by circles were obtained at times at which the first, accelerating, stage of the reaction was occuring and show that the rate of the reaction can be interpreted using Eq. (4.52). The data shown as crosses are from later stages of the reaction and show clearly that a change of mechanism occurs when the slope of the line in Fig. 4.19 changes after about 3.5 minutes. During the reactions at 130°C, 140°C and 150°C the end of the first stage of the reaction occurred at 10.5, 5.0 and 3.55 minutes and formation of metal powder at 17.1, 13.0 and 7.0 minutes respectively. The change in slope of the line for the second stage of the reaction occurred shortly before that time.

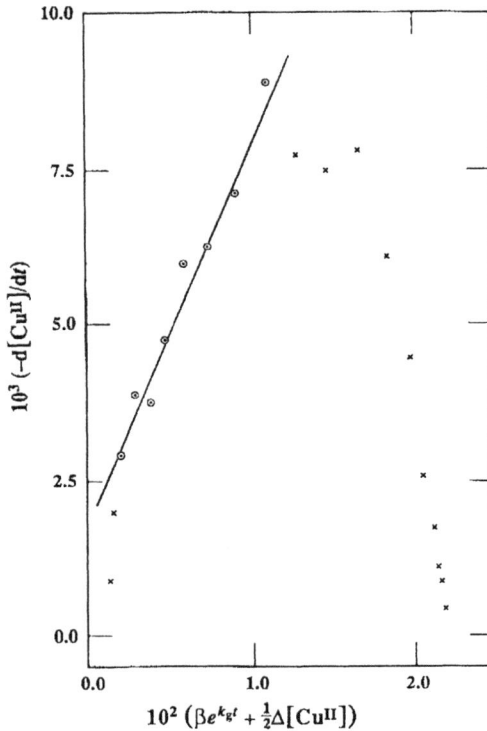

Fig. 4.21.   Reduction of $Cu(CH_3COO)_2$ by H_2. Plot of rate Eq. (4.52). Conditions as for Fig. 4.19; 140°C. Crosses relate to data from the later part of the reaction for which the rate equation does not apply.

The rate constants k_o, determined from the slopes of the lines fitted to the points by linear regression analysis, are given in Table 4.6. The error values are one standard deviation of all the lines that could be passed through the points considered. The activation energy of the reaction is 1.906 ± 0.013 kJ mol^{-1}. The fit of the points to a straight line is very good, giving further support to the model for the reaction on which Eq. (4.52) is based.

Table 4.6.

Temperature°C	130	140	150
Pressure lb in^{-2}	400	400	400
Rate constant k_o	0.2351 ± 0.0137	0.6194 ± 0.0412	1.5170 ± 0.0897

Stage 2 — Decaying Part of The Reaction

The second stage of the reaction was studied using constant values of hydrogen partial pressure during experiments and recording the optical density of the solution as reaction proceeded. The rate of decrease in Cu^{II} concentration under the same conditions of solution composition, initial p_{H_2} and temperature as were used to obtain the results for stage 1 of the reaction shown on Figs. 4.19 and 4.21 is shown in Fig. 4.20. The linear portion of the $\ln[Cu^{II}]$-t curve starts at the time when the linear first part of Fig. 4.19 ends, 3.5 min. The reaction is then first-order with respect to the Cu^{II} concentration until the value of this falls to about 6×10^{-3} M, after which the rate of reaction decreases progressively. In another set of experiments in which the total acetate concentration was 4.93 M buffered at pH 3.68, copper metal was formed after 7.50 minutes of reaction at 150°C, the shape of the curve being similar to that in Fig. 4.20.

In a series of reduction experiments on solutions containing 0.159 M Cu^{II} initially and 0.242 M total acetate at temperature between 110°C and 150°C the general shapes of the $\ln[Cu^{II}]$-t curves were similar to those with the higher acetate concentrations, Fig. 4.22 being typical, but no solid reaction product

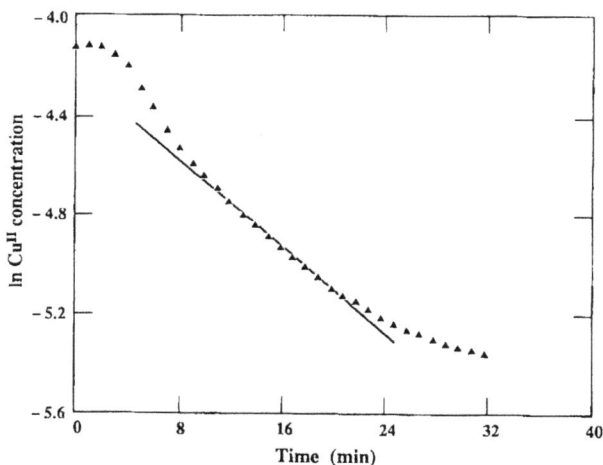

Fig. 4.22. Reduction of $Cu(CH_3COO)_2$ by H_2. Change of $\ln[Cu^{II}]$ with time. Conditions: 130°C; p_{H_2} 650 psi (4.49 MPa); initial $[Cu^{II}]$, 0.0159 M; total $[CH_3COO^-]$, 0.242 M. Typical curve for conditions where no insoluble reaction product formed at the reaction temperature but Cu_2O precipitated while cooling to ambient temperature.

formed at the reaction temperature. When the solution was allowed to cool Cu_2O precipitated.

Discussion of the Reaction Mechanism

The accelerating stage of the reduction reaction, seen in Figs. 4.20 and 4.22 can be explained by the increase in concentration of molecular H_2 in the solution from zero after H_2 gas is admitted to the reactor. During the second stage the reaction is first-order with respect to $[Cu^{II}]$ and it is reasonable to suppose that since p_{H_2} in the reactor is kept constant the concentration of molecular H_2 in the solution is constant. The nature of the reduced Cu^I species in the solution is not established by these experiments.

Between the mid 1950's and the early 1970's, a number of studies were carried out on the reduction of Cu^{II} salts in solution by H_2 and by CO. Thus work was in two categories, one concerned with controlling the reactions so as produce saleable copper metal; the other directed towards elucidation of the mechanisms of the reactions taking place. In this latter case the behaviour of copper as a hydrogenation catalyst was also considered and the mechanism proposed in Ref. 142 was

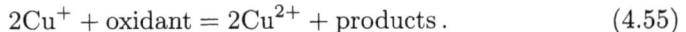

$$Cu^{2+} + H_2 = CuH^+ + H^+ \qquad (4.53)$$

$$CuH^+ + Cu^{2+} = 2Cu^+ + H^+ \qquad (4.54)$$

$$2Cu^+ + \text{oxidant} = 2Cu^{2+} + \text{products}. \qquad (4.55)$$

This applies only to solutions in which Cu^{2+} and Cu^+ are not complexed. The rate of the reaction in which copper metal was formed was studied between 150°C and 175°C in perchlorate solution, to check the proposed mechanism and rate Eq. (4.55) being replaced by

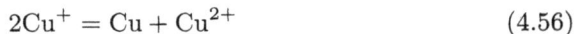

$$2Cu^+ = Cu + Cu^{2+} \qquad (4.56)$$

the simple disproportionation reaction of Cu^+.[143] The rate equations based on these mechanisms were found not to hold, however, at 250°C using O_2 as the oxidant[144] or at higher temperatures in the range 160°C to 200°C using dichromate ion as the oxidant.[145] The difficulty was that the dependence of the rate on acid concentration was smaller than that predicted. It seems probable that the mechanisms outlined above are valid for the reactions where an oxidising agent is present in sufficient quantity to prevent the build up of a significant concentration of Cu^+ ions. An alternative mechanism has been proposed which involves the reduction of Cu^{2+} by Cu to form $2Cu^+$ ions followed by formation of CuH as an intermediate species.[146]

The reaction forming copper metal was studied between 160°C and 200°C in perchlorate solutions under conditions selected to permit the building up of the Cu^I concentration, which was measured before it could change by reaction or disproportionation during cooling.[147] If the initial Cu^{II} concentration was greater than 0.03 M, the Cu^I concentration increased to a limiting value and then began to fall due to disproportionation at the reaction temperature. In more dilute solutions the limiting value was not reached and there was no evidence of metal formation at that temperature. Metallic copper deposited from these dilute solutions as they cooled.

During the reduction of copper (II) acetate solution, when copper metal precipitated in the reactor in the spectrophotometer at the reaction temperature,[141] a simultaneous increase in the concentration of Cu^{II} species was observed. This was presumably due to the disproportionation reaction (4.56). Quantitative determination of the increase was not possible because of the presence of the suspended metal, of very small particle size. However, by measuring the optical density due to the turbidity at a wavelength at which Cu^{II} did not absorb and subtracting this from the value at which $[Cu^{II}]$ was measured, an approximate estimate was obtained. The concentration of Cu^{II} ions formed while metal was being precipitated showed that much less than half of the total quantity of copper in the solution which had been reduced reacted according to Eq. (4.56). The rest of the metal may have been produced by the reaction

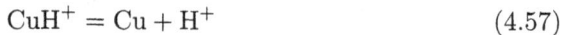

$$CuH^+ = Cu + H^+ \qquad (4.57)$$

if hydrogenated Cu^I ions do actually exist in the solution. In that case it would appear that in the early stages of the reaction CuH^+ reacts rapidly with Cu^{II} species (4.54), whereas in the later stages CuH^+ accumulates in the solution.

Strong evidence for the presence of the species CuH^+ or some similar molecule was provided by the use of D_2 instead of H_2 to reduce Cu^{II} in sulphate and perchlorate solutions in H_2O and D_2O.[148] The work was concerned with supporting the rate equation proposed in an earlier paper[149] and used the same equipment as was used in the kinetic investigation. During the reduction reactions, gas and solution samples were taken to measure the extent of both the reduction of Cu^{II} and of the formation of HD in the gas. Deuterium exchange occurred in every experiment except one in which copper ions were not present in the solution. The reactions involved were assumed to be

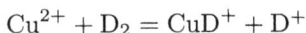

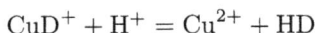

$$Cu^{2+} + D_2 = CuD^+ + D^+$$

$$CuD^+ + H^+ = Cu^{2+} + HD$$

and similarly when H_2 and D^+ from D_2O were used. The kinetics of the reduction of Cu^{II} in sulphate solutions had been investigated previously[143,146,147] and the reactions proposed were essentially (4.53)–(4.55), in the latter case the oxidant being $CuSO_4(aq)$. Additional equivalent reactions involving this uncharged solute species were also written. The reaction is considerably faster than in perchlorate solutions. In dilute solutions the rate of formation of metal is second-order with respect to $[Cu^{2+}]$ and is first-order at higher concentrations.

4.4.5.   *Reduction Reactions between Metal Salts and Carbon Monoxide in Hydrometallurgy*

Presently, metals are not produced commercially by reduction of aqueous solutions of metal salts by carbon monoxide but a considerable amount of work was carried out to investigate the possibility of doing so. The only cases of possible interest were the reduction of silver,[150] copper,[151–153] nickel[154] and cobalt.[155] Copper is reduced to a lower valency state but metal is not produced, nickel forms the carbonyl $Ni(CO)_4$, silver gives the metal and cobalt is reduced to metal in an ammoniacal Co^{II} solution in the presence of precipitated cobalt sulphide, but only to a carbonyl species in the absence of the sulphide catalyst. The papers listed give references to earlier work.

The kinetic studies indicated that in most cases a carbon monoxide insertion complex of the type

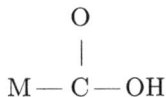

$$\begin{array}{c} O \\ | \\ M - C - OH \end{array}$$

formed as an intermediate in the reactions. In the case of cobalt the reaction to form metal proceeds in two steps, having the stoichiometries

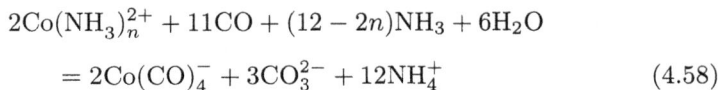

$$2Co(NH_3)_n^{2+} + 11CO + (12 - 2n)NH_3 + 6H_2O$$

$$= 2Co(CO)_4^- + 3CO_3^{2-} + 12NH_4^+ \tag{4.58}$$

and

$$9Co(NH_3)_n^{2+} + 2Co(CO)_4^- + (32 - 9n)NH_3 + 16H_2O$$

$$= 11Co^0 + 8CO_3^{2-} + 32NH_4^+ . \tag{4.59}$$

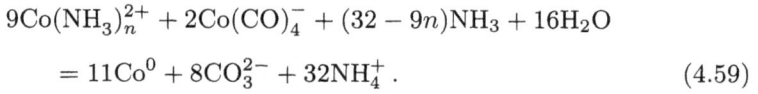

Reaction (4.59) occurs only in the presence of CoS_x produced by adding a little $(NH_4)_2S$ to the system, the solid being probably γ-Co_6S_5 or possibly Co_9S_8.

Chapter 5

Chemistry of Leaching Processes

5.1. Introduction

Some industrially important leaching processes used in large scale hydrometallurgical plants are listed in Table 5.1. Methods of dissolving metals such as tungsten, the rare earth elements, Zr/Hf, Nb/Ta, Be etc. from their ores are used on a much smaller scale. As indicated in the table, in some cases an ore mineral is not itself leached but is decomposed by a chemical process before the metal of value is taken into an aqueous solution. In the case of zinc concentrates an oxidising roast is used in the electrolytic process to convert the sphalerite, ZnS, to a calcine containing ZnO with some ferrite, $ZnFe_2O_4$, formed by reaction with the iron present. This elegant process was first used commercially in 1916 and the roasting step was necessary because the conditions required to dissolve the zinc by direct leaching of the sulphide in sulphuric acid are rigorous and could not be achieved on a commercial plant at that time, whereas ZnO dissolves readily. Direct leaching of the concentrates is also being used commercially now because pressure hydrometallurgy has become accepted technology after being used successfully since the 1950's for the production of nickel from its sulphide concentrates and subsequently in many other applications.

All other hydrometallurgical processes for recovery of nickel from sulphide ores first produce a matte by smelting. The Ni/Cu ratio in this is determined by

Table 5.1.

Feed	Leaching solution	Oxidant	Temperature	Pressure	Equipment
Oxidised Cu ore	Dilute H_2SO_4 [1]	None	Ambient	Atmospheric	Vats, tanks or heaps
Cu sulphides; low grade ores and tailings	Dilute H_2SO_4	$Fe_2(SO_4)_3$ + bacteria	Ambient	Atmospheric	Heaps of ore or *in situ*
Zinc calcine ($ZnO + ZnFe_2O_4$)	Dilute H_2SO_4 [2]	None	Ambient and hot [2]	Atmospheric	Agitated tanks
ZnS concentrates	Return anolyte + H_2SO_4 [3]	Air	About 150°C	pO_2 207 kPa	Autoclaves
Ni sulphide concentrates	NH_3 + $(NH_4)_2SO_4$	Air	94°C	794 kPa	Autoclaves
Ni matte	Return anolyte [4]	Air	Ambient and hot [4]	Atmospheric and elevated [4]	Agitated tanks and autoclave
Ni matte	Concentrated metal chlorides	$Cu^{II} \cdot Cu^{I} + Cl_2$	Ambient	Atmospheric	Agitated tanks
Ni/Cu/Pt group metals matte	Return anolyte [5]	Air	132–143°C [5]	0.93 MPa [5]	Autoclaves
Precipitated Ni/Co sulphide	Water	Air	150–170°C	1.5–4.8 MPa	Autoclaves
Ni laterite ore	H_2SO_4 [6]	None	240–260°C	4.2 MPa	Pressurised vertical tanks
Reduced Ni laterite ore	NH_3 + $(NH_4)_2CO_3$	Air	90°C	Atmospheric	Agitated tanks
Uranium ore	Dilute H_2SO_4	$Fe^{III} + NaClO_3$ or MnO_2 or H_2SO_5	Ambient	Atmospheric	Agitated tanks
Uranium ore	$Na_2CO_3 + NaHCO_3$	O_2	90°C or ambient [7]	Atmospheric	Agitated tanks or *in situ* [7]
Rich uranium ore	H_2SO_4	Air; O_2	Ambient	Atmospheric 650 kPa O_2	Agitated tanks; Autoclaves

Table 5.1. (*Continued*)

Feed	Leaching solution	Oxidant	Temperature	Pressure	Equipment
Gold ore	NaCN	Air	Ambient	Atmospheric	Agitated tanks
Refractory gold ore[8]	Water	Air or $Fe_2(SO_4)_3$ + bacteria	$\sim 200°C$ or ambient	High or atmospheric	Autoclaves or heaps
Bauxite ("Al_2O_3")	NaOH	None	$104\text{-}200 + °C$	Water vapour pressure	Autoclaves or pipeline reactor
Scheelite or wolframite	Na_2CO_3	None	$190\text{-}225°C$	$1\text{-}2 - 2\text{-}6$ MPa	Autoclaves

(1) Return anolyte from tankhouse or fresh acid usually produced from smelter gases.

(2) ZnO is dissolved in dilute H_2SO_4 at ambient temperature. $ZuFe_2O_4$ in the residue is dissolved in return anolyte to which more H_2SO_4 is added, at 95°C.

(3) Total free H_2SO_4 around 180 g L^{-1}.

(4) Return anolyte; free H_2SO_4 45 g L^{-1}. Four stages of countercurrent leaching, 3 at ambient temperature and pressure and a final hot pressure leach.

(5) Return anolyte contains 80–100 g L^{-1} free H_2SO_4. A sequence of three distinct stages of leaching is used, first to leach out Ni and not Cu, the second to leach out as much as possible of the Ni, Co and Cu remaining; the third to remove essentially all Ni, Co and Cu from the PMs concentrate.

(6) Concentration not specified. Required leaching time (1–2 hours) was stated to be a function of temperature, acid concentration and ore composition. 98% H_2SO_4 feed into first tank.

(7) Carbonate leaching of uranium has been used for recovery from ores which consumed much acid. Agitated tanks were used with O_2 at 90°C. The system has also been used for *in situ* leaching of U from porous sandstone deposits.

(8) Gold is held within particles of sulphides, e.g. FeS_2, FeAsS etc. which are oxidised by air in a hot slurry, or by bacterial leaching, so releasing Au for recovery by leaching with NaCl.

the ratio in the concentrates but the metal to sulphur ratio is controlled during smelting so as to achieve the required results in the leaching process used. The composition of the matte produced in the Outokumpu process is typically $NiCu_{0.26}S_{0.18}$, leaching being with sulphuric acid and air as oxidant. In the Falconbridge chlorine leach process the matte contains much more sulphur, its composition being around $NiCu_{0.81}S_{1.08}$, and leaching is carried out in a boiling concentrated solution of the chlorides of the metals with Cl_2 as the oxidant. The nickel ores which are smelted contain very small amounts of platinum group metals and gold; however these account for an important fraction of the value of the metals for which the ore is worked. The Falconbridge matte contains about 1200 ppm of precious metals and this represents practically quantitative extraction of them from the ore into the sulphide phase. During the successive stages of leaching to recover the nickel and copper the PMs remain in the insoluble residue which finally is treated as a PMs concentrate from which the individual pure metals are produced. Only nickel sulphide ores which contain negligible amounts of precious metals are leached directly for recovery of nickel.

A significant proportion of the primary copper metal produced now is obtained by leaching followed by solvent extraction and electrowinnning (the SXEW process), much of it by leaching oxide ore with sulphuric acid, this being produced from smelter gas. The oxide was present as a cap over the sulphide ore, formed by weathering, and formerly was not mined or was stacked because only limited amounts of it could be fed into the smelters with the sulphide concentrates. Heaps of low-grade sulphide ores and dumps of waste rock are being used also to produce copper-containing solutions suitable for solvent extraction or cementation for recovery of the metal. The copper sulphide minerals are oxidised by acidic solutions containing $Fe_2(SO_4)_3$, formed by oxidation of iron sulphides by certain bacteria. A number of processes involving the leaching of copper sulphide concentrates which were developed during the 1960's and 1970's are not shown in Table 5.1 because none of them has yet been used commercially on a full scale plant. They contain some interesting chemistry which may be used commercially in future. Work on these processes commenced before solvent extraction of copper from leach liquors was commercially accepted technology. Thus the leaching steps could now be considered again for producing solutions to be treated by solvent extraction and electrowinning from either sulphate or chloride solution.

5.2. Leaching of Oxides and Silicates

5.2.1. *Reactions of Oxides not Involving Oxidation or Reduction*

In many cases metals can be leached from their oxides without a change of valency by dissolution in aqueous acid or alkali, the conditions necessary being controlled by their solubility products. Examples are: dissolution of ZnO, $ZnFe_2O_4$, $Cu(OH)_2$ and UO_3 in acids; and Al_2O_3, $CaWO_4$ and UO_3 in alkaline solutions (Table 5.1). When a change of valency is necessary before dissolution can occur, the value of E in the system must be adjusted, as well as the pH, by addition of an oxidising or reducing agent to produce values outside the region of thermodynamic stability of the solid but within the stability range of the metal ion required to be in solution. Examples are Cu_2O in acidic and UO_2 in acidic or alkaline solutions, an oxidant being required; and MnO_2 in acid with, for example, SO_2 causing reduction to Mn^{II} and dissolution of the oxide.

The first step in the leaching of a solid must be wetting of the surface, a process in which interaction between atoms at the solid surface and water molecules takes place, resulting in the formation of M–OH groups. This is referred to as hydroxylation of the surface. It can be observed by infra-red spectroscopy and the early work using this technique has been reviewed[156] as also has the general behaviour of metal oxides during leaching reactions.[157] The chemically bound –OH groups behave in a similar manner to those in liquid water, becoming associated with H^+ or OH^- ions, depending on the pH of the solution, forming $-OH_2^+$ or $-(OH)_2^-$ groups on the surface. Thus the hydroxylated surface acts as a cation or anion exchanger but the pK value for the equilibrium between $-OH_2^+$ and $-(OH)_2^-$ will not be the same as pK_w for water itself due to the interaction of the bound –OH groups with the solid, see also Sec. 3.11.4. The wetted solid surface has a charge, therefore, which is treated theoretically in terms of the zeta potential in studies of electrokinetic phenomena. In the case of simple hydroxylation of metal oxides the pH of the solution at which there is no net charge on the oxide surface, the zero point of charge or isoelectric point, is different for each solid due to this interaction. The electric charge on the solid surface at other pH values is neutralised by the presence of an excess of ions of opposite charge over ions having the same sign of charge as the surface, in the water close to the solid–water interface. Migration of ions due to this electric field was one cause of mass transport considered in the calculation of the rate of dissolution of zinc calcine in dilute acid solutions[121] discussed in Sec. 4.3.3.

When dissolution of an oxide in aqueous acid takes place the reaction is frequently anisotropic, that is solid is lost more rapidly from some crystal planes than from others. Crystal defects cause an enhanced leaching rate which is also found very frequently along grain boundaries. Such behaviour is the basis for etching procedures used in metallography and its occurrence during leaching of metal oxides has been reviewed.[157] In the case of single crystals of α-Fe_2O_3 dissolving in hydrochloric acid the rate varied with orientation over a tenfold range, the basal plane (0001) being attacked most rapidly and (1011) most slowly.[158]

Comparison of the rates at which different metal oxides dissolve in $HClO_4$, HCl, HNO_3 and H_2SO_4 did not show any consistent order in the leaching efficiency of the acids. However measurements showing that α-Fe_2O_3 dissolved some ten times more rapidly in HCl than in HNO_3 led to a correlation between increasing "absolute" rates of leaching in different acids with increasing equilibrium constants for the formation of complexes between Fe^{3+} and the anions of the acids.[159] A similar correlation has been made in a number of other systems but no explanation was found for the effects of changes in the concentration of acids on the rates of leaching of different oxides.[157] Further measurements[160] of the rates of leaching of Fe_2O_3, Al_2O_3, Cu_2O and CuO in $HClO_4$, HCl and H_2SO_4, over concentration ranges which gave convenient rates of dissolution, were interpreted in terms of a series of equilibria, which were not themselves rate-controlling, involving protonation and adsorption on the solid surface. The equilibrium constants were included in a rate equation which was based on the assumption that the rate-determining steps for the dissolution of the oxides were either the formation of activated metal complexes at the surface of the solid, or desorption of these complexed surface species. It was used as a fitting equation to include all of the experimental results obtained for the leaching rates of the oxides in $HClO_4$ and HCl, using four rate constants in the equation. The "absolute" rates of leaching of the oxides studied were very different from one another and this could not be explained in terms of the model used.[160] A significant advance has been made in the understanding of the factors which control the "absolute" rate of leaching of a simple oxide, ZnO, in aqueous acids, based on the ionic diffusivity and ionic mobility of each chemical species taking part in the reaction, together with the equilibrium constants and activity coefficients for the equlibria,[121] see Sec. 4.3.3.   The diffusivity and mobility values control the rate of mass transport to and from the reacting surface and in consequence the model is applicable only when the leaching rate is not controlled by forced diffusion. That is to say applicable only when the rate is independent of the conditions of agitation.

5.2.2. *Reactions of Oxides Involving Oxidation*

The industrially important minerals of some metals contain iron also, and frequently other metals as impurities. These include chromite, (Fe, Mg) (Cr, Al, Fe)$_2$O$_4$ with amall quantities of Mn, Zn, Ti etc. ilmenite, FeTiO$_3$; the columbite-tantalite series, (Fe, Mn) (Nb, Ta)$_2$O$_6$ the Fe–Mn and Nb–Ta ratios being continuously variable; and wolframite (Fe, Mn)WO$_4$. These are very stable, unreactive substances and, except in the case of wolframite, require very severe conditions in order to decompose them before recovering their metal values. In general attack by alkali is preferred, in order to reject the iron as early in the process as possible, and very concentrated solutions of alkali at high temperatures, or fusion conditions are used. Oxidation is also necessary in order to oxidise the Fe^{II} and so destroy the crystal lattice. In the case of chromite oxidation of Cr^{III} is necessary also, to form soluble chromate. Some minerals of this type behave in an unusual manner when leached in aqueous alkali solutions and for satisfactory recovery of their metal values require careful selection of conditions used, which often varies with the source of the mineral concentrates. This is the case with wolframite for example.

For the production of chromium chemicals the mineral chromite is fused with alkali and oxidised with oxygen to form chromate, which is treated further. More recently attention has been given to the possible advantages of treating ferrochrome instead of the mineral, because it is more reactive. Since the alloy is produced by reducing chromite with carbon and is not porous this increase in reactivity is presumably due to the difference in behaviour of an oxide and a metal containing the same elements. Freshly precipitated Cr(OH)$_3$ dissolves to some extent in fairly concentrated solutions of alkali on standing for a month or two, forming the alkali metal chromite, MCrO$_2$; in air this oxidises to the chromate.[161] Cr$_2$O$_3$ dissolves fairly readily in 2.5 M NaOH solution at 210°C under an air pressure of about 146 lb in^{-2} (1 MPa).[162] Treatment of this oxide or of chromite mineral under such conditions was suggested[163] as an alternative to roasting for the production of chromate and reactions of this type were claimed in patents.[164] The rate of reaction of chromite mineral was said to be slow and use of chromium oxide or ferrochrome was preferred.

Alkaline Oxidative Leaching of Chromite

The kinetics of dissolution of Cr_2O_3 and of chromite mineral were studied[165] using 2.5, 5, and 7.5 m sodium hydroxide solutions over the temperature range 190–280°C, with p_{O_2} in most cases 212 lb in^{-2} (1.46 MPa). Under oxygen-free N_2, heating Cr_2O_3 in 5 m NaOH at 210°C and at 260°C for 3 hours gave solutions containing 0.10 g Cr_2O_3 l^{-1}. After standing in air or under N_2 for several days the chromium precipitated as a greenish lilac flocculant solid. No chromate was detected in the supernatent liquids.

The dependence of the rate of dissolution of the oxide in the presence of O_2 on the quantity of oxide used, sodium hydroxide activity, and p_{O_2} was interpreted as follows. Both homogeneous oxidation of the chromite ions formed by dissolution of the solid and heterogeneous oxidation of chromium at the surface of the solid can occur, the former being most noticeable at the higher temperatures used. The apparent activation energy in the initial stages of the reaction is about 80 kJ mol^{-1} in the range 190–260°C. The observed dependence of the rate on NaOH activity using 2.5, 5, and 7.5 m solutions was 1.75. Plotting log rate against log concentration did not give a straight line. The observed dependency on {NaOH} is more likely to reflect the change in the nature of the reaction medium than to have an explanation based on conventional chemical kinetics.

Chromite mineral was heated at 280°C in 7.5 m NaOH for 5 hours under oxygen-free N_2. No chromium could be detected in the cold solution. This did contain a small amount of iron in solution or colloidal suspension which separated out as a brown flocculant solid after a few days. The washed and dried residual solid had a tarnished appearance.

When leached in 5 and 7.5 m NaOH with p_{O_2} 212 lb in^{-2} (1.46 MPa) at 200°C, no significant quantity of chromium was leached from chromite. Between 220°C and 280°C the amount of chromate formed increased with increasing temperature but the concentrations were only about 10% of those obtained using Cr_2O_3 under similar conditions. When 10 to 15% of the chromium present in the mineral had dissolved a little red oxide of iron was present, some as very small particles in suspension, some firmly attached to small areas on the mineral surface which, after washing and drying, was dull black, as distinct from the highly reflecting black particles of the untreated mineral. When ferrochrome was leached under similar conditions dissolution of chromium was much faster than with chromite and large amounts of red iron oxide were produced.

Leached particles of chromite were mounted in epoxy resin (Aruldite) and the disc was polished to produce sections through particles at its surface. Optical microscopy showed that a coating up to 5 μm thick was present around each particle, having a slightly different colour and reflectivity compared with the unaltered chromite inside the coating. The coating was examined using electron microprobe analysis. The results presented in 1963[166] are believed to be the first use of this techhnique to investigate the behaviour of minerals when leached. There was a diffusion gradient of both chromium and iron in the coating; the concentration of iron increased from the value in the unaltered chromite to a maximum close to, or at, the edge of the section through the particle. The chromium concentration decreased within the same region.

The irregular shapes of the mineral particles produced by grinding caused some difficulty in making accurate observations through layers of solid 5 μm thick forming the outside of the particles. To overcome this a well formed octahedral crystal of chromite from Sierra Leone, of side about 1 cm, was cut across the central square plane and one of the pyramids which resulted was leached in 7.5 m NaOH solution for 3 hours at 260°C, with 500 lb in^{-2} (3.45 MPa) air pressure (at 20°C). The leached and unleached pyramids were mounted to expose the square faces at the surface of the disc, and the faces were polished. Electron microprobe analysis showed that the leached pyramid had a surface coating about 10–24 μm thick. The unleached specimen did not, but the outermost surface of the solid was iron-rich compared with the bulk of the solid. This was presumably due to weathering of the crystal.[4] The coating on the leached pyramid had well defined diffusion gradients of both iron and chromium, the edge of the crystal being clearly defined by the sharp fall in the response to the iron content. The results presented in 1963 were in the form of traces on a cathode ray screen. Results obtained using the same mounted specimen in a modern instrument in 1992 are shown in Fig. 5.1.

The observations were interpreted as follows.[166] The inverse spinel lattice of chromite (Sec. 1.2.3) is a close packed oxide structure with iron and chromium ions in some of the holes. Removal of chromium from the surface of the solid by the leaching reaction results in a deficiency of positive charge which is compensated by oxidation of some iron to Fe^{III}. Chromium ions diffuse down the activity gradient to the surface of the solid where further reaction can take place. Some iron also diffuses towards the surface and maintains what may be regarded as a magnetite structure containing chromium. The rate of leaching of chromium from chromite is, according to this model, controlled after an

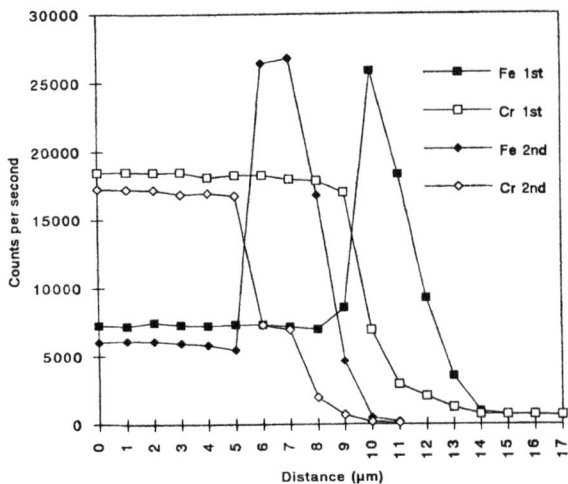

Fig. 5.1.   Electron microprobe analyses of 2 sections through the surface layer of a chromite crystal leached for 3 hours in 7.5 m NaOH at 260°C with 3.45 MPa air pressure (at 25°C).

initial period during which a thin film of solid product is formed, by the rates of diffusion of the metal atoms in the spinel oxide structure. The rate of oxidation of magnetite is slow under such conditions and the thickness of the layer of product increases until eventually it acts as a completely protective layer and leaching of chromium from the solid ceases. The extent to which leaching of particulate chromite occurs before this happens is greater the higher the alkali concentration or the temperature. If residual solid which is no longer reacting with the alkali is removed from the solution, washed and digested with HCl the layer of iron-rich solid dissolves and the residue will then react with alkali and oxygen until a new layer of product has built up.

The red oxide of iron which is formed after a significant quantity of chromium has been leached from the chromite is probably formed because in order to maintain the mass balance as $FeCr_2O_4$ forms Fe_3O_4, oxide ions must be lost by the solid in addition to chromium. Thus some of the surface regions of the particles must be destroyed, releasing iron from them.

Alkaline Oxidative Leaching of Columbite

Columbite is the name used for minerals at the niobium-rich end of the unlimited solid solution series $(Fe, Mn)Nb_2O_6$–$(Fe, Mn)Ta_2O_6$. It reacts with sodium hydroxide solutions to give a surface layer of sodium metaniobate, $NaNbO_3$, which is very insoluble. Reaction then ceases. With potassium hydroxide solutions however soluble niobate is produced and at 220°C in the presence of an oxidising agent such as oxygen, reaction proceeds to an extent which depends on the concentration of the alkali and then stops.[166] When the leach residue was digested with dilute mineral acid for a few minutes, washed, and leached again with the potassium hydroxide solution at 220°C, it again formed soluble niobate before the reaction slowed and finally stopped again.

A piece of columbite, not a single crystal, was cut and the freshly formed faces were polished. One half was leached in 3.5 m potassium hydroxide at 260°C, the air pressure at 21°C being 205 lb in^{-2} (1.41 MPa). After mounting and repolishing the previously polished faces the specimens were examined using electron microprobe analysis. The results[167] confirmed those obtained using particulate mineral[166] but were more accurate, being based on counts

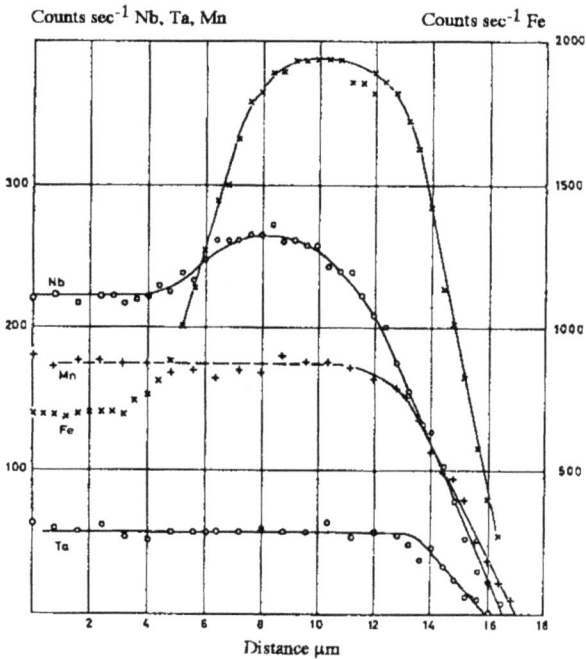

Fig. 5.2. Electron microprobe analyses through the surface layer of a polished face of columbite leached in 3.5 m KOH at 260°C with 1.41 MPa air pressure (at 21°C).

made when the electron beam was kept stationary at points on the sample, Fig. 5.2. No tantalum appeared in the leach solution. The unleached specimen showed no surface layer.

At 280–300°C, the reaction of columbite with 7.5 m potassium hydroxide solution is rapid and proceeds almost to completion. When the reaction was terminated before this occurred partially leached particulate residues showed patches of iron-rich outer layer which was never thicker than 2 μm, other regions of the surface having no coating. Thus fast reaction prevents the buildup of a protective layer, the iron forming a dispersed, red, finely-divided hydrated oxide. Chromite does not react as rapidly with 7.5 m sodium hydroxide as columbite does with 7.5 m KOH and reaction does not go to completion at 300°C. However, as was stated above, reaction of chromite continues further towards completion before it stops the higher the temperature or the greater the NaOH concentration.

It can be tentatively concluded from this behaviour of chromite and columbite that when a mineral has a composition corresponding to a particular ratio of two metal oxides which can form a continuous series of solid solutions over at least some range of composition in which the proportion of one metal is low, a region of varying composition can occur as that metal is leached out. If the crystal structure does not break down when the alteration occurs, the layer of product reduces the rate of leaching because diffusion of metal ions through it must take place. The effects of alkali concentration on the leaching rate must, on this basis, be due to the structure of the product layer, and hence diffusion rates through it, this structure being affected by the alkali concentration. Values of the critical increment of energy for columbite leaching vary with alkali concentration; within the range 150–270°C being for 5 m, 6.25 m and 7.5 m 160, 105 and 80 kJ mol^{-1}. Measurements at the lower concentrations could be made only at higher temperatures where appreciable reaction occurred and the values must be taken as being very approximate.[165] They are very high even for diffusion in a solid and the temperature of leaching, as well as alkali concentration, doubtless affects the structure of the solid. The product layer formed on columbite by leaching in the region of 300°C disintegrates.

Alkaline Oxidative Leaching of Wolframite

Wolframite, $(Fe, Mn)WO_4$ and scheelite, $CaWO_4$ are the two industrially important minerals used for the production of tungsten and its compounds. The extractive metallurgy of tungsten consists of the chemical treatment of an ore concentrate to obtain an intermediate product from which tungsten metal can be made.

Wolframite is decomposed industrially by roasting at 800–900°C with sodium carbonate, by reacting with 40–50% sodium hydroxide solution at 110–130°C in air at atmospheric or slightly higher pressure, or by leaching in 10–18% sodium carbonate solution at 190–225°C. Problems with the formation of coatings of products make it necessary to leach solid of small particle size, and attrition is used in some cases to break off the material. Production of magnetite from wolframite has been reported.

In a laboratory study[166] using a sample of wolframite which visually showed no sign of surface oxidation the leaching rate with 0.5 M sodium hydroxide solution at temperatures from 100° to 150°C increased with increasing partial pressure of oxygen from practically zero in deoxygenated solution under an atmosphere of nitrogen, to a maximum rate at about 3 lb in^{-2} (21 kPa) oxygen partial pressure measured at 20°C. The dissolution of tungsten did not proceed to completion however. When the oxygen partial pressure was raised to 50 lb in^{-2} (345 kPa) much less tungsten dissolved before reaction ceased than at the lower pressure, under similar conditions of alkali concentration and temperature. For example using 0.5 M NaOH at 150°C for 2 hours the percentage of WO_3 dissolved fell from 71% at 1.5 lb in^{-2} (10.4 kPa) to 32% at 50 lb in^{-2} (345 kPa) oxygen.

A slightly weathered sample of wolframite did react with 0.5 M sodium hydroxide solution in the absence of oxygen, at temperatures from 100° to 150°C. In the presence of oxygen up to 10 lb in^{-2} (69 kPa) (at 20°C) under these conditions the rate of dissolution of tungsten was initially fast but then decreased as a surface coating of a product formed. When sufficient complexing agent to chelate the iron in the leach solution was present, complete decomposition of the wolframite occurred. Diaminoethane-tetra-acetic acid was used. This substance is oxidised to oxalic acid and subsequently to carbonate under oxygen at elevated temperatures.

Microscopic examination of polished sections of partially leached wolframite showed that extensive penetration into the particles along grain boundaries or faults occurred in many samples. The residue was a porous solid shown by electron microprobe analysis to contain iron and manganese in the same ratio as in the original wolframite but little or no tungsten. Similar material occurred around the outside of each particle but in no case was a diffusion gradient found at the phase interface. Thus the mechanism of the reaction by which wolframite forms iron oxide on leaching is different from that which occurs with chromite and columbite. A considerable amount of a highly reflecting solid was also found in the layers of product around leached wolframite. This was dense, strongly magnetic and is believed to be magnetite. However, no lines other than those of wolframite were found in X-ray powder photographs. The dense product occurred in bands within the porous solid and more was present in specimens leached at the higher oxygen pressures than at lower pressures. This is probably the reason for less reaction of wolframite occurring at the higher pressures, the dense solid forming a more effective barrier against diffusion of oxygen and hydroxide ions to the underlying wolframite. Three photomicrographs of leached wolframite particles and one set of electron microprobe area images for W, Fe and Mn are reproduced in Ref. 166

The difference in behaviour of chromite and columbite minerals when Cr and Nb are leached out on the one hand, and of wolframite when W is removed from the solid on the other, can be attributed to two factors. Whereas the first two have Cr and Nb in close packed oxide structures, the tungsten atoms in wolframite have four tetrahedrally coordinated oxygen atoms around them, forming WO_4 units. Also, whereas Cr–Fe and Nb–Fe can replace one another in a lattice giving regions of solid having a range of composition, this is not the case in wolframite, in which it is Fe and Mn which can form the solid solution range.

The formation of a protective layer based on the crystal structure of the reacting solid must be distinguished from other solid products formed by different mechanisms during leaching. Thus when pentlandite $[(Ni, Fe)_9S_8]$, chalcopyrite ($CuFeS_2$), and nickeliferous pyrrhotite $[(Fe, Ni)_{1-x}S]$, are leached· with aqueous ammonia a pseudomorphic hydrated iron (III) oxide is formed. That is, the particles of sulphide mineral retain their original size and shape while the reaction takes place.[168,169] Transport of reactants and soluble products takes place through the solid product. Since dissolved ammonia and oxygen molecules diffuse to the reacting surface from the bulk of the solution and nickel ammine and thiosulphate ions move in the other direction, this transport cannot occur within the crystal lattice of the sulphide minerals. The hydrated iron oxide is porous.

The porous iron and manganese hydrated oxides and carbonates formed during the leaching of wolframite in alkaline solutions do not form pseudomorphic particles but tend to remain where they are produced within the particles of the friable mineral. During the decomposition of scheelite, $CaWO_4$, by leaching in sodium carbonate solutions in autoclaves insoluble protective layers of

$Na_2CO_3 \cdot CaCO_3$ and $Na_2CO_3 \cdot 2CaCO_3$ form on the particles and attrition is commonly used to break off the solid. During the leaching of uranium ores in alkali carbonate solutions the rate of dissolution of UO_2 at high oxidation potentials is limited to the rate at which a film of UO_2CO_3, produced by the oxidation reaction, is removed by chemical dissolution. A major problem in the leaching of gold ores can be caused by the formation of a passive film of oxide which hinders, or prevents the dissolution of the metal in cyanide solutions. Such passive films on metals are in many cases effective in protecting the metal from further oxidation, or corrosion. The formation of a layer of a solid on a reacting solid substance is a very common occurrence. It can be helpful, or a hindrance. In order to avoid difficulty it is often useful to understand the reason for the production of the compound and to change conditions to prevent it forming.

5.2.3. *Electrochemical Model for the Oxidation Leaching of Uranium Dioxide*

The minerals of uranium contain U^{VI}, U^{IV} or the metal in both oxidation states. Low-grade ores are treated under very mild conditions to minimise costs and extensive investigations were carried out in order to optimise the conditions used with individual ores. Systematic studies to identify the factors which control the kinetics of the oxidative leaching reaction were undertaken using synthetic uranium dioxide[170–172] as well as uraninite $(U^{IV}_{(1-x)} U^{VI}_x)O_{(2+x)}$ concentrates[173] because of the ill-defined chemical nature of most minerals containing U^{IV}. The major factors which influence the reactions and their effects on the rate of leaching of UO_2 were established and it is accepted that oxidation of U^{IV} to U^{VI} takes place before a uranium ion transfers to the aqueous phase. This is fundamentally different from a reaction in which a uranium atom in the U^{IV} oxidation state leaves the solid lattice and is oxidised as a U^{IV} salt in the aqueous phase. No satisfactory explanation was found for the difference between the leaching process in acidic and in alkali carbonate solutions, or for the differences between the effects of different oxidising agents on the kinetics of leaching of UO_2. In particular it was shown that the rates of dissolution of uraninite and UO_2 do not correlate with the standard redox potentials ($E°$ values) of the oxidants used.[173,174] Thus the rate of oxidation of U^{IV} in solid UO_2 is not controlled solely by the thermodynamic driving force provided by the difference between the value of the equilibrium potential E_{el}

for the half cell reaction for the oxidation of UO_2

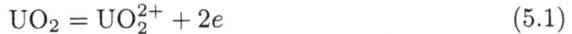

$$UO_2 = UO_2^{2+} + 2e \qquad (5.1)$$

and E_{e2} for the reduction of an oxidant M^{z+} in the solution, written as

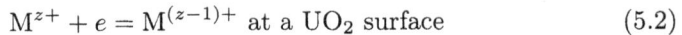

$$M^{z+} + e = M^{(z-1)+} \text{ at a } UO_2 \text{ surface} \qquad (5.2)$$

where E_{e2} depends on both the value of E° for the reversible redox couple M^{z+}, $M^{(z-1)+}$ and on the ratio of the activities of the two ions. This is the driving force shown as existing by predominance area diagrams for the U–H_2O and the M–H_2O systems. No consideration is given to the kinetics of the half-cell Reactions (5.1) and (5.2) in the thermodynamic analysis.

The electrochemical approach to chemical reactions involving electron transfer does take into account the kinetics of this transfer, which can be measured as the current density flowing while the reaction takes place, see Sec. 4.2.5. Application of this procedure to the leaching of UO_2 has shown that certain experimental observations can be satisfactorily explained on the basis that the mechanism of the reaction is electrochemical.[175,176] These papers appear to have been the first in which an electrochemical mechanism has been proved to control the kinetics of an oxidative leaching reaction.

The reaction is the sum of (5.1) and (5.2), balancing the charge

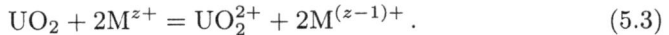

$$UO_2 + 2M^{z+} = UO_2^{2+} + 2M^{(z-1)+}. \qquad (5.3)$$

Proof that the mechanism of Reaction (5.3) is electrochemical requires a demonstration that it is in all respects equivalent to the component half-cell Reactions (5.1) and (5.2). In addition it must be shown that the oxidation of UO_2 follows the same reaction path whether the potential ϕ of the UO_2 surface is applied from outside the aqueous systems by a potentiostat, or from within it by a chemical oxidant in solution. The potential is that existing between the surface of the UO_2 and a reference electrode situated in the solution beyond the electrical double layer. It should be noted that a proof such as that described here only holds good for a solid which is an electronic conductor.

Leaching experiments were carried out[175,176] using experimental methods essentially similar to those described in Sec. 4.2.5 except that an electrode consisting of an 0.4 cm cube of a single crystal of high purity UO_2, mounted in an epoxy resin block with a connector wire through the back, was placed on the bottom of the leaching vessel with the polished face at which reaction occurred directly underneath the stirrer. The oxygen to uranium ratio in the oxide was 2.03 and because of its relatively high resistivity, 1000 ohm cm, potentials applied by the potentiostat were corrected manually for potential drop across the UO_2 electrode after the resistance between it and the platinum auxiliary electrode in the solution had been measured.

The electrochemical treatment of the kinetics of leaching of UO_2 considers the relationship between the rate of transfer of charge and the potential for the half-cell Reactions (5.1) and (5.2). The rate of transfer of charge is measured as the current density when the total potential is applied by a potentiostat in the absence of a chemical oxidant in the solution, and in separate experiments by the specific rate of dissolution of uranium when the same total potential is applied in the presence of a chemical oxidant in the solution. The rate of appearance of uranium in solution is half the rate of formation of $M^{(z-1)+}$ from M^{z+} in the case of the oxidant considered.

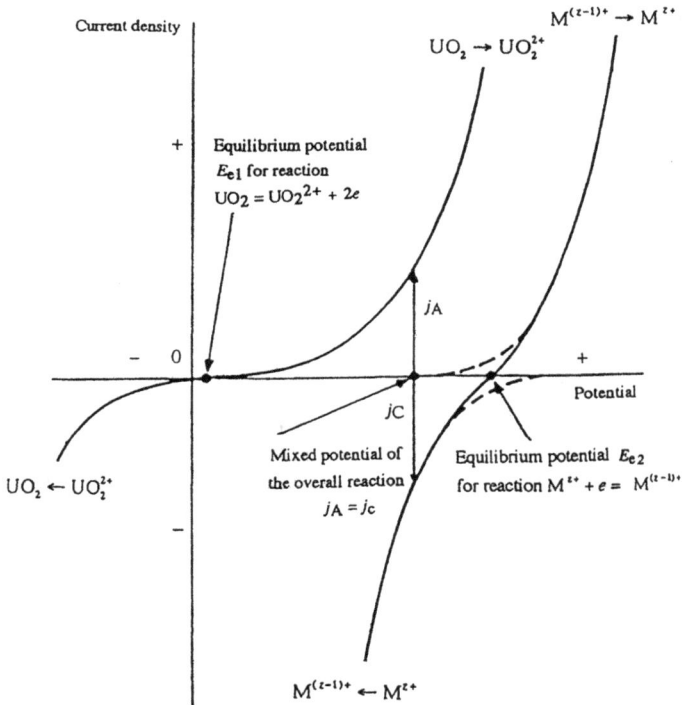

Fig. 5.3. Electrochemical treatment of leaching of UO_2 in terms of current density-potential curves for two half-cell reactions.

The relationships between the current density and potential for the two half-cell reactions are represented diagramatically in Fig. 5.3. If the solution contains the oxidant the equilibrium potential E_{e2} which is set up is determined by the value of $E^o_{M^{z+}.M^{(z-1)+}}$ and by the ratio of the activities of the two ions. Since the ions are at equilibrium the current density is zero. If a reducing potential is applied either by a potentiostat or by addition of a reducing agent to the system some M^{z+} will be reduced and the rate at which this occurs will be indicated either by the measured current density or by the rate of oxidation of the reducing agent. If the reducing agent is a flat surface of UO_2 of constant, known surface area, at the instant when the UO_2 surface touches the solution containing the oxidant it takes up the equilibrium potential E_{e1} for reaction (5.1). The system is now thermodynamically unstable and there is a driving force towards equilibrium given by the potential difference $E_{e2} - E_{e1}$. Since the surface of the UO_2 has cathodic areas and anodic areas electrochemical cells are set up on the surface and current is forced to flow. A steady state is reached when the anodic and cathodic currents caused by oxidation of UO_2 and reduction of the oxidant are balanced by each other. The potential at which this is achieved is known as the *mixed*, or *rest* potential. The difference between the mixed potential and the equilibrium potential of the UO_2, U^{VI} couple represents the driving force of the leaching reaction and the current density j_A is equivalent to the specific rate of leaching of UO_2.

Two sets of experiments were carried out in acidic solution.[175] In the first set chemical oxidants in solution were used to apply a series of overpotentials to the UO_2 surface and the rates of leaching of UO_2 were measured by taking and analysing samples of the solutions. In the second set no chemical oxidant was present in the solution; the same overpotentials were applied to the UO_2 surface by means of a potentiostat and the currents due to the anodic dissolution of UO_2 were measured. If the rates of leaching (quantity of UO_2 per unit area per unit time) from the first set of experiments are equal to the rates of anodic dissolution, in the second set of experiments, calculated from the current using the laws of electrolysis, the leaching of UO_2 in the presence of an oxidant can be said to follow an electrochemical mechanism. This assumes 100% current efficiency for the anodic oxidation of UO_2 which was confirmed by analysis of solutions produced by anodic oxidation using known amounts of current. The results obtained using oxidants, and a curve drawn from results of anodic dissolution are shown in Fig. 5.4. It was concluded that the leaching of UO_2 by chemical oxidants takes place by an electrochemical mechanism in acidic solutions.

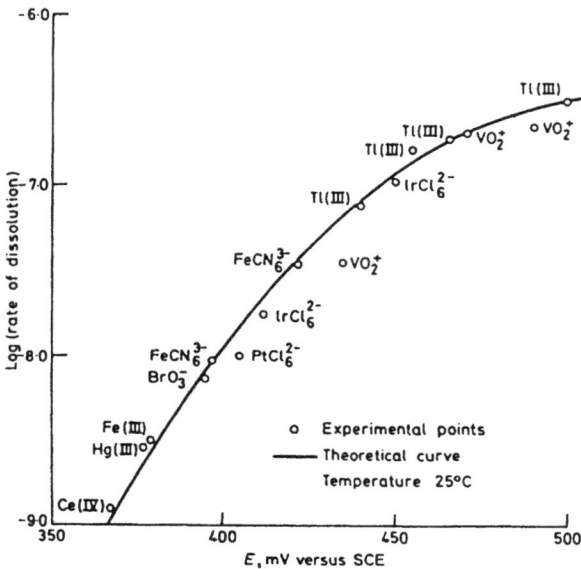

Fig. 5.4.   Comparison of leaching rates of UO_2 in acid perchlorate solutions when oxidation potentials were applied using dissolved oxidants (chemical leaching) and the same overpotentials were applied to the reaching UO_2 electrode from an external source in the absence of a chemical oxidant (electrochemical leaching).[175]

Rates of Reactions Having Electrochemical Mechanisms

The Tafel equation.   The rate of an electron transfer reaction which has an electrochemical mechanism, between a solid which is an electronic conductor and an aqueous solution, is determined by the value of the overpotential η. In 1905, Tafel showed that the relationship between η and the rate of a reaction taking place continuously, measured as the current density j amperes per cm^2, when the potential is applied from an external source is given by the Tafel equation

$$\eta = a + b \log j \tag{5.4a}$$

with fair accuracy over a considerable range of current density. In many cases the rate of evolution of hydrogen at metal electrodes was studied. a and b are constants for the reaction and b may be written

$$b = 2.303RT/\beta F. \tag{5.4b}$$

The value of β is often about 0.5 and b about 0.12 V, but values of b have been found between 0.02 (β about 3) and 0.3 (β about 0.2).

Since

$$\ln j = (\eta - a)(\beta F/RT)$$

$$j = \exp[-\beta F(a - \eta)/RT] = \exp[-(w - \beta\eta F)/RT].$$

(5.5)

The form of this equation, derived to fit experimental observations, suggests that the rate of the reaction is controlled by a rate-determining step having an activation energy $(w - \beta\eta F)$ kJ mol^{-1}. At zero overpotential the value is $w = \beta a F$. This argument indicates that the value of β for a reaction determines how the input of electrical energy (ηF) affects the rate of the reaction. β is known as the symmetry factor and represents the fraction of the applied overpotential which is used in providing a driving force for the reaction. This overpotential, and driving force, can also be provided, as the redox potential, by an oxidising agent present in the solution. The value of β depends not only on the composition of the solid at which the reaction is taking place but also on the properties of the surface.

The Butler–Volmer equation.   This expresses the relationship between the rate of electron transfer across the interface between the solid and solution and, (i) the thermodynamic driving force for the reaction; (ii) the nature and properties of the solid surface; and (iii) the characteristics of the electric field. The general form of the relationship between the reaction rate and the thermodynamic driving force expressed as a potential is shown in Fig. 5.3 for the oxidation of uranium dioxide in an acidic oxidising solution of a metal M^{z+}. The fact that the chemical reaction has an electrochemical mechanism means that the rate of the forward reaction $\vec{\nu}_{el}$, in which the UO_2 donates 2 electrons to the M^{z+} ions, is influenced by the presence of the electric field, produced either from an outside source or, as in this case, by the oxidation potential created by the $M^{z+}/M^{(z-1)+}$ couple. If the solid reacting were not an electronic conductor the field would have no effect on the reaction rate and the rate of the forward reaction $\vec{\nu}_c$ would be controlled entirely by chemical factors.

The rate of the chemically controlled reaction is given by

$$\vec{\nu}_c = \vec{k}_c\, c_{ox}^n$$

where $\vec{k}_c$ is the rate constant, c_{ox} the concentration of the dissolved oxidant and n the order of the reaction with respect to the oxidant. The simplest case will be considered in which $n = 1$ and a single electron is transferred in one step to complete one occurrence of the overall reaction. The rate constant is given by

$$\vec{k}_c = A \exp(-E_a/RT)$$

where E_a kJ per mol is the activation energy of the reaction and A is the preexponential factor. It is convenient to represent E_a in an alternative form, as $\Delta G_c^{0\neq}$, the standard free energy of activation of the forward reaction. This is the change in free energy required for the oxidant ion, for example Fe^{3+}, to reach the summit of the energy barrier when there is no electric field acting on the ion in its motion.

In the presence of the electric field caused by the Fe^{III}/Fe^{II} redox couple the total free energy of activation of the forward reaction between UO_2 and Fe^{III} is equal to the chemical free energy of activation plus an electrical contribution $\beta F \Delta\phi$ where $\Delta\phi$ is the potential difference between the surface of the reacting solid and a reference electrode situated in the solution beyond the electrical double layer present at the solid–liquid interface. Thus according to the discussion of the Tafel equation given above

$$\Delta G_{el}^{0\neq} = \Delta G_c^{0\neq} + \beta F \Delta\phi$$

and the rate $\vec{\nu}_{el}$ of the forward reaction under the influence of the electric field can be written

$$\vec{\nu}_{el} = \vec{k}_c \, c_{ox} \exp(-\beta F \Delta\phi/RT).$$

The rate is the number of moles of oxidant ion which react with 1 cm^2 of the surface of the solid per second. Each ion accepts one electron and the current density is therefore

$$\vec{j} = F \, \vec{k}_c \, c_{ox} \exp(-\beta F \Delta\phi/RT) \text{ amperes cm}^{-2}. \tag{5.6}$$

Many oxidation-reduction reactions are reversible and as the concentrations of the products of the forward reaction increase, or they are added for some purpose, the effects of the reverse or back reaction must be taken into account. This is the reaction between Fe^{2+} and U^{6+} for example. Since the activation energy of the forward reaction is lowered by $\beta F \Delta\phi$ due to the presence of the electric field, that of the back reaction is raised by $(1 - \beta)F\Delta\phi$. When the forward and the back reactions are taking place simultaneously the resulting exchange current density is given by the difference between the two rates

$$j = \overleftarrow{j} - \overrightarrow{j} = F\,\overleftarrow{k}_c\,c_{red}\exp[(1-\beta)F\Delta\phi/RT]$$
$$- F\,\overrightarrow{k}_c\,c_{ox}\exp[-\beta F\Delta\phi/RT]$$

where c_{red} is the concentration of the reducing agent and $\Delta\phi$ is the nonequilibrium value of the potential across the interface which produced the value j of the current density. If the equation is written for $j = \overrightarrow{j} - \overleftarrow{j}$ this changes the sign of j.

The value of $\Delta\phi$ can be separated into ϕ_e, the equilibrium potential at which $j = 0$, shown in Fig. 5.3 as the mixed potential of the overall reaction, at which the anodic and cathodic currents are equal, and the additional potential η. The separate components of j give

$$j = [F\,\overleftarrow{k}_c\,c_{red}\exp(1-\beta)F\Delta\phi_e/RT]\{\exp(1-\beta)F\eta/RT\}$$
$$- [F\,\overrightarrow{k}_c\,c_{ox}\exp-\beta F\Delta\phi_e/RT]\{\exp-\beta F\eta/RT\}\,.$$

The two terms inside the square brackets form the expression for the equilibrium exchange current density j_o and it is convenient to write

$$j = j_o[\exp\{(1-\beta)F\eta/RT\} - \exp\{-\beta F\eta/RT\}]\,. \tag{5.7}$$

This is the Butler–Volmer equation. The first term represents the current density $\overleftarrow{j}$ of the reduction, cathodic reaction, and the second term, the anodic, oxidation current density $\overrightarrow{j}$. The exchange current density j_o describes the kinetic properties of the cathodic and the anodic reactions at equilibrium, when they have the same rate. Thus j_o is related to the rate constant k_0 for the reaction through the equation

$$j_o = Fk_o[M^{(z-1)+}]\exp(1-\beta)[M^{z+}]\exp-\beta \tag{5.8}$$

if one electron is exchanged and the order of reaction of M^{z+} and $M^{(z-1)+}$ is 1 in each case.

It can be seen from Fig. 5.3 that when j is plotted against potential the curve obtained resembles that of a hyperbolic sine function, $[\exp x - \exp(-x)]/2 = \sinh x$. If the value of β is 0.5, then from (5.7)

$$j = j_o[\exp(\eta F/2RT) - \exp(-\eta F/2RT)] = 2j_o\sinh\eta F/2RT\,. \tag{5.9}$$

This value of β corresponds to the presence of a symmetrical j versus η curve. Other values of β do not, so that β is referred to as the symmetry factor. There are two limits to the shapes of the curves.

(i) *The high overpotential approximation.* If the reaction proceeds so that j is given by Eq. (5.7) with $\beta = 0.5$, the value of η being large

$$\exp F\eta/2RT \gg \exp -\eta F/2RT$$

and $\exp -F\eta/2RT$ approaches zero. Then from (5.9)

$$2 \sinh \eta F/2RT \approx \exp \eta F/2RT \quad \text{and} \quad j \approx j_o \exp \eta F/2RT .$$

The same conclusion can be drawn directly from (5.7). As η increases $\exp -F\eta/RT$ decreases until it can be neglected relative to $\exp F\eta/RT$. The high field approximation of the Butler–Volmer equation is therefore

$$j \approx j_o \exp(1 - \beta)F\eta/RT$$

and

$$\eta = -[2.303RT/(1 - \beta)F] \log j_o + [2.303RT/(1 - \beta)F] \log j . \tag{5.10}$$

This high field approximation is valid at values of η greater than 0.10 V with an error less than $+2\%$, and less $+1\%$ above 0.20 V. If η is plotted against $\log j$ in accordance with Eq. 5.4(a), a "Tafel line" is obtained and extrapolation of the linear portion to zero overpotential gives the value of the exchange current density.

(ii) *The low overpotential approximation.* When η is small $\eta F/2RT$ is very much smaller than 1 so than the approximation $\sinh \eta F/2RT \approx \eta F/2RT$ can be used. For this special case the Butler–Volmer equation becomes a linear relationship

$$j = j_o \eta F/RT . \tag{5.11}$$

If it is assumed that $\eta F/RT$ must be less than 0.2 for the low field approximation to be regarded as acceptable then η must be less than about 0.01 V for a one-electron one-step transfer reaction. This region of "ohmic behaviour" can be seen in Fig. 5.5(a).

The equation for multistep reactions. In deriving the equations given above it has been assumed that the reactions considered involve the transfer

Oxidation

Current density
$+j$Am^{-2}

$j_A = [j_o \exp(\alpha_A nF\eta / RT)]$

Overpotential ηV

40

20

-0.2   -0.1   0   0.1   0.2

$-j$Am^{-2}

-20

$j_c = [-j_o \exp(\alpha_c nF\eta / RT)]$

-40

Reduction

(a)

2

$\log j$ (Am^{-2})

1

Overpotential ηV

0

-0.2   -0.1   0.1   0.2

j_o

-1

(b)

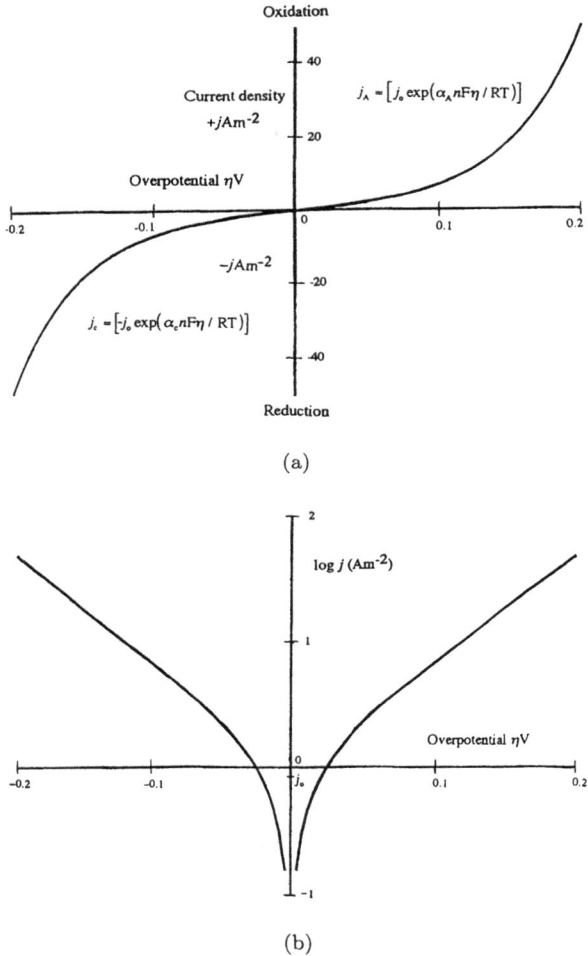

Fig. 5.5.   (a) Diagramatic plot of the Butler–Volmer equation for $\beta = 0.5$. (b) Semiloga-rithmic plot of the Butler–Volmer equation for $\beta = 0.5$.

of one electron in a single step. Many reactions however take place in several steps, for example the oxidation of U^{VI} to U^{VI} in the reaction of uranium dioxide with ferric sulphate. It is generally accepted that only 1 electron can be transferred in a step. An intermediate product can also be formed; for example the reduction of H^+ during the electrolysis of an acid using a metal cathode forms hydrogen atoms which remain associated with the metal until, in a second stage, reaction occurs between pairs of them to form H_2 molecules. Thus in this multistep reaction the electron transfer step must occur twice before a hydrogen molecule can form. This is indicated by stating that the stoichiometric number v of the reaction scheme is 2. There is a v — to — one correspondence between the occurrence of this step and completion of the overall reaction.

In the general case the overall reaction may require the completion of a sequence of several consecutive steps, each of which involves the transfer of one electron. Every chemical reaction in the sequence has an energy barrier associated with it and the one which has the highest free energy of the activated state compared with the initial state is the rate-determining step. Thus the rate equation for a multistage reaction contains one value for the activation energy. However each species except the first in the sequence is formed by a charge transfer reaction so that its concentration depends on the potential. The rate of the overall reaction, therefore, is that constant value achieved when the individual steps are in a state of kinetic equilibrium, when the forward and backward rates are equal.

An analysis of this situation should be sought in textbooks on electrochemistry. The result is the most general form of the Butler–Volmer equation for a reaction in which n electrons are transferred overall

$$j = j_0[\exp\{\overleftarrow{\alpha}\ F\eta/RT\} - \exp\{-\ \overrightarrow{\alpha}\ F\eta/RT\}] \tag{5.12}$$

which is valid for a multistep reaction in which there may be electron transfers in steps other than the rate-determining step, and in which this step may have to take place v times for a single occurrence of the overall reaction. The equation has the same form as the Butler–Volmer equation for a one-step single electron transfer reaction

$$j = j_0[\exp\{(1 - \beta)F\eta/RT\} - \exp\{-\beta F\eta/RT\}]. \tag{5.7}$$

The transfer coefficients $\overleftarrow{\alpha}$ and $\overrightarrow{\alpha}$ play the same role in a multielectron multistep reaction as the symmetry factor does in a single step single electron transfer reaction. They represent the fraction of the applied overpotential which is utilised in providing a driving force for the reaction. The values of α are determined in part by the total number n of 1-electron steps in the overall reaction. Hence the value of n is not included in the exponential terms of Eq. (5.12). It does however remain in the value of j_0.

The Butler–Volmer equation is plotted diagramatically in Fig. 5.5(a) for $\beta = 0.5$. It gives a symmetrical line through the origin, positive values of j arising from an oxidation reaction and negative values from reduction. The semilogarithmic plot in Fig. 5.5(b) shows the linear relationship between $\log j$ and the overpotential [Eq. (5.10)] of the high overpotential approximation. Extrapolation of the straight lines to $\eta = 0$ gives the value of j_0. The linear relationship of the low overpotential approximation is shown in Fig. 5.5(a). At high anodic overpotential ($\eta > 0.1$ V) the effect of the cathodic reaction can be neglected and Eq. (5.12) becomes

$$j_A = j_0 \exp(\alpha_A F\eta/RT)$$

$$\log j_A = \log j_0 + (\alpha_A F\eta/2.303RT).$$

At high cathodic overpotential ($\eta < -0.1$ V) the effect of the anodic reaction can be neglected and (5.12) becomes

$$-j_c = j_0 \exp(-\alpha_c F\eta/RT)$$

$$\log(-j_c) = \log j_0 - (\alpha_c F\eta/2.303RT).$$

Thus for a reaction between two distinct reactants the Butler–Volmer equation can be written

$$j = j_0[\exp(\alpha_A F\eta/RT) - \exp(\alpha_c F\eta/RT)]. \tag{5.13}$$

When this is used in the investigation of a leaching reaction, of UO_2 for example, each of the half-cell Reactions (5.1) and (5.2) is considered separately. The current density overpotential relationships are represented diagramatically in Fig. 5.3, which shows that UO_2 is oxidised at positive values of η giving positive values of current density. The electrons are transferred to M^{z+}, the reduction reaction producing a negative current density. The Butler–Volmer equation also takes into account the two reverse reactions, oxidation of $M^{(z-1)+}$ to M^{z+} and reduction of UO_2^{2+} to UO_2. The curve for the oxidation of UO_2 is determined experimentally by applying potentials from an external source across a UO_2 anode in a solution of the required composition, and an inert electrode. Reduction of UO_2^{2+} ions in a solution onto a cathode requires a high overpotential before deposition of UO_2 occurs, so that anodic oxidation of UO_2 can be regarded as an irreversible process. Thus the term in the Butler–Volmer equation which describes the cathodic reduction of UO_2^{2+} can be neglected.

The rate of an oxidation reaction depends on the concentration of the oxidant, Eq. (5.6), and this is introduced into the Butler–Volmer equation through the value of the exchange current density j_o. Modification of Eq. (5.8) for a reaction involving n electrons and an oxidant and reductant having orders of reaction p and q respectively gives

$$j_o = nFk_o c_{ox}^p (\exp -\beta) c_{red}^q (\exp 1 - \beta).$$

The value of k_o, the rate constant for the overall reaction producing the exchange current density, is achieved by a combination of values of k for the separate reactions and the concentrations of the reactants. Then from (5.13) if $p = q = 1$

$$j = [nFk_o c_{ox} \exp(\alpha_A F\eta/RT)] - [nFk_o c_{red} \exp(\alpha_c F\eta/RT)].$$

For ease of manipulation such equations were simplified[175,176] by using

$$b = 2.303 RT/\beta F. \tag{5.4}$$

In order to indicate that the reactions may have several steps the authors replace β by $f(\beta)$. b is the slope of the line obtained by plotting $\log j$ against η and $1/b = \alpha F/2.303RT$. Hence

$$j = \left[nFk_A c_{ox} \exp \frac{2.303\eta}{b_A} \right] - \left[nFk_C c_{red} \exp -\frac{2.303\eta}{b_C} \right]. \tag{5.14}$$

The rate constants k_A and k_C are those for the chemically controlled reactions and so are defined for the reactions taking place at zero potential against a reference electrode situated in the solution outside the electrical double layer present at the surface of the reacting solid. The "concentration" of the reacting solid is related to the area of its reacting surface. If the order of reaction of the dissolved solute is to be determined the value of the potential E should be used rather than the overpotential (see Appendix A).

Considering as an example the oxidation of UO_2 using a solution of iron (III) sulphate, the Butler–Volmer equation describing the anodic oxidation of the UO_2 is

$$j = 2Fk_{UO_2} \exp 2.303E/b_{UO_2} \tag{5.15}$$

since j is expressed in amperes cm^{-2} and c_{UO_2} is unity therefore, and the reaction is an irreversible process. In a leaching operation the ratio Fe(III) to Fe(II) is kept high and so the partial anodic current due to the anodic oxidation of Fe(II) is negligible compared with that due to the anodic oxidation of UO_2. Under these conditions the back reaction involving oxidation of Fe(II) can be neglected. The reduction reaction of Fe(III) at a UO_2 surface has been shown to be first-order with respect to the concentration of iron (III) and zero order of iron (II). Then the Butler–Volmer equation for the cathodic reduction of iron (III) at a UO_2 surface is

$$-j = F k_{Fe(III)} [\text{Fe(III)}] \exp(-2.303 E / b_{Fe(III)}) . \tag{5.16}$$

At the mixed potential, the currents are equal and by combining the two equations a value for the mixed potential E_M can be obtained

$$E_M = \frac{b_{UO_2} \cdot b_{Fe(III)}}{[b_{UO_2} + b_{Fe(III)}]} \log \frac{F k_{Fe(III)} [\text{Fe(III)}]}{2 F k_{UO_2}} . \tag{5.17}$$

Substitution of Eq. (5.17) into (5.15) gives the following expression

$$j = (2 F k_{UO_2})^{\varphi_1} (F k_{Fe(III)} [\text{Fe(III)}])^{\varphi_2} \tag{5.18}$$

where

$$\varphi_1 = b_{UO_2} / (b_{UO_2} + b_{Fe(III)}) \quad \text{and} \quad \varphi_2 = b_{Fe(III)} / (b_{UO_2} + b_{Fe(III)}) .$$

From Eq. (5.18), it is predicted that the exponent of the concentration of iron (III) in the reaction is equal to

$$b_{Fe(III)} / (b_{UO_2} + b_{Fe(III)}) .$$

This and some other specific examples of situations occurring during industrial leaching processes are described below.

Leaching of Uranium Dioxide in Sodium Carbonate Solution[176]

The reaction in which UO_2 is leached in a sodium carbonate solution is written

$$UO_2 + 3 CO_3^{2-} + X = UO_2(CO_3)_3^{4-} + X^{2-}$$

where X is an oxidant. Experiments were carried out in 0.2 M Na_2CO_3–0.2 M $NaHCO_3$ buffer solution at 45°C using as oxidants $Fe(CN)_6^{3-}$; MnO_4^-; Hg^{II} (saturated); oxygen; and air. The electrochemical potential of the UO_2 surface with respect to a saturated calomel electrode (SCE: $E = -0.2420$ V) in the solution was measured in each experiment and the rates of leaching of the uranium were measured by chemical analysis. In other experiments no oxidant was present in the solution but potentials were applied between the UO_2 surface and the electrode, the current density being used to measure the rate of leaching. The results indicated that the reaction in each case has an electrochemical mechanism, being similar to those obtained in acidic solutions, Fig. 5.4.

The equilibrium potential E_e of the half-cell reaction

$$UO_2 + 3CO_3^{2-} = UO_2(CO_3)_3^{4-} + 2e$$

at pH 10 was calculated from free energy data to be -0.34 V versus SCE. The equilibrium potential of the cathodic half-cell reactions used, at pH 10 and 25°C, in alkali carbonate solutions are given in Table 5.2. If the only factor controlling the effectiveness of the oxidising system for leaching UO_2 were the differences between the equilibrium potentials of the two half-cell reactions, then the order of the rate of leaching would be:

$$H_2O_2 > O_2 > Fe(CN)_6^{3-} > Cu(NH_3)_4^{2+} .$$

It is found by experiment to be

$$Fe(CN)_6^{3-} > H_2O_2 > Cu(NH_3)_4^{2+} > O_2 .$$

At the time this work was carried out (about 1970) the only published data on the kinetics of leaching of UO_2 and U_3O_8 in alkaline carbonate solutions

Table 5.2. Equilibrium potentials of some cathodic half cell reactions at pH 10.0, 25°C.

Reaction	Conditions	E_e, V versus SCE
$O_2 + 4H^+ + 4e = 2H_2O$	With $p_{O_2} = 1$ atm (approximately 10^{-3} M)	0.40 (calculated)
$Fe(CN)_6^{3-} + e = Fe(CN)_6^{4-}$	10^{-2} M $Fe(CN)_6^{3-}$ + 10^{-4} M $Fe(CN)_6^{4-}$	0.33 (measured)
$Cu(NH_3)_4^{2+} + e$ $= Cu(NH_3)_2^+ + 2NH_3$	6×10^{-3} M $Cu^{(II)}$ (O_2 saturated)	< 0.15 (measured)
$H_2O_2 + 2e = 2OH^-$ $HO_2^- + H_2O + 2e = 3OH^-$	5×10^{-3} M H_2O_2	0.94 (calculated)

related to the use of oxygen, which was known to be a very poor oxidant for this purpose at ambient temperatures and atmospheric pressure. In each of the three investigations it was found that the rate of dissolution of uranium was proportional to the square root of the partial pressure of the gas. This had been taken to indicate that each molecule of oxygen adsorbed on the solid dissociated to form two species ("oxygen atoms") which separately oxidised U^{IV} to U^{VI}. Experiments were carried out therefore to determine how the rate of leaching of UO_2 depends on the concentration of each of the oxidants listed in Table 5.2. The results are shown as a logarithmic plot in Fig. 5.6, the exponents of the oxidant concentrations, i.e. the order of reaction of each oxidant, being: $Fe(CN)_6^{3-}$, 0.57; H_2O_2, 0.50; O_2, 0.60; 0.41 with UO_2 from a different source; $Cu(NH_3)_4^{2+}$, 0.50. In each case the rate is approximately

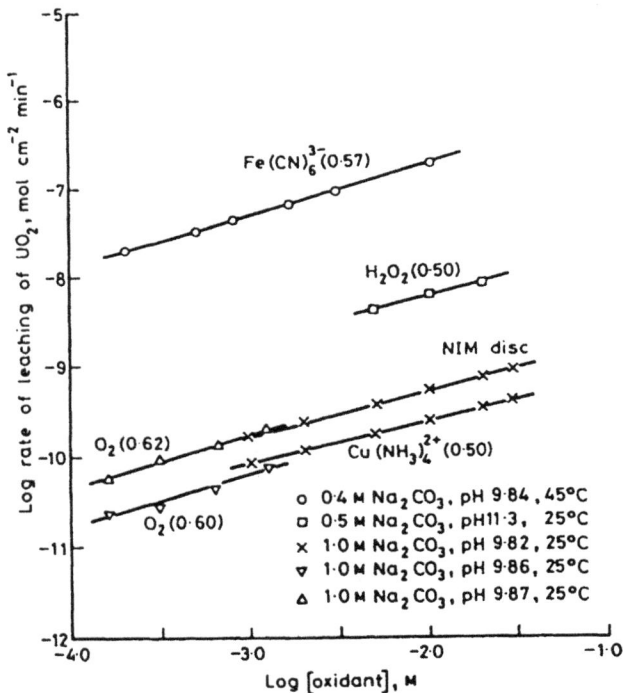

Fig. 5.6.   Variation of rate of leaching of UO_2 in sodium carbonate solutions with oxidant concentration. The value of the exponent of the oxidant concentration is shown on each line.[176]

proportional to the square root of the concentration of the oxidant. Although it might be argued that dissociation of hydrogen peroxide in the rate-determining step could give a square root dependence, such a suggestion could not be supported for $Fe(CN)_6^{3-}$ or $Cu(NH_3)_4^{2+}$. It seemed probable therefore that the behaviour was due to the electrochemical mechanism of the reaction.

The anodic characteristics of UO_2 were determined by studying its anodic oxidation in solutions of sodium carbonate of concentrations in the range 10^{-3} M to 1.5 M at pH 9.82 and 25°C. In order to provide substantial ionic strength in all solutions, each was 1.0 M in sodium perchlorate. The results showed that at low applied potentials the current density changed "relatively" linearly with E, but at values greater than about 0.4 V the current became independent of the potential. The exact value at which this occurred depended on the pH and the carbonate concentration. This behaviour was attributed to the formation of a film of a solid which was identified as UO_2CO_3, produced by oxidative hydrolysis of an intermediate solid. At high potentials the rate of dissolution of uranium is controlled by the rate of leaching of this film, which is independent of the electrical potential, no electron transfer being involved. Since the film is an electrical insulator the rate of reaction of a UO_2 surface with an oxidant in the solution decreases while it is forming, and is considerably slower once a coherent layer of the uranyl carbonate has been produced. This was demonstrated in separate experiments in which a ring disc electrode was used to measure the rate of anodic oxidation of hydrogen peroxide and of ferrocyanide at a UO_2 surface as a function of the potential applied. In both cases the current density increased as the potential applied was increased and reached a maximum value, after which increasing the potential produced a lower current density until zero current indicated that the solid product completely covered the surface of the UO_2.

The rate of reduction of oxygen at a UO_2 surface was studied using known concentrations of the gas in the range 10^{-3} to 10^{-5} M and sodium carbonate solutions with pH 9.5, at 25°C. Cathodic potentials of -0.20, -0.30, -0.60 and -0.75 V were used, the cathodic current being measured in each case. The results showed clearly that the reaction is first-order with respect to oxygen concentration and therefore that dissociation of oxygen molecules on the surface of the UO_2 does not occur. Consequently the order of the leaching reaction with respect to oxidant concentration shown in Fig. 5.6 cannot be explained purely in terms of the cathodic half-cell reaction. The kinetics of the anodic half-cell reaction must also be taken into account.

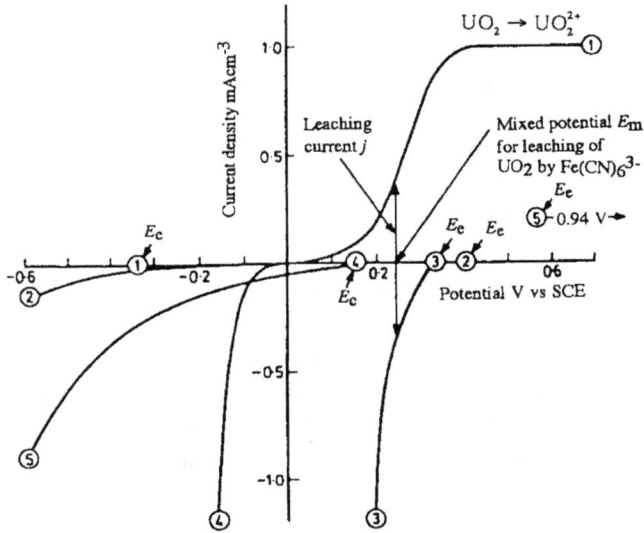

Fig. 5.7.   Current density-potential curves for (1) the anodic oxidation of UO_2 and the cathodic reduction of (2) oxygen; (3) $Fe(CN)_6^{3-}$; (4) $Cu(NH_3)_4^{2+}$ and (5) hydrogen peroxide in carbonate solutions at pH 10 and 25°C.[176]

This required measurement of current density potential curves for (i) the anodic oxidation of UO_2; and the cathodic reduction at a UO_2 surface of (ii) oxygen; (iii) $Fe(CN)_6^{3-}$; (iv) $Cu(NH_3)_4^{2+}$; and (v) hydrogen peroxide. These measurements were made in carbonate solutions at pH 10, and 25°C. The results are shown in Fig. 5.7. While leaching is taking place the UO_2 takes up the potential at which the currents produced by the anodic and cathodic half-cell reactions are equal, because equivalent amounts of oxidant and reducing agent react. The oxidant which reacts with the UO_2 most rapidly is, therefore, the one which produces the most anodic mixed potential during the reaction. Its equilibrium potential E_e is important because it determines the maximum overpotential that can be applied to the UO_2 surface. However the efficiency with which this driving force for the reaction can be utilised depends on the relative positions and shapes of the two half-cell current density potential curves. The fact that the value of E_e for the oxidation of UO_2 in carbonate solutions at pH 10 is -0.34 V versus SCE and that the reaction is practically irreversible means that the efficiency of the reaction will be low. If the reduction of an oxidant at a UO_2 surface is also highly irreversible, as is the case for oxygen, then the rate of leaching of UO_2 by oxidation with O_2 in carbonate solutions is also slow. This is shown clearly in Fig. 5.7 by the positions of the curves for the two reactions. The fact that a reaction is said, in this context, to be highly irreversible means that is very slow. Thus Fig. 5.7 provides information concerning both the thermodynamics and the kinetics of the individual half-cell reactions which are represented. The cathodic reduction of the $Fe(CN)_6^{3-}$ ion at the UO_2 surface is very reversible (i.e. fast) so that although it has approximately the same value of E_e (0.33 V) as does oxygen ($E_e = 0.40$ V versus SCE) the theoretical argument predicts that it should be much more efficient as an oxidant for leaching UO_2 in carbonate solutions, as is found to be the case. This does not mean of course that it would be considered for use in a process.

The rate of dissolution of UO_2 in terms of current was formulated above for the case of iron (III) as oxidant, in Eq. (5.18). In the general case for an oxidant M this becomes

$$j = (2Fk_{UO_2})^{\varphi_1}(n_M Fk_M[M])^{\varphi_2} \tag{5.19}$$

where

$$\varphi_1 = b_{UO_2}/(b_M + b_{UO_2}) \quad \text{and} \quad \varphi_2 = b_M/(b_{UO_2} + b_M)$$

b_{UO_2} and b_M are the Tafel slopes for the anodic oxidation of UO_2 and the cathodic reduction of the oxidant at the UO_2 surface, [M] is the concentration of the oxidant and k is defined as the overall rate constant of the reaction at the potential of zero versus the reference electrode used; in this case a saturated calomel electrode. From Eq. (5.19), it is predicted that the exponent of the oxidant concentration is equal to $b_M/(b_{UO_2} + b_M)$. The values obtained using the experimental data shown in Fig. 5.7, and those predicted from electrochemical measurements of b_M and b_{UO_2}, are given in Table 5.3.[176] Agreement is good except in the case of hydrogen peroxide.

Table 5.3. Comparison of experimental and predicted exponents of oxidant concentration for leaching of UO_2.

Oxidant	b_{UO_2}	b_M	Values of exponents Predicted	Experimental
O_2	0.073	0.142	0.66	0.61
$Fe(CN)_6^{3-}$	0.073	0.081	0.53	0.57
H_2O_2	0.073	0.52	0.88	0.45
$Cu(NH_3)_4^{2+}$	0.073	0.071	0.49	0.50

This lack of agreement is caused by additional chemical reactions which can occur in the H_2O_2–UO_2^{2+} system. Thus at the potentials required for the leaching of UO_2 hydrogen peroxide can be simultaneously oxidised to oxygen and reduced to water at a UO_2 surface

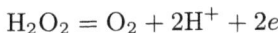

$$H_2O_2 = O_2 + 2H^+ + 2e$$

$$H_2O_2 + 2e = 2OH^-\ .$$

At any peroxide concentration the proportion of the peroxide contributing to the leaching reaction depends on the electrochemical parameters of these two half-cell reactions. (Evolution of oxygen is frequently observed as a side reaction when hydrogen peroxide is used as an oxidant.) Also the rate of anodic oxidation of UO_2 is increased in the presence of H_2O_2. Thus in 0.5 M sodium carbonate solution, pH 11.3, at 25°C, at potential 0.3–0.5 V versus SCE the anodic current is increased from about 420 μA to about 950 μA in the presence of 1.67×10^{-2} M peroxide.[176] This cathodic shift is considered to be probably due to the reactions involving the formation of the peroxy-complexes of the uranyl carbonate ion

$$UO_2(CO_3)_3^{4-} + H_2O_2 = UO_2(CO_3)_2O_2H^{3-} + HCO_3^- \qquad (5.20)$$

$$UO_2(CO_3)_3^{4-} + HO_2^- = UO_2(CO_3)_2O_2H^{3-} + CO_3^{2-}\ . \qquad (5.21)$$

For the equilibrium $HO_2^- + H^+ = H_2O_2$ the logarithm for the association constant lies between -11 and -12 at 25°C.[46] Thus at pH values below pH 10 Reaction (5.20) predominates and above pH 13 Reaction (5.21) does so. Using other thermodynamic data the authors conclude that if the ion HO_2^- is substituted for one of the carbonate ions in the uranyl tricarbonate complex according to Eqs. (5.20) and (5.21), then depending on the pH of the solution a cathodic shift of between 0.065 and 0.114 V would be expected. The observed shift at low currents is approximately 0.096 V. Thus the proposed mechanism is reasonable.

Previous work on the leaching of UO_2 in carbonate solutions using oxygen at elevated temperatures and pressures had shown that the rate increases with increasing carbonate concentrations to a limiting value above which no further increase occurs. Similar trends were observed at atmospheric temperatures and pressure and it was found that the cathodic reduction of oxygen at a UO_2 surface was not affected by the carbonate concentration. It was expected therefore that the observed behaviour is due to the effect of carbonate on the anodic oxidation of UO_2. Leaching of this by oxygen at atmospheric pressures occurs in the range of potential -0.10 V to $+0.05$ V versus SCE and electrochemical measurements were carried out at 0.05 V in equimolar $NaHCO_3$–Na_2CO_3 solutions with total carbonate concentrations 0.1 to 0.5 M, at 25°C. The results were compared with those obtained using chemical leaching with oxygen and good qualitative agreement was found.

Leaching data in the literature for 7 atm oxygen pressure at 93°C agreed similarly with electrochemical measurements made at 0.40 V versus SCE. This potential gave the best quantitative agreement but the pattern of the results overall support the suggestion that the theory developed by the authors for the leaching of UO_2 at ambient temperature and pressure is also applicable to the conditions used for pressure leaching.

The experimental procedures used are also able to provide commercially useful information of other kinds; for example on the effects of additives to leaching solutions. Results are given for the effects of concentrations of cyanide, phosphate, sulphide and dodecylbenzene sulphonic acid from 10^{-4} to 5×10^{-3} M, on the anodic oxidation at 0.4 V of UO_2, and the cathodic reduction at -0.6 V of oxygen at a UO_2 surface in 0.4 M sodium carbonate solution at pH 9.82 and 25°C. Sulphate additions from 0.05 to 0.20 M were also investigated.

The results showed that phosphate, sulphate and sulphide have a negligible effect on the anodic oxidation of UO_2 and the cathodic reduction of oxygen at a UO_2 surface. It was expected therefore that the presence of these substances would not seriously affect the rate of leaching of UO_2. Cyanide does, however, produce a significant decrease in the current due to the cathodic reduction of oxygen. This would be disadvantageous to the use of a leaching process for the simultaneous extraction of gold and uranium which had been the subject of intensive investigation in South Africa at that time.

It had been claimed that the addition of surface active agents has a beneficial effect on the recovery of uranium from its ores by the alkaline carbonate process. The addition of dodecylbenzene sulphonate was found to inhibit both half-cell reactions however. Further semiquantitative tests with other surface active reagents showed no increase in the rate of either half-cell reaction, except for a small increase in the presence of less than 0.2 g L^{-1} cetylpyridinium bromide. Since surface active reagents do not accelerate the chemical reactions any improvement they may make to the rate of leaching uranium ores must be due to changes in the gross physical condition of the surface of the particles.

Leaching of Uranium Dioxide in Sulphuric Acid Solutions with Iron (III)[175]

Equation (5.16) relates to the situation where the partial anodic current due to the anodic oxidation of Fe(II) is negligible compared with that due to the anodic oxidation of UO_2. Therefore, the back reaction involving oxidation of Fe(II) is ignored.

Experimental results for the rate of leaching of a sintered pellet of UO_2 in 0.5 M sulphuric acid at constant concentration of iron (III), against the concentration of iron (II) are shown in Fig. 5.8, as a logarithmic plot. With very low concentrations of Fe(II) the rate of leaching of UO_2 is independent of the concentration of Fe(II) but depends on the concentration of Fe(III) to the power of 0.73. A similar fractional order had been reported previously by other workers for UO_2, uraninite and U_3O_8. Equation (5.18) gives the exponent of the concentration of Fe^{III} as $\phi_2 = b_{Fe(III)}/(b_{UO_2} + b_{Fe(III)})$. For the sample of UO_2 used the relevant values are $b_{UO_2} = 0.049$ V and $b_{Fe(III)} = 0.158$ V, giving

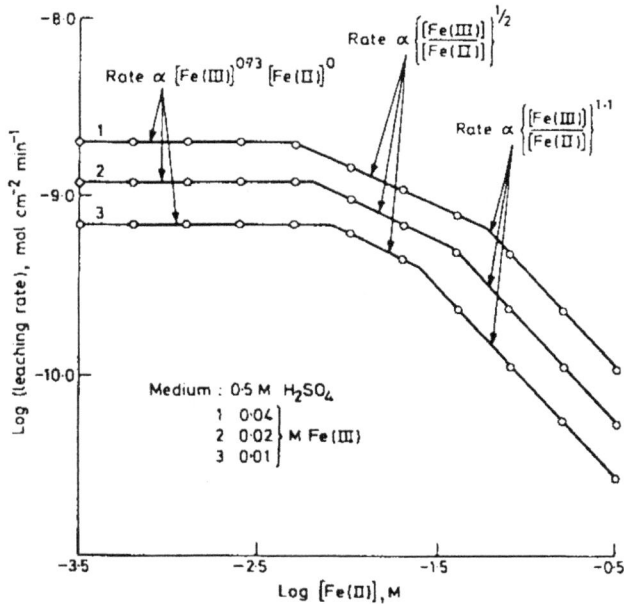

Fig. 5.8.   Effect of concentration of Fe(II) on rate of leaching of UO_2 by Fe(III) in sulphuric acid at 25°C.[175]

an exponent of 0.76. Using a different sample of UO_2 the relevant values are $b_{UO_2} = 0.049$ V and $b_{Fe(III)} = 0.088$ V, giving an exponent of 0.64.

At higher concentrations of iron (II) the rate of leaching of UO_2 decreases as the concentration of Fe(II) in the solution is increased. The anodic oxidation of Fe(II) at a UO_2 surface has been shown to be first-order and zero-order with respect to the concentrations of Fe(II) and Fe(III) respectively. Then the full Butler–Volmer equation for the second half-cell reaction can be written

$$j = F k_{Fe(II)}[\text{Fe(II)}] \exp\left[\frac{2.303E}{b_{Fe(II)}}\right] - F k_{Fe(III)}[\text{Fe(III)}] \exp - \left[\frac{2.303E}{b_{Fe(III)}}\right].$$

(5.22)

Under steady state conditions the net current is zero and the mixed potential can be calculated by combining this equation with that for the anodic oxidation of UO_2 (5.15). There is no exact analytical solution to the resulting equation and as an approximation the authors make the assumption that the Tafel slopes b_{UO_2}, $b_{Fe(III)}$ and $b_{Fe(II)}$ are all equal. The solution for E_M then becomes

$$E_M = \frac{b}{2} \log \left[\frac{k_{Fe(III)}[\text{Fe(III)}]}{2k_{UO_2} + k_{Fe(II)}[\text{Fe(II)}]}\right]$$

where b has replaced b_{UO_2}, etc. If this equation is substituted into (5.15), the final result is obtained

$$j = 2F k_{UO_2} \left[\frac{k_{Fe(III)}[\text{Fe(III)}]}{2k_{UO_2} + k_{Fe(II)}[\text{Fe(II)}]}\right]^{1/2}$$

When $k_{Fe(II)}[\text{Fe(II)}] \gg 2k_{UO_2}$ the rate of leaching of UO_2 should be proportional to the square root of the ratio of the concentrations of Fe(III) and Fe(II) in the leaching solution.

The third situation considered by the authors is that in which the partial anodic current due to the anodic oxidation of UO_2 is small in comparison with that due to the anodic oxidation of Fe(II) under steady state conditions. This appears to be of academic interest only. The final result is

$$j = 2F k_{UO_2} \left[\frac{k_{Fe(III)}[\text{Fe(III)}]}{k_{Fe(II)}[\text{Fe(II)}]}\right]^{\varphi_3}$$

where $\varphi_3 = b_{Fe(II)}b_{Fe(III)}/[b_{Fe(II)} + b_{Fe(III)}]b_{UO_2}$.

A diagramatic interpretation of the effect of increasing the concentration of iron (II) on the leaching rate of UO_2 is given in Fig. 5.9.[175] Under steady

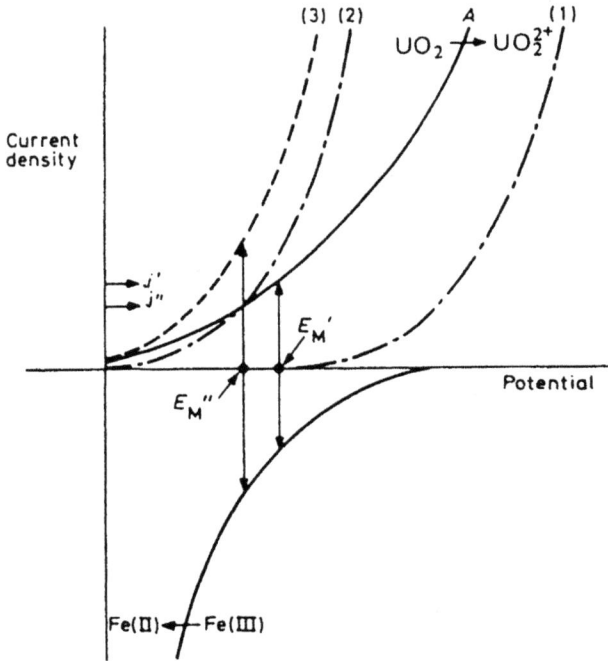

Fig. 5.9.   Effect of increasing the concentration of iron (II) on the leaching rate of UO_2.[175] See text.

state conditions the total anodic and total cathodic currents must balance one another, no matter what the reactions may be which produce them. Curve 1 represents the anodic oxidation of Fe(II) when the iron (II) concentration is low and its effect can be neglected. Curve 2 is for the anodic oxidation of Fe(II) when its concentration is very high. If it is assumed that there is no interaction between Fe(II) and UO_2 the overall anodic current is the algebraic sum of the two partial currents. As the concentration of Fe(III) is unchanged the only way in which this increase in the total anodic current can be compensated for is by a decrease in the potential at which the steady state condition is achieved (Curve 3). Thus the overpotential for the anodic dissolution of UO_2 is decreased and the rate of leaching is lowered. The mathematical analysis given above indicates that the rate of leaching of the solid should depend approximately on the square root of the Fe(III) to Fe(II) ratio, which is the result shown in Fig. 5.8.

At very high concentrations of Fe(II) relative to that of UO_2 in the leach liquid the potential of the UO_2 surface, and therefore the mixed potential, is determined solely by the equilibrium potential of the Fe(III)–Fe(II) couple. Substitution of the relevant values ($b_{Fe(II)} = 0.095$ V) in the expression for φ_3 gives the exponent 1.2, in good agreement with the experimental value, 1.1.

Effect of temperature. The activation energy of each half-cell reaction depends in part on the potential. At a potential of 0.425 V, which is typical of the conditions used for the leaching of UO_2 by acid ferric sulphate solutions, the activation energy for the anodic oxidation of UO_2 is 63.5 kJ mol^{-1} and that for the reduction of Fe(III) at a UO_2 surface is 59.1 kJ mol^{-1}. The activation energy of the chemical leaching reaction is the arithmetic mean of these values, 61.3 kJ mol^{-1}. A typical value measured by experiment is 66.4 kJ mol^{-1}.

In alkaline carbonate solutions activation energies of 55.6 kJ mol^{-1} and between 46 and 63 kJ mol^{-1} have been measured using oxygen as the oxidant, the value depending on the partial pressure of oxygen. From electrochemical data the activation energies at a potential of zero versus SCE for the anodic oxidation of UO_2 and the cathodic reduction of oxygen at a UO_2 surface have been calculated as 59.5 and 57.6 kJ mol^{-1} respectively. The arithmetic mean of these values, 58.5 kJ mol^{-1} agrees with the experimental values.

5.2.4. *Leaching of Silicates*

Treatment of solid silicates with acids can result in (i) no reaction or, (ii) leaching of cations from the solid, leaving a siliceous residue, or (iii) complete destruction of the silicate structure and dissolution of the metals which were incorporated in the structure and also of the silica. The soluble form of silica is usually represented as $Si(OH)_4$ and its apparent solubility in the presence of amorphous solid silica is about 0.1 g SiO_2 L^{-1} at pH values below about 9. This is the form which rapidly reacts under appropriate conditions to give the yellow silicomolybdate ion $[SiMo_{12}O_{40}]^{4-}$ when treated with ammonium molybdate in acidic solution. In the analytical test for silicate the silicomolybdate ion is reduced with stannous chloride or sodium stannite, or with benzidine to give "molybdenum blue" (Phosphates and arsenates also form heteropoly molybdates and interfere with the test). In alkaline solutions the solubility of amorphous silica increases rapidly due to the formation of species such as $SiO_2(OH)_2^{2-}$; $SiO(OH)_3^-$; $Si_4O_6(OH)_6^{2-}$ etc. During reactions between solid silicates and acids, much higher concentrations of silica may pass into solution than 0.1 g L^{-1} SiO_2. Solutions containing more than about 1 g L^{-1} SiO_2 can form a colloidal sol; a silica gel; or a filterable precipitate. If a sol or gel is produced at any stage during a hydrometallurgical process it can make removal of the silica from the system impossible, or at best very difficult. The behaviour of silica in this respect is not fully understood, but it has been summarised and reviewed.[178] Silicate minerals are not widely used as sources of metals unless adequate alternatives are not available, but they are frequently present in low-grade ores which may be treated using a hydromet-allurgical route. In this case it is desirable to know which silicates are liable to behave as "gelatinising" silicate minerals, which react with acidic leach liquors giving extensive dissolution of silica, and why they do so. These matters have also been well reviewed.[179−181]

Solid silicates containing discrete silicate anions, SiO_4^{4-} in orthosilicates; $Si_2O_7^{6-}$ in pyrosilicates; $Si_3O_9^{6-}$, $Si_4O_{12}^{8-}$, $Si_6O_{18}^{12-}$ in the ring silicates, dissolve in acidic solutions without decomposition of the anions and dissolution of the cations and anions cannot take place independently. This is sometimes described as congruent dissolution. Chain, sheet and framework silicates must suffer breakdown of the silicate structure before silica can pass into solution as $Si(OH)_4$ or small polymeric fragments. Subsequent polymerisation of the dissolved silica species can then occur. Thus one major factor controlling the reactivity of silicate minerals when leached in acid solutions is the strength of the chemical bonds in the silicate framework.

It was pointed out in Sec. 1.2.2 that the group AlO_4^{5-} is also tetrahedral and has almost the same size as SiO_4^{4-}, making it possible for AlO_4 to replace some SiO_4 groups in a silicate structure, forming the aluminosilicates. These are generally classified as silicate minerals. The Si–O bond energy is $13\,012$ to $13\,146 \text{ kJ mol}^{-1}$ and the Al–O bond energy $7201–7858 \text{ kJ mol}^{-1}$ so that the strength of a silicate framework is significantly lowered, in thermodynamic terms, as Al replaces Si in the structure. The influence of the metal cations present on the reactivity of silicates is dealt with extensively by Terry.[179] It has been estimated empirically[181] that if the Al to Si ratio exceeds 2 to 3 in framework silicates the silicate structure will be largely broken up on acid attack, and so will probably lead to gelatinisation.

Since the metal–oxygen bond in silicates appears to be very important in controlling their reactions with acids it might be expected that the dissolution of cations from silicates would show similar behaviour to their dissolution from oxides. It would be of interest therefore to compare the fundamental chemical reaction rates of a given metal from an oxide and a silicate. Terry has examined published rate data for the dissolution in acids of MgO, Mg_2SiO_4; ZnO, Zn_2SiO_4 and $ZnFe_2O_4$.[182] Since the specific rates are to be compared the surface area per unit mass of the reacting solids must also be known, and the method of measuring this must be known in each case to ensure comparability of the nature of the surface whose area was measured. The extent to which diffusional resistance, in small pores for example, contributes to the dissolution rate must also be taken into account. Analysis of the data available showed that at 25°C, the rates of chemical dissolution of the metals from oxides were significantly faster than from the silicates. Possible explanations for this were discussed using the current ideas on the mechanisms for the dissolution of metal oxides considered in Sec. 5.2.1.

Chemical Treatment of Silicates in Industrial Processes

Probably the largest tonnage of silicate ore which is treated by a hydrometallurgical process for recovery of a metal is of nickel laterites. Such ores contain silica largely in the form of magnesium silicates which have been precipitated from ground waters during the weathering of rocks; and these are readily decomposed by mineral acids. NIckel occurs in the ore body in two separate strata, in the upper region closest to the surface in the limonite zone, which is primarily goethite, FeOOH, and in the lower depths in the garnierite zone, which is largely hydrated magnesium silicates. The garnierite ore is dried and reduced at high temperature to form nickel and cobalt metal while minimising the reduction of iron. The nickel and cobalt are then leached in ammoniacal ammonium carbonate solution with air as oxidant to form the nickel and cobalt ammines. By this means the two metals are dissolved with minimal dissolution of iron although some magnesium does pass into solution. The silica content of the ore causes no problems. Limonite ores are also being treated by a hydrometallurgical process in which the wet ore is digested with sulphuric acid at 240–260°C. At this temperature the iron does not dissolve; it forms goethite, hematite and jarosite. Again, after such rigorous treatment the silica present causes no difficulties during solid–liquid separation.

Zircon, $ZrSiO_4$, is an example where the metal is part of the silicate structure. The most technically advanced process used to recover the metal values is that in which the mineral is dissolved and the solution is treated by solvent extraction to separate the zirconium from the hafnium which is always present with it in zircon and baddeleyite (ZrO_2). A brief outline of the procedures which have been proposed for dissolving the metals present in the silicate is given here.

The silicate structure can be destroyed or damaged by pretreatment before attacking the mineral with acid. In the extreme case the SiO_2 can be driven off by heating a mixture of zircon and carbon in air in an electric arc furnace, forming a carbide–nitride of the metal. Zircon can be sintered or melted with CaO, $CaSO_4$ or $CaCl_2$ to make it more amenable to acid attack, although a simpler method would appear to be to heat it at 1000°C and quench it in water before acid treatment. Sulphuric and hydrochloric acids have been used. The presence of sufficient fluoride in an acid liquor will dissolve the SiO_2 as silicofluoride or, under suitable conditions, drive off SiF_4.

Generally, treatment of zircon with an alkali is the preferred route as long as the presence of some alkali in ZrO_2 produced without further treatment steps can be tolerated. It can be advantageous when the oxide is to be used in the glass, enamel or porcelain industries. In an early procedure finely ground zircon was added to molten sodium hydroxide in the ratio 1 to 6. The rate of addition had to be controlled because of the violent frothing which is described as being "very troublesome, filling the surrounding atmosphere with caustic spray which is very obnoxious". Because of this, reaction with soda ash (anhydrous Na_2CO_3) is sometimes preferred as an industrial process in spite of the higher temperature necessary and the relative slowness of the reaction. The product of the alkali fusion contains most of the zirconium as sodium zirconate and most of the silicon as sodium silicate. After cooling and crushing the solid is leached with water. Sodium zirconate tends to hydrolyse and unless the wash liquor contains sufficient alkali a gelatinous hydroxide forms which is difficult to filter.

In the period around 1960, there was considerable interest in the use of alkalis to decompose silicate minerals containing metals of actual or potential interest to the nuclear industry, including zircon and beryl, $Be_3Al_2Si_6O_{18}$. The published information was largely concerned with the fact that treatment under certain conditions would decompose the mineral, and information concerning the effects of temperature and alkali concentration on leaching rates was lacking. In order to establish the general behaviour of SiO_2 on leaching in alkali solutions at elevated temperatures work was carried out on finely ground quartz which had passed through a 200-mesh sieve (minus 76 μm) and been subjected to a standard decantation treatment to remove particles smaller than about 10 μm.[183] The quartz was washed with 2 M HCl to remove all iron from the surface. In 1 M NaOH at 150°C the rate of dissolution of SiO_2 was first-order with respect to the surface area of quartz (weight of quartz put into the autoclave), the rate of appearance of SiO_2 in the solution being constant while the first 30% of the quartz dissolved. After this the rate decreased in the usual way. This behaviour was observed under conditions where the time required to dissolve this proportion of the quartz varied from less than 1 hour to more than 120 hours. The simple explanation proposed was that even if the crystal faces dissolve uniformly, remaining smooth, the surface area decreases much less rapidly than the mass of the solid. In fact however many of the faces were etched and pitted by the reaction, increasing the specific surface area of the solid.

The apparent activation energy for the reaction is in 1 M NaOH, 150–170°C, 73 kJ mol^{-1} in 7.5 M NaOH, 125–160°C, 94.5 kJ mol^{-1}.

The data for 1 M NaOH at 150°C, 162°C and 170°C gave a good straight line on plotting log rate against $1/T$ but the point for 100°C was well off the line. The rate result for 100°C was reproducible using the autoclave, necessary for the higher temperatures, or a stirred reactor under a reflux condenser at atmospheric pressure. Thus it appears that there is a change in reaction mechanism between 100°C and 150°C in 1 M NaOH. In 7.5 M NaOH the rate for 100°C lies not so far off the line joining the points for 125°C, 150°C and 160°C. It seems from this and from data obtained using other silicate minerals that when the rate of dissolution is very slow because of low temperature or low

Table 5.4.

Size analysis of quartz

Size μm	+60	−60 + 50	−50 + 40	−40 + 30	−30 + 20	−20 + 10
Weight %	17.5	23.2	17.8	16.1	14.3	11.1

Typical experimental data, 40 g, quartz, 100°C 1 M NaOH

Duration of run, hours	0.25	0.50	0.75	1.0	1.5	2.0	3.5
SiO_2 in solution g L^{-1}	8.14	9.85	12.42	19.27	23.55	24.62	27.00

Rates of reaction of quartz

NaOH M	1.0	1.0	1.0	1.0	2.0	7.5	7.5	7.5	7.5
T°C	100	100	162	170	150	100	125	150	160
$10^6 k$	0.012	0.67	1.16	1.73	0.767	0.0495	0.644	4.65	9.78

reagent concentration the rate is much slower than would be predicted by extrapolation from conditions where normal behaviour for the system is observed. There appears to be some sort of lower limit below which reactivity is depressed. In Table 5.4, data are given for the rates of dissolution, 10^6 k, in g moles SiO_2 sec^{-1} g^{-1} of quartz, calculated from data for the first 30% of the reaction. The size analysis for the quartz used and typical experimental data are also included.

Zircon

As was stated above, fusion of zircon with NaOH or Na_2CO_3 forms sodium zirconate or sodium silicate which dissolve when the cooled product is leached with water. When leached with concentrated NaOH solutions at high temperatures, however, zircon produces a solution containing sodium silicate and a small amount of zirconium.[183] The rest of the zirconium, equivalent to the silicate in solution, appears as a new, unidentified, crystalline phase. In the early stages of the reaction this deposits on parts of the surface of the zircon particles, reducing the surface area of the reacting mineral and making estimation of specific reaction rates impossible. However, some measured rates are given in Table 5.5, 10^8 k, g moles SiO_2 sec^{-1} g^{-1} of zircon.

The rate at 190°C in 10 M alkali is very low and the fact that log k does not lie on the line joining the other four points on plotting log k against $1/T$ is probably due to experimental error in the analysis. Below 175°C, decomposition of zircon is barely detectable; in 7.5 M NaOH the rate is extremely low below about 200°C.

Table 5.5.

Size analysis of zircon

Size μm	> 211	−211 + 152	−152 + 124	−124 + 104	−104 + 89	−89 + 76	−76 + 53	−53
Weight %	0	0.20	8.58	7.22	35.76	17.33	19.18	11.73

Rates of reaction of zircon

NaOH M	10	10	10	10	10	7.5	7.5	7.5
T°C	190	230	250	270	300	230	270	300
10^8 k	0.16	1.32	4.5	10.2	40	1.05	7.08	20.9
E_A	7.5 M NaOH			230–300°C			104 kJ mol^{-1}	
E_A	10 M NaOH			230–300°C			117 kJ mol^{-1}	

Beryl

Beryl ($3BeO \cdot Al_2O_3 \cdot 6SiO_2$ in the representation as distinct oxides) was found to dissolve rapidly in 5 M NaOH at 250°C, but as the concentration of silicate and aluminate in the solution increased reprecipitation of much of the beryllium occurred as an apparently amorphous solid.[183] When only 10 g of beryl was digested with 1 litre of 5 M NaOH for 3 hours at 250°C and the liquid was discharged rapidly through a cooled condenser tube a clear colourless solution was obtained and no solid was left in the autoclave. On standing, a solid was deposited from the solution. It contained almost all the beryllium, some of the aluminium and a little of the silica which had been present in the beryl. The beryllium and aluminium content of the solid readily dissolved in dilute HCl.

When the experiment was repeated using 100 g of beryl a slurry of suspended solid was discharged from the autoclave at 250°C. Similar behaviour was found after leaching at 265°C and 300°C. A number of the solid products were washed with water and dried to constant weight at 65°C. A typical analysis was: loss on ignition, 0.61; BeO, 9.8; Al_2O_3, 12.0; SiO_2, 46.3%; residue occluded alkali. The composition of the beryl was: BeO, 12.7; Al_2O_3 17.9; SiO_2, 65.2%. As long as sufficient mineral was present in the autoclave to produce a precipitate at the temperature of leaching the solid product from the autoclave contained BeO and SiO_2 in approximately the same ratio as in the mineral, but less Al_2O_3. X-ray powder photographs from the solid products discharged showed either no lines or weak lines of beryl, attributed to unreacted mineral.

5.3. Leaching of Sulphides

5.3.1. *Thermodynamics of the Metastable $S–H_2O$ System*

The sulphur–water E-pH diagram for 298.15 K shown in Fig. 3.2 includes all of the relevant substances which are thermodynamically stable in some area within the stability limits for water. It was pointed out that a number of other sulphur-containing species exist but do not appear in the diagram because they are metastable. The relationships between them can be shown however if sulphuric acid and the ions derived from it are excluded from the calculations. All of the known species which would appear in such a diagram have recently been included in the preparation of a metastable predominance area diagram[184] using the MTDATA program.[185,186] The equations given have

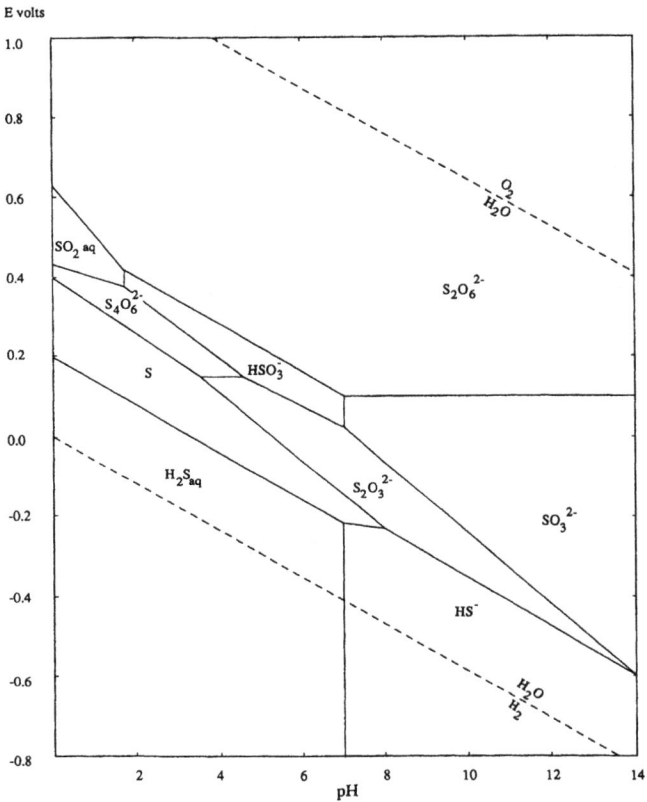

Fig. 5.10.   Equilibrium diagram for the metastable S–H$_2$O system, 298.15 K.

been used to draw Fig. 5.10. The following conclusions were reached by Kelsall and Thompson.

(i) Peroxymonosulphuric and peroxydisulphuric acids and their anions lie above the region of stability of water.

(ii) If all species containing S^{VI} are excluded from consideration, in the ranges pH 0 to 14, $E = +1$ to -1 volt the following species appear: H$_2$S; HS$^-$; S; SO$_2$(aq), see below; HSO$_3^-$, bisulphite; SO$_3^{2-}$, sulphite; S$_2$O$_3^{2-}$, thiosulphate; S$_2$O$_6^{2-}$, dithionate; S$_4$O$_6^{2-}$ tetrathionate.

(iii) Dithionite ion, S$_2$O$_4^{2-}$, has no area of stability even in the metastable diagram; it would lie in the region of stability of H$_2$S and HS$^-$.

(iv) If all species containing sulphur in an oxidation state greater than zero are excluded then polysulphide species S_n^{2-} with $n = 2, 3, 4$ or 5 have narrow areas of predominance at high pH values.

The thermodynamic data used are given in Table 5.6. Sulphur dioxide is soluble in water; the solution possesses acidic properties and has generally been described as containing sulphurous acid, H_2SO_3. However this substance is either not present in such solutions or is present only in extremely small amounts. Thus in Table 5.6 the ΔG° value is given for $SO_2(aq)$ and this equals the sum of the values of SO_2 gas and for liquid water. The only polythionate species considered is $S_4O_6^{2-}$ because ΔG° values for others are not included in the set of data used. The equations for the lines drawn in Fig. 5.10 are given in Table 5.7.

Table 5.6. Thermodynamic data used to draw Fig. 5.10, metastable diagram for the S–H_2O system.

Species	$H_2S(aq)$	HS^-	$SO_2(aq)$	HSO_3^-	SO_3^{2-}	$S_2O_3^{2-}$	$S_2O_4^{2-}$
ΔG° kJ	-27.87	12.05	-300.708	-527.81	-486.2	-518.8	-600.4

$S_2O_6^{2-}$		$S_4O_6^{2-}$		H_2O
-966		-1022.2		-237.178

Data source: S. Zhdanov. In: Standard Potentials in Aqueous Solution. Ed. A.J. Bard, R. Parsons and J. Jordan. Dekker, New York (1985) pp. 93–111.

Table 5.7. Equations for the lines in the metastable predominance area diagram for the S–H_2O system, 298 K.

$$H_2S(aq) = H^+ + HS^-$$
$$\log K_1 = -6.99 = \log\{HS^-\} - \log\{H_2S(aq)\} - pH \tag{1}$$

$$SO_2(g) = SO_2(aq)$$
$$\log\{SO_2(aq)\} = 0.09 + \log p_{SO_2}(g) \tag{2}$$

$$HSO_3^- = H^+ + SO_3^{2-}$$
$$\log k_2 = -7.22 = \log\{SO_3^{2-}\} - \log\{HSO_3^-\} - pH \tag{3}$$

$$4SO_2(aq) + 4H^+ + 6e = S_4O_6^{2-} + 2H_2O$$
$$E = 0.5074 - 0.0394\ pH + 0.0394\log\{SO_2(aq)\} - 0.0098\log\{S_4O_6^{2-}\} \tag{4}$$

$$S_4O_6^{2-} + 12H^+ + 10e = 4S + 6H_2O$$
$$E = 0.4155 - 0.071\ pH + 0.0059\log\{S_4O_6^{2-}\} \tag{5}$$

Table 5.7. (*Continued*)

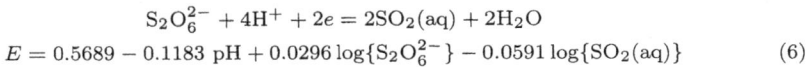

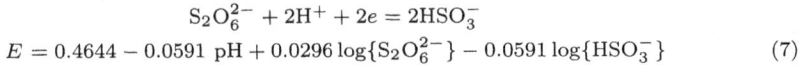

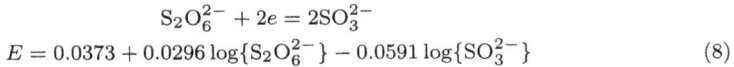

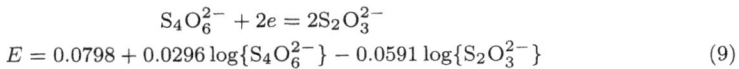

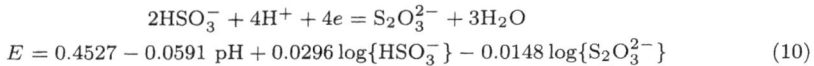

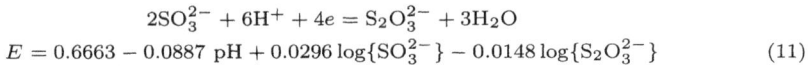

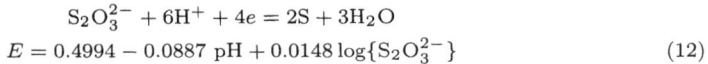

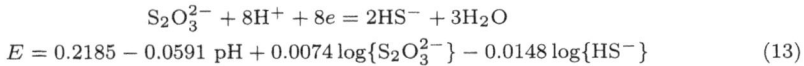

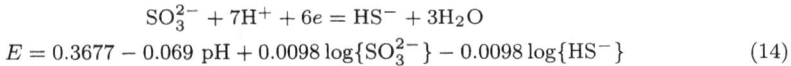

$$S_2O_6^{2-} + 4H^+ + 2e = 2SO_2(aq) + 2H_2O$$

$$E = 0.5689 - 0.1183 \text{ pH} + 0.0296 \log\{S_2O_6^{2-}\} - 0.0591 \log\{SO_2(aq)\} \tag{6}$$

$$S_2O_6^{2-} + 2H^+ + 2e = 2HSO_3^-$$

$$E = 0.4644 - 0.0591 \text{ pH} + 0.0296 \log\{S_2O_6^{2-}\} - 0.0591 \log\{HSO_3^-\} \tag{7}$$

$$S_2O_6^{2-} + 2e = 2SO_3^{2-}$$

$$E = 0.0373 + 0.0296 \log\{S_2O_6^{2-}\} - 0.0591 \log\{SO_3^{2-}\} \tag{8}$$

$$S_4O_6^{2-} + 2e = 2S_2O_3^{2-}$$

$$E = 0.0798 + 0.0296 \log\{S_4O_6^{2-}\} - 0.0591 \log\{S_2O_3^{2-}\} \tag{9}$$

$$2HSO_3^- + 4H^+ + 4e = S_2O_3^{2-} + 3H_2O$$

$$E = 0.4527 - 0.0591 \text{ pH} + 0.0296 \log\{HSO_3^-\} - 0.0148 \log\{S_2O_3^{2-}\} \tag{10}$$

$$2SO_3^{2-} + 6H^+ + 4e = S_2O_3^{2-} + 3H_2O$$

$$E = 0.6663 - 0.0887 \text{ pH} + 0.0296 \log\{SO_3^{2-}\} - 0.0148 \log\{S_2O_3^{2-}\} \tag{11}$$

$$S_2O_3^{2-} + 6H^+ + 4e = 2S + 3H_2O$$

$$E = 0.4994 - 0.0887 \text{ pH} + 0.0148 \log\{S_2O_3^{2-}\} \tag{12}$$

$$S_2O_3^{2-} + 8H^+ + 8e = 2HS^- + 3H_2O$$

$$E = 0.2185 - 0.0591 \text{ pH} + 0.0074 \log\{S_2O_3^{2-}\} - 0.0148 \log\{HS^-\} \tag{13}$$

$$SO_3^{2-} + 7H^+ + 6e = HS^- + 3H_2O$$

$$E = 0.3677 - 0.069 \text{ pH} + 0.0098 \log\{SO_3^{2-}\} - 0.0098 \log\{HS^-\} \tag{14}$$

5.3.2. *Leaching of Iron-Containing Sulphides in Alkaline Solutions*

When sulphide minerals are leached under acidic oxidising conditions the products of the oxidation of the S^{2-} ions present in the solid are elemental sulphur or sulphate ions. However studies of the surfaces of some sulphide minerals which have been kept wet in the air at slightly alkaline pH values have indicated the presence of thiosulphate groups. Leaching under alkaline oxidising conditions can result in the formation of metastable oxyacids of sulphur. Leaching of pentlandite concentrates, $(\text{Fe, Co, Ni})_9\text{S}_8$, in aqueous ammonia at temperatures between about $80°C$ and $95°C$ with air as oxidant is used to dissolve the nickel and cobalt content in the Sherritt–Gordon process. Elemental sulphur is not formed under the alkaline conditions and only part of the sulphur present in the leach liquor is in the form of sulphate. In the early stages of the reaction thiosulphate is produced at the same time as sulphate. Later, polythionate is formed together with increasing amounts of sulphate and sulphamate $(H_2NSO_3^-)$, the latter formed by the oxidation of sulphur species in the presence of ammonia. In batch leaching experiments which are allowed to continue long enough the thiosulphate and polythionate disappear, leaving only sulphate and sulphamate as the products of the oxidation of the sulphur present in the pentlandite. Under industrial conditions leaching is carried out as a continuous process and this does not occur. Sulphite is not produced by the leaching reactions.

The rate of oxidation of thiosulphate to thionates, sulphate and sulphamate is increased by the presence in the solution of copper (II) which acts as an oxidation catalyst. Chalcopyrite is present in most pentlandite concentrates and is oxidised under the leaching conditions used, the copper forming soluble complexes with the ammonia present. The rate of oxidation of $S_2O_3^{2-}$ and $S_xO_6^{2-}$ is also increased greatly if the temperature is raised from about 50°C to the boiling point, about 104°C. At the start of a batch leaching experiment, or as fresh concentrate slurry is introduced into the continuous autoclave leaching system, the rate of reaction of the mineral is so rapid that availability of oxygen for the oxidation of S^{2-} to SO_4^{2-} is restricted. This leads to the formation of the thiosulphate, Fig. 5.10. The $S_2O_3^{2-}$ ion has a structure which is similar to that of SO_4^{2-}, with one oxygen replaced by sulphur. Thus it has a S–S bond which can act as the beginning of the sulphur chains present in the polythionates. Oxidation of cold thiosulphate solutions by hydrogen peroxide is a standard method of preparing polythionates, and when pure ammoniacal ammonium thiosulphate solution is oxidised in the laboratory under oxygen pressure a mixture of sulphate and polythionate is formed.

$$2(NH_4)_2S_2O_3 + 2O_2 = 2(NH_4)_2S_3O_6 + (NH_4)_2SO_4 .$$

Thus the sequence of reactions which occur during the leaching of pentlandite in ammoniacal solutions is consistent with thermodynamic predictions in that sulphide forms thiosulphate in alkaline conditions if oxidation through to sulphate cannot occur. Subsequent oxidation of the thiosulphate follows the usual chemical route.

Pyrite present in the pentlandite concentrates does not react under the oxidising ammoniacal conditions used in the leaching step and any nickel and cobalt present in the pyrite are not dissolved. The pentlandite, chalcopyrite and pyrrhotite react to produce soluble salts of nickel, cobalt and copper in such a way that the iron present in each mineral particle is converted to hydrated iron oxide *in situ*, with the result that when leaching is complete the particles consist of hydrated iron oxide pseudomorphic with the original mineral[168] as was stated in Sec. 5.2.2 when contrasting this behaviour with the formation of protective layers of product (on wolframite).

The lack of reactivity of pyrite in the ammoniacal system is in marked contrast with its behaviour when leached in dilute sodium hydroxide or carbonate using oxygen as oxidising agent. The sulphide forms sulphate directly with no intermediate formation of the lower oxyacids and the iron remains as a coating which has been reported[187] to be largely amorphous, although X-ray powder photographs showed lines due to ferric oxide and magnetite. The report of the possible formation of the latter substance led to a further study of the reaction.[167,188] Two kinds of pyrite were used, a relatively impure highly disseminated ore (Rio Tinto) made up of small grains of pyrite and other minerals, cemented together, and well formed crystals of high purity. The rate curves obtained using ground, sized samples in 0.5 M sodium hydroxide with oxygen partial pressure 10 to 120 psi (6.8–82 kPa) at temperatures between 70°C and 120°C showed some unusual features. With the pure pyrite smooth curves were obtained under all conditions, reactions proceeding to completion. The disseminated pyrite behaved in the same way at 120°C and higher temperatures, but at 60°C the reaction slowed down and then almost stopped after a time that dependent on a number of factors. At 100°C, the reaction rate decreased steadily and then, after some time, rose again.

Partly leached samples of the two kinds of pyrite were quite different in appearance. The disseminated material had a coating of product whereas the particles of pure pyrite looked clean, although dull compared to the original surface. A cubic crystal of the pure pyrite, of side about 1 cm, was selected as having areas of one face that were optically flat. It was treated with 0.5 M sodium hydroxide solution in a gently stirred autoclave under 1 atmosphere air pressure at 90°C for one hour and was then washed and dried in air. The flat areas were examined by electron diffraction and it was reported that they had a thin coating of maghemite, γ-Fe_2O_3, which was positively distinguished from magnetite, and there was evidence of a preferred orientation. Another crystal of pyrite was treated with alkali under the same conditions for five minutes and was then examined by electron diffraction. Some maghemite was found in patches but other unidentified phases were present on the surface. Neither these nor maghemite were observed on pyrite crystals which had not been treated with alkali. It was concluded that the reaction proceeded via some unidentified crystalline phases, formed as very thin films, to give maghemite as the end product. It has been shown recently[189] that at pH 9.2, when a pyrite electrode is oxidised the voltammograms suggest that the initial product on the surface is an iron-deficient layer in the pyrite lattice, represented by $Fe_{1-x}S_2$, plus sulphate. The alternative possibility that the initial oxidation product involves the formation of elemental sulphur was not supported by X-ray photoelectric spectroscopy measurements.

X-ray powder photographs obtained from small particles of pure pyrite that had been slightly leached with alkali showed no lines other than those caused by pyrite. Similarly treated disseminated pyrite showed lines consistent with the presence of γ-Fe_2O_3. From the powder photographs it was not possible to distinguish between magnetite and maghemite, but since the latter was reported to have been proved conclusively by electron diffraction to be formed on a pure pyrite crystal it was assumed to be present on the pyrite grains cemented together in the leached ground ore.

Partly leached particles of pure pyrite were mounted and polished to provide sections through them. No outer coating of solid reaction product was seen except at a few positions where a crack penetrating into the particle was filled with a solid that extended just around the outer face. A few regularly shaped holes in the crystals were also found to contain a solid shown to be rich in iron but free from sulphur. These voids were formed by preferential attack along defects in the pyrite crystal other than cracks caused during grinding.

Partly leached particles of the disseminated pyrite, which had been mounted and very carefully polished so as to minimise the pulling away of soft material, were quite different in appearance. Attack had occurred preferentially along the boundaries between the individual grains of pyrite making up the particles, and a large amount of reaction product was held in these spaces, which were so close together that a thick layer of the product was bound all around the particles. When the pyrite had a well defined shape the solid produced during leaching could be seen to have the same shape. This is characterisitic behaviour when a solid state reaction occurs. It can be clearly seen in the photomicrograph shown as Fig. 4(a) of Ref. 167, which is incorrectly labelled.

Two kinds of solid could be seen in addition to the residual pyrite, macroscopic pieces of a brown material which could accept a polished surface, and a finely divided porous material having the same colour, which looked as though the massive product had shattered. In no case was the latter in contact with residual pyrite, but it seems certain that the massive solid must have been produced by replacement of sulphide in a pyrite crystal by oxide ions, the thin crystalline film observed by electron diffraction steadily increasing in thickness. The massive product was found by electron microprobe analysis to contain $70 \pm 2\%$ Fe and no detectable sulphur. This is consistent with maghemite, which contains 69.9% Fe, and was almost certainly present. However Fe_3O_4 contains 72.3% Fe.

The X-ray diffraction pattern of maghemite is so similar in most respects to that of magnetite that its formation as a product of leaching processes may have sometimes been overlooked, but it certainly is not often produced. It is of interest therefore to try to explain why it is formed from pyrite under very mild conditions of leaching. If the S_2^{2-} groups in pyrite are replaced by O^{2-} ions occupying the same positions in the lattice the product is FeO, wustite, the change in volume being about 40%. Wustite is unstable at low temperatures and under oxidising conditions changes to Fe_3O_4 and then to γ-Fe_2O_3 by a series of topotactic reactions, Sec. 1.2.3, Fig. 1.3. It seems likely that this is the course that the reaction takes when the pyrite is leached relatively slowly.

The change in lattice dimensions which accompanies the change from sulphide to oxide causes the latter to break away from a perfect pyrite crystal when the layer is very thin, except when reaction has occurred along faults or grain boundaries. In such cases it is retained in the solid and builds up in thickness. The disseminated pyrite holds a great deal of oxide in position by the same mechanisms and the oxide builds up in thickness before the strain causes shattering at the interface. The extent of this build-up depends on the conditions of leaching.

The irregularities in the rate curves of the disseminated pyrite are readily explained on this model. At low temperatures, around 60°C, nonporous massive layers of oxide form and protect the pyrite from further attack. At intermediate temperatures the layers of maghemite build up and then shatter so that diffusion occurs through the pores and reaction becomes more rapid again. At 120°C and higher temperatures the initial reaction is so fast that no massive protective layer of oxide is formed.

Any diffusion gradient between pyrite and oxide would be largely within the shattered solid, possibly extending very slightly into the pyrite. Electron microprobe analyses[167.] traversing across the boundaries between pyrite and oxide product indicated that the composition changes from Fe, 52; S, 48.5; to

Fe, 54; S, 12.5 percent over a distance of approximately 1.5 μm. Counts at a point 1 μm back into the residual pyrite gave Fe, 49; S, 50.5 percent and at a point 1 μm further into the oxide layer Fe, 62; S, 2.5 percent. There was no indication of discontinuity between the phases.

When pyrrhotite was leached in sodium hydroxide or carbonate solutions using oxygen as oxidant under a wide range of conditions thiosulphate was produced together with its oxidation products, thionates and sulphate. The solid which formed was a porous hydrated iron oxide. No evidence was found suggesting that pyrrhotite could react with alkali solutions to form pieces of solid product similar to those formed on pyrite. Pyrhhotite has the NiAs crystal structure and it is not possible to form an oxide of iron from it by replacement of sulphide by oxide ions.

5.3.3. *Oxidative Leaching of Chalcocite in Acidic Solutions*

The main result of a study[111] of the kinetics of the reaction between synthetic Cu_2S and iron (III) chloride solutions were given in Sec. 4.2.4 where the system was used to illustrate the experimental procedures which can be used with particulate solids. Two distinct stages of the reaction were observed in the lower range of the temperatures used, Fig. 4.5. In the first, part of the copper content of the chalcocite dissolved giving covellite, CuS, as the only solid product at the end of the first leaching stage. During the second stage the CuS was decomposed with the formation of elemental orthorhombic sulphur, the copper passing into solution. The mechanism of the change of crystalline Cu_2S to crystalline CuS particles of similar size and shape is clearly of interest.

During the period around 1950, the possibility was investigated of electrorefining nickel and copper, using anodes of their sulphides which could be cast from mattes. This was achieved in the case of nickel, using its most nickel-rich sulphide Ni_3S_2, Sec. 3.7.2. Anodes of Cu_2S disintegrate during electrolysis however, making use of this compound impracticable. Because of the potential saving of energy compared with electrorefining metal anodes, a considerable amount of work has been carried out on the electroleaching of chalcocite in the hope of finding a way of overcoming this difficulty. The major matter of interest is the nature of the changes which occur in the solid as copper is progressively removed from it. It would be expected that these are similar to those which occur during the leaching process using a chemical instead of an electrochemical oxidation reaction.

The phase diagram of the copper–sulphur system has been studied by a number of workers and the solid state phase equilibria in the Cu_2S–CuS region have been determined by X-ray diffraction methods.[190-196]. The phases stable at atmospheric temperature and pressure, and their ranges of stoichiometry have been stated to be as follows:[195] chalcocite, $Cu_{1.993-2.000}S$; djurleite, $Cu_{1.934-1.965}S$; covellite, $Cu_{1.000}S$; and anilite, $Cu_{1.75}S$. However digenite, with composition around $Cu_{1.8}S$, has been accepted as a stable phase in the system since the first studies were carried out. In addition copper-rich covellite has been found as a natural mineral and given the name blaubleibender covellite. Two other minerals have more recently been discovered in the same area of copper mineralisation and given the name spionkopite, $Cu_{1.4}S$, and yarrowite, $Cu_{1.125}S$. Geerite, $Cu_{1.60}S$ was discovered as a mineral which occurs in epitactic replacement of sphalerite.

Chalcocite. The low temperature form, chalcocite-III, $Cu_{1.993-2.000}S$, was originally referred to a pseudo-orthorhombic cell[190] but was later[197] shown to have true symmetry not higher than monoclinic. The proposed structure has 48 Cu_2S in the unit cell with hexagonal close packed sulphur atoms. All copper atoms are 3-coordinated, mostly in or near positions of planar coordination in the array of the sulphur atoms. At 104°C, this chalcocite-III is transformed to chalcocite-II, with hexagonal symmetry, and at 470°C to a cubic high temperature form with cubic close packed sulphur atoms.

Djurleite. Djurle[190] reported the composition of djurleite to be $Cu_{1.96}S$ and stated that there were three forms: a low symmetry form now called djurleite-III; a tetragonal form, probably metastable, djurleite-II; and a cubic high temperature phase formed from this at 100°C. The extremely complex structure of djurleite-III is not known but it has 128 sulphur atoms in the unit cell.[198] In the metastable tetragonal form the sulphur atoms are in almost perfect cubic close packing.

Covellite. According to Roseboom[191] there is no deviation of stoichiometry from $Cu_{1.000}S$ and the solid is stable at temperature up to 507°C, where it transforms to high digenite (see below) and a sulphur-rich liquid. Covellite has a unique structure.[199] One third of the copper atoms have three sulphur atoms as closest neighbours with Cu–S distance 2.19 Å, at the corners of a triangle. The remainder have four sulphur atoms arranged tetrahedrally (Cu–S 2.32 Å). Two thirds of the sulphur atoms are also bound together as S_2 groups like those in pyrite. Represented in terms of conventional valencies the formula is $Cu_4^I Cu_2^{II}(S_2)_2 S_2$, but see also the general conclusions at the end of this discussion of the structures of the individual minerals.

Digenite. At room temperature the composition range of digenite is $Cu_{1.765-1.79}S$. The low copper limit is not affected by temperature, but the copper-rich limit changes to $Cu_{1.83}S$ at 83°C, at which temperature it transforms to high digenite. Digenite of composition $Cu_{1.8}S$ is unstable below 50°C. Digenite is now not regarded by some[195] as a phase in the pure copper–sulphur system having been claimed to be metastable[196] in accordance with the statement that it occurs in nature only when stabilised by the presence of an appropriate amount of iron in the lattice.[192] It would be very unusual to find a copper sulphide mineral in an iron free environment, and oxidative leaching of copper sulphide deposits is usually caused in nature and in extractive processes by the action of Fe^{III} as the oxidant. Digenite is an important phase when studying the leaching of chalcocite because its formation represents the major change in the crystal structure of the solid which occurs during the process, from hcp to ccp sulphur. The sulphur atoms in digenite are in cubic close packing with the copper atoms situated along the diagonals of the cube. The subcell dimension is $a = 5.555$ Å at composition $Cu_{1.765}S$ and $a = 5.571$ Å at composition $Cu_{1.79}S^{200}$ showing the expansion of the lattice structure as more copper atoms are packed into it.

Blaubleibender covellite. $Cu_{1.05-1.10}S$ or $Cu_{1.05-1.4}S$. This is the accepted name for naturally occurring copper-rich covellite present in the oxidation zone of ore deposits where it occurs as an intermediate phase in the alteration of copper– and copper–iron sulphides to normal covellite. Roseboom[191] did not observe a solid in the composition range given for blaubleibender covellite in his work on the phase diagram, and considered that it remained to be proved whether such material is more stable (thermodynamically stable) than a mixture of covellite and sulphur-rich digenite. Moh[201,202] considered the material to be best represented as $Cu_{1+x}S$ where x is 0.05–0.10, and reported that above 157°C it is transformed to a mixture of normal covellite and digenite. Frenzel[203,204] found that blaubleibender covellite formed by leaching chalcocite could have a copper excess as high as $Cu_{1.4}S$.

Two methods are commonly said to be used for the synthesis of "blue remaining covellite". (i) Acidic oxidising leaching of copper– or copper–iron sulphides in aqueous solutions. (ii) sulphidising digenite or chalcocite using reducing conditions, under aqueous or organic solvents. The solid is removed from the solution when the required composition has been reached. The success of these methods would indicate that if dissolution and reprecipitation of new solid particles does not occur, which seems to be the case, and if progressive alteration in the composition of the product occurs, then either copper ions can be removed from solid copper– or copper–iron sulphides, or sulphide ions can be added to solid digenite or chalcocite, forming the same material. The mechanism by which this can be achieved from a copper–iron sulphide is obscure.

Anilite. $Cu_{1.75}S$. This mineral was first described in 1969.[205] Its crystal structure is based on a cubic closed packed array of sulphur atoms with the copper atoms arranged in an ordered fashion in the interstices.[206] It is an unstable solid which forms low digenite when ground and doubt was expressed as to its stability by Cook[193,194] and other workers later. It was synthesised to provide pure samples for determination of its X-ray powder diffraction pattern by heating copper wire (99.975%) and sulphur (99.999%) at 450°C in a sealed evacuated silica tube for a sufficiently long time to reach equilibrium (24–120 hours) and quenching the solid in cold water.[195]

Electrochemical measurements[196] indicate that anilite is a thermodynamically stable phase with a composition range $Cu_{1.750\pm0.003}S$ and with a possible solid solution field within these limits. Difficulties were encountered in determining the solid solution range owing to poor nucleation of anilite. The solid breaks down to form covellite plus high digenite at 75 ± 3°C. It is worth noting that the copper used to synthesise the anilite was specified as containing not more than 100 ppm iron and it was accepted by Potter[195,196] that low digenite is metastable in the absence of iron. High digenite is formed when low digenite is heated to a temperature which depends on the composition of the digenite but lies between 76°C and 83°C. The reflections due to the superstructure in low digenite disappear and the remaining reflections can all be indexed by use of a cubic cell of side about 5.56 Å. The inversion is readily reversed and high digenite cannot be quenched to room temperature. It has not been observed as a mineral. Thus interconversion of the phases anilite and low digenite must be considered as being possible when the solids contain only very low concentrations of iron.

Geerite. $Cu_{1.6}S$. This was found occurring as an epitactic replacement of sphalerite.[207] In consequence its structure is based on the ZnS lattice on which the crystals grew. It has cubic close packed sulphur atoms and lattice dimensions close to those of the host sphalerite, but its structure has not been determined. The relationship between the formation of a geerite-type structure on the surface of sphalerite particles due to activating them with copper sulphate before froth flotation has been investigated.[208]

Spionkopite. $Cu_{1.393}S$ or $Cu_{1.40}S$; and *yarrowite*; $Cu_{1.125}S$. These minerals were found in the deposits where blaubleibender covellite was originally discovered and they occur as replacement of other sulphides, being formed by selective replacement of copper.[209] Their crystal structures are not yet known but their well developed subcells resemble the unit cells of covellite. They are believed therefore to possess a combination of hexagonal close packed and covalently bonded sulphur atoms.

General conclusions. The physical properties of the copper sulphides, for example their metallic lustre; intrinsic semiconductivity; the ductility of some of them; the close proximity of copper atoms in the lattices of some of the solids; and the variety of configurations of the bonds from the copper atoms, indicate that these are semimetallic phases. The triangular and tetrahedral coordination of the copper atoms in CuS indicate that all copper is present as Cu(I). The solid is diamagnetic, is a metallic conductor of electricity and becomes superconductive below 1.62 K. Thus covellite is not Cu(II) sulphide; the conventional valency formula given above in the description of the mineral is not applicable and a more appropriate theoretical model must be used in discussions involving its electronic structure.

Leaching Studies. The two distinct stages of the leaching of chalcocite in acidic solutions of iron (III) sulphate reported by Sullivan[112] were also observed in acidic sulphate solutions between 100°C and 200°C when oxygen under pressure was used as the oxidant.[211] The apparent activation energies were 27.6 and 7.5 kJ mol^{-1} for stages 1 and 2, the low value for the latter being attributed to coating of the mineral particles by liquid sulphur formed by the reaction.

The rotating disc technique was used with synthetic minerals to study the kinetics of leaching covellite,[212] and of digenite and chalcocite[213] in acidic iron (III) sulphate solutions. Partially leached minerals were examined and it was concluded that chalcocite formed digenite, followed by blaubleibender covellite and stoichiometric covellite. At temperatures between 25°C and 60°C almost all of the chalcocite or digenite had altered before any sulphur appeared, but at higher temperatures both stages occurred simultaneously after reaction had been taking place for a short time. E_A for the reaction during the first stage was approximately 25 kJ mol^{-1}. When chalcocite was leached in dilute sulphuric acid with oxygen in the temperature range 29.6°C to 67°C the final reaction product in each experiment was covellite.[214] Thus in this system (oxygen as oxidant) stage 2 does not occur below 67°C but it does occur between 100°C and 200°C.[211]

Electroleaching of chalcocite was reported to occur with formation of covellite as an intermediate step,[215] and a surface layer of this was said to form almost immediately after starting electrolysis using a chalcocite anode.[216] Sections through electroleached anodes of white metal (a copper matte containing about 77% Cu) showed that a dark blue product had formed and it was concluded[217] that digenite was an intermediate product as well as covellite. In 1969, the results were reported[218] of an investigation in which partially electroleached chalcocite particles were studied using X-ray powder diffraction methods. It was concluded that the reaction proceeded to blaubleibender covellite via the intermediate phases djurleite and digenite. Under certain conditions a polymorph of digenite, tentatively referred to as Cu_7S_4 was obtained as an intermediate product. Anodic treatment of digenite gave blaubleibender covellite by way of several unidentified solids.

Also in 1969, it was reported in a review[167] that X-ray powder diffraction measurements on partially leached particulate synthetic chalcocite indicated the presence of phases with compositions far outside the normally accepted ranges of nonstoichiometry of chalcocite, djurleite, digenite, blaubleibender covellite, and covellite, the copper sulphides known to exist at that time. The material studied had been produced during the kinetic investigation outlined in Sec. 4.2.6 and had overall compositions in the range $Cu_{1.823}S$ and $Cu_{0.974}S$. It had been concluded[219] that during the first stage of the leaching of chalcocite it was transformed rapidly to CuS via the intermediate phase digenite, the copper being removed from the nonstoichiometric copper sulphides having the structures of the Cu_2S and $Cu_{1.8}S$ phases by diffusion of copper in the lattices. X-ray powder diffraction photographs of partially leached chalcocite particles indicated that either the Cu_2S structure is changed progressively to the $Cu_{1.8}S$ structure or else there is an easy transformation from the chalcocite to the digenite structure (hcp to ccp sulphur) when the limit of nonstoichiometry in the chalcocite phase structure is reached. This occurs when the overall composition of the particles of leached residue is about $Cu_{1.85}S$. The nonstoichiometric digenite phase structure loses copper by diffusion until the composition reaches about $Cu_{1.5}S$ when a further structural change occurs. Eventually the continuing loss of copper by diffusion through the lattice and dissolution from the surface of the solid produces stoichiometric CuS. This does not have the lattice structure of the naturally occurring mineral covellite.

Because of the unexpected nature of these results in 1966, additional experimental work was carried out.[111] The data can now be reinterpreted using data published on the natural minerals in the Cu–S system described since then. This appears to be the most structurally complex system on which leaching studies have been carried out and it provides good examples of the ways in which X-ray diffraction data can be used.

Examination of partially leached chalcocite by optical microscopy and electron microprobe. Approximately cubic pieces of fully dense synthetic chalcocite of side 0.7 to 1 cm were cut from a block and one face of each was ground flat and lightly polished. Each block was leached in acidic iron (III) chloride solution under one of a selected set of concentrations and temperatures, below 80°C. The partially leached blocks were cut perpendicular to the previously polished face and the newly cut faces were polished and examined to observe the leached solid extending inwards from the flat surface which had been in contact with the solution. The behaviour reported previously[213] was confirmed. The solid close to the leached surface was very porous, the sulphur present always being in pores and holes of the residual mineral, never forming a continuous layer. Residual covellite in this outer region of the block had a fibrous appearance. The pores of the unleached surface were in preferred crystallographic directions and the orientation of the individual crystal grains determined the angle of inclination of the pores to the cube surface. The porosity decreased through the CuS phase and the solid between this and the unleached Cu_2S phase in the centre of the cube, which was compact.

Proceeding from the edge towards the centre of a cube which had been leached for 20 hours at 40°C, the residual mineral showed the complete anisotropic colour sequence of covellite, crystals of different orientation, produced because of twinning in the chalcocite, appearing in contrasting colours. The colours due to covellite gradually disappeared on passing towards the un-attacked chalcocite. A sharp line of demarkation separated this from a blue phase which was the first indication of a solid which had been formed by re-moval of copper from the chalcocite. The blue phase became darker as more copper was removed and is probably the $Cu_{1.8}S$ phase identified by X-ray diffraction. Within a field of view of the blue phase a wide range of depth of blue was apparent, indicating different rates of removal of copper by diffusion along different crystallographic planes.

Electron microprobe examination. Point counts across a large flat surface of the massive synthetic chalcocite showed exact uniformity of compo-sition. A rotating disc of the fully dense chalcocite, 2.5 cm in diameter and approximately 1.5 cm in length was mounted in araldite resin and the exposed face was ground flat. It was leached for 20 hours at 80°C using a shaft speed of 1500 rpm. The disc was cut perpendicular to the leached surface and the face was polished. Calibration curves for electron microprobe analysis were constructed using spectrographically pure copper rod, and the Cu_2S prepared from the pure elements for the 79.86% Cu point.

Copper and sulphur scans approximately 200 μm apart were taken from the leached surface, perpendicular to it. The copper line showed a gradual decrease in copper content of the solid from 80% in unleached solid to approximately 65 wt % Cu over a distance of approximately 3.5 mm, corresponding to the change from Cu_2S to CuS. The electron beam width was 1 μm and there was no evidence of the existence of two phases of different composition being present in close proximity. Within the solid of composition Cu_2S, that is in unaltered chalcocite, there were occasional sharp falls and rises in the copper counts and specimen current at grain boundaries and defects. These were much less frequent in the material of composition between Cu_2S and CuS, and when they were present in this they were much less pronounced. The portion of the scan indicating removal of copper from the solid containing the CuS phase was rough and uneven due to the poor polishing properties of the leach residue which was porous and contained elemental sulphur. However there were indications that the copper content of solid fell from about 65 to about 57 wt % over a distance of about 0.1 mm, remained constant over a distance of about 0.15 mm after which the beam was responding to the outer surface of the specimen. The sulphur scan followed the same pattern as that for copper but showed many more marked fluctuations in counting rate.

Copper scans were taken along the long axes and across the breadths of lamellae of the porous structure close to the leached surface of the disc. A scanning distance across the breadth of the parallel lamellae of 0.6 mm showed eleven minima indicating a copper content of about 57 wt % Cu and eleven maxima at about 68 wt %. Scans along the length of lamellae close to the surface of the disc and perpendicular to the surface showed a steady fall from about 67 to about 62 wt % Cu as the surface was approached. This indicates that copper diffuses (i) from the solid copper sulphide into the pores of the structure and subsequently through the pores into the bulk of the leach solution and (ii) along the length of the lamellae of the porous structure towards the disc surface.

No detectable amounts of iron were found in the solid structure. Traces were detected in some pores and grain boundaries, no doubt originating from leach solution not washed out.

X-ray powder diffraction measurements.   Synthetic fully dense chalcocite prepared from spectrographic quality copper and sulphur, of particle size 150–300 μm, was leached in $FeCl_3$–HCl solutions under conditions and for times to produce residual particles of required compositions, based on the earlier kinetic studies. The overall composition of each batch of residue was (i) calculated from the quantity of copper present in the leach solution when the solid was removed from it and (ii) by chemical analysis of a sample by dissolution in HNO_3 + Br_2 and use of the atomic absorption method, checked in some cases by gravimetric analysis using quinaldinic acid. Agreement was satisfactory.

The X-ray powder diffraction measurements of King[219] were made using a Debye–Scherrer camera with either CuK_α or CoK_α radiation, intensities of lines being measured with a microdensitometer. In the second set of measurements a Guinier camera was used with CoK_α radiation and relative line intensities were estimated visually. The X-ray reflections gave narrow lines with no sign of line broadening or diffuseness. Values of d were measured for all lines observed in the patterns obtained with the partially leached chalcocite particles and all of the values with d below 3.5 Å are given in Table 5.8, taken from Refs. 111 and 219. Values for the chalcocite used were measured with the Guinier camera and are included in the table so that changes occurring during the very early stages of leaching are reliably indicated. For purposes of discussion here, published d-spacings and relative line intensities of relevant

Table 5.8. Diffraction data for partially leached samples of synthetic chalcocite.

Cu2S	used	Cu1.966	S	Cu1.949	S	Cu1.891	S	Cu1.854	S	Cu1.823	S	Cu1.800	S	Cu1.793	S	Cu1.776	S
d Å	I	d Å	I	d Å	I	d Å	I	d Å	I	d Å	I	d Å	I	d Å	I	d Å	I
3.733*	w	3.401	w	3.417	w	3.407	m	3.473	2.5	3.390	5	3.396	5	3.204	s	3.223	20
3.596	w	3.363*	ew	3.379*	ew	3.376*	vw	3.396	5	3.223	20	3.217	20	2.831	ew	3.045	5
3.314	vw	3.314	vw	3.333*	vw	3.321*	vw	3.217	5	3.031	10	3.040	10	2.778	s	2.782	20
3.273	w	3.273	m	3.294*	m	3.296*	w	3.031	5	2.882	5	2.892	5	2.383	ew	2.396	15
3.185	w	3.188	m	3.214	m	3.219	m	2.906	2.5	2.844	5	2.835	5	2.152	ew	2.164	5
3.154*	w	3.161*	m	3.171*	m	3.208*	w	2.838	2.5	2.791	20	2.782	20	1.963	vs	1.966	100
3.054	w	3.112*	vw	3.126*	vw	3.174*	w	2.795	2.5	2.396	20	2.396	20	1.872	ew	1.874	15
2.954	m	2.950	m	3.057	m	3.117*	w	2.709	5	1.966	100	1.962	100	1.765	ew	1.691	5
2.931*	vw	2.933*	ew	3.008*	ew	3.094*	ew	2.673	5	1.872	20	1.872	20			1.678	15
2.885	ew	2.899	ew	2.976*	ew	2.976*	ew	2.573	5	1.695	5	1.693	5			1.138	5
2.869*	ew	2.869*	vw	2.966	vw	2.957	w	2.531	5	1.681	15	1.679	15			1.072	10
2.823	ew	2.838*	ew	2.912*	ew	2.908	w	2.483	5	1.138	5	1.138	5				
2.762*	ew	2.820	ew	2.882	w	2.878	w	2.403	40	1.071	10	1.071	10				
2.726	m	2.770*	ew	2.854*	ew	2.850	w	1.966	100								
2.665	w	2.727	m	2.836*	w	2.820	w	1.874	70								
2.644*	ew	2.692*	ew	2.802	ew	2.793	vw	1.696	15								
		2.665	m	2.778*	ew	2.745*	ew	1.682	15								
2.617*	w	2.623*	w	2.741*	w	2.715	vw	1.649	10								
2.558	ew	2.601*	ew	2.701	vw	2.697*	vw	1.518	5								
2.527	m	2.558	vw	2.692*	vw	2.666	m	1.277	10								
2.473	m	2.527	m	2.672	m	2.629*	vw	1.075	5								
2.400	vs	2.475	m	2.631*	m	2.605*	vw										
2.330	m	2.398	vs	2.612*	vs	2.569	vw										
2.306*	ew	2.380*	ew	2.570	ew	2.524	w										
2.264*	ew	2.328	w	2.535	w	2.481	w										
2.240*	w	2.298*	vw	2.484	vw	2.439*	m										
2.208	m	2.267*	ew	2.446*	ew	2.420*	ew										
2.181*	ew	2.240*	w	2.422*	w	2.395	s										
2.119*	w	2.209	m	2.401	m	2.388*	m										
		2.182*	ew	2.388*	ew	2.362*	ew										

Table 5.8. (Continued)

Cu_2S d Å	used I	$Cu_{1.966}$ d Å	S I	$Cu_{1.949}$ d Å	S I	$Cu_{1.891}$ d Å	S I	$Cu_{1.854}$ d Å	S I	$Cu_{1.823}$ d Å	S I	$Cu_{1.800}$ d Å	S I	$Cu_{1.793}$ d Å	S I	$Cu_{1.776}$ d Å	S I
2.026*	ew	2.119*	vw	2.336	w	2.324	vw										
2.010*	w	2.074*	ew	2.307*	w	2.293*	vw										
1.973	vs	2.008*	ew	2.276*	ew	2.271*	ew										
1.950*	vw	1.972	s	2.248*	ew	2.240*	ew										
1.910*	vw	1.959	m	2.214*	m	2.204*	ew										
1.878	vs	1.910*	vw	2.191	vw	2.192	ew										
1.797*	vw	1.895*	ew	2.173*	vw	2.169*	ew										
1.782*	vw	1.881*	vs	2.159*	vw	2.110*	ew										
		1.872	vs	2.120*	vw	2.074*	w										
		1.796*	ew	2.070*	ew	2.025*	ew										
		1.785*	ew	2.015*	ew	1.988*	ew										
				1.992*	ew	1.966	vs										

$Cu_{1.728}$ d Å	used I	$Cu_{1.586}$ d Å	S I	$Cu_{1.532}$ d Å	S I	$Cu_{1.460}$ d Å	S I	$Cu_{1.306}$ d Å	S I	$Cu_{1.176}$ d Å	S I	$Cu_{1.042}$ d Å	S I	$Cu_{0.974}$ d Å	S I	$Cu_{0.790}$ d Å	S I
3.217	35	3.214	30	3.223	20	3.217	20	3.091	40	3.287	2.5	3.281	5	3.287	10	3.867	5
3.045	5	3.087	5	3.091	5	3.091	30	2.896	30	3.240	2.5	3.240	5	3.246	10	3.287	10
2.778	35	3.051	5	3.002	5	2.906	5	2.766	30	3.082	50	3.135	5	3.077	50	3.240	10
2.169	5	2.778	30	2.774	30	2.774	35	1.960	30	2.874	40	3.077	50	2.865	40	3.118	5
1.962	100	2.174	5	2.174	5	2.174	2.5	1.907	100	2.757	20	2.869	40	2.757	30	3.077	40
1.673	25	1.962	100	1.960	100	1.962	100	1.809	15	1.902	100	2.753	20	1.897	100	2.860	30
1.387	10	1.921	20	1.921	40	1.915	60	1.590	20	1.803	10	1.900	100	1.801	10	2.757	20
1.132	15	1.812	2.5	1.812	5	1.811	5			1.731	5	1.801	10	1.735	5	1.896	100
1.069	10	1.672	20	1.673	30	1.672	5			1.611	5	1.732	10	1.610	5	1.798	10
		1.500	2.5	1.600	5	1.601	5			1.585	30	1.614	10	1.581	20	1.732	5
		1.386	2.5	1.386	5					1.232	5	1.584	30	1.228	5	1.611	5
		1.316	2.5	1.316	5					1.098	5	1.228	5	1.095	5	1.578	20
		1.131	10	1.132	5							1.096	5			1.229	5
		1.068	5													1.094	4

Table 5.9. Published diffraction data for some copper sulphide minerals.

Djurleite-III[220]		Digenite[221]		Anilite[222]		Geerite[223]		Spionkopite[224]		Yarrowite[225]		Covellite[226]	
$Cu_{1.93}$ d Å	S I	$Cu_{1.8}$ d Å	S I	$Cu_{1.75}$ d Å	S I	$Cu_{1.60}$ d Å	S I	$Cu_{1.40}$ d Å	S I	$Cu_{1.125}$ d Å	S I	Cu d Å	S I
3.39	35	3.35	4	3.36	20	3.128	100	3.408	8	3.215	8	3.285	14
3.35	14	3.21	35	3.32	17	2.712	10	3.278	15	3.061	55	3.220	30
3.28	14	3.01	14	3.20	57	1.918	50	3.223	5	2.767	35	3.048	65
3.19	16	2.779	45	2.77	65	1.870	10	3.076	85	2.292	12	2.813	100
3.11	6	2.553	4	2.75	6	1.683	10	2.964	8	2.180	8	2.724	55
3.04	18	2.141	10	2.69	14	1.637	30	2.926	5	2.120	8	2.317	10
3.02	16	1.967	100	2.59	29	1.576	10	2.777	30	2.075	8	2.097	6
2.892	16	1.814	2	2.54	31	1.247	10	2.592	5	1.961	5	2.043	8
2.875	14	1.752	6	2.39	10	1.109	20	2.483	5	1.899	100	1.902	25
2.836	16	1.678	20	2.16	39			2.386	20	1.827	5	1.896	75
2.819	16	1.607	4	2.13	15			2.297	25	1.783	8	1.735	35
2.785	6	1.392	10	2.05	5			2.185	8	1.731	15	1.634	4
2.710	6			1.956	100			2.123	8	1.687	8	1.609	8
2.693	12			1.873	10			2.060	8	1.670	8	1.572	16
2.656	16			1.847	3			1.965	12	1.639	5	1.556	35
2.605	8			1.677	35			1.910	100	1.613	12	1.463	6
2.561	14							1.820	30	1.586	12	1.390	6
2.519	16							1.738	12	1.572	15	1.354	8
2.474	12							1.684	8	1.534	5	1.343	6
2.389	85							1.622	20	1.504	5	1.280	10
2.288	6							1.573	12	1.472	8	1.227	6
2.144	6							1.473	15	1.441	5	1.210	10
2.067	8							1.424	8	1.421	5	1.0998	8
2.047	10							1.406	8	1.385	5	1.0946	10
1.966	40							1.386	8	1.360	8	1.0607	10
1.959	95							1.361	8	1.314	5	1.0155	8
1.925	8							1.321	12	1.279	5		
1.870	100							1.241	12	1.232	8		
								1.151	5	1.210	5		
								1.104	12	1.163	5		
								1.089	12	1.097	12		
								1.065	8	1.073	5		
								1.032	8	1.063	5		
										1.028	5		
										1.022	5		

copper sulphide phases described above are given in Table 5.9. Where alternative sets of data are available, the one containing the number of *d*-values most nearly corresponding to that obtainable with the cameras used with the leached chalcocite samples has been selected. This avoids the inclusion of large numbers of lines of low intensity such as those given in Ref. 195, which are not generally relevant to the discussion.

Published X-ray powder diffraction data have been compiled in a data file under the auspices of the Joint Committe on Chemical Analysis by Powder Diffraction Methods (JCPDS) now superseded by the International Centre for Diffraction Data, Swarthmore, Pa., USA. The data are published in sets, the set number being followed by the number of the file card. This contains the reference to the sources of the diffraction data and other information relating to the interpretation of it. *d*-spacings, relative intensities and the assignments of the lines are given.

Discussion of the X-ray diffraction data.   The purpose of the investigation was to determine the phase changes which occur in the chalcocite particles during the leaching process. Such data can also be used to study in some detail the alterations in lattice dimensions and orientations which take place.

d-spacings of leached samples and of copper sulphide minerals given in Tables 5.8 and 5.9 are plotted in Fig. 5.11. For reasons of clarity the diffraction data for djurleite and some values for samples and other minerals are omitted. These are indicated in the tables by an asterisk. It is clear from the figure and from the number of *d*-spacing values in the list for each solid in the table that a very substantial alteration occurs in the lattice type as the composition of the solid approaches $Cu_{1.80}S$ from the copper-rich side. The large number of reflections resulting from the complex hexagonal close packed sulphur structures of chalcocite and the djurleite phases is replaced by the much smaller number from the simple cubic close packed sulphur structure of digenite. Analysis of the phase changes can therefore be carried out separately for solids within the two composition ranges $Cu_{2.0}S$–$Cu_{1.8}S$ and $Cu_{1.8}S$–$Cu_{1.0}S$.

$Cu_{2.0}S$–$Cu_{1.80}S$.   The mineral phases stable at room temperature within this range of composition are chalcocite-III; djurleite-III; the tetragonal form djurleite-II (probably metastable); and the digenite solid solution which may be thermodynamically metastable at such temperatures but is formed nevertheless by oxidative leaching of chalcocite or djurleite.

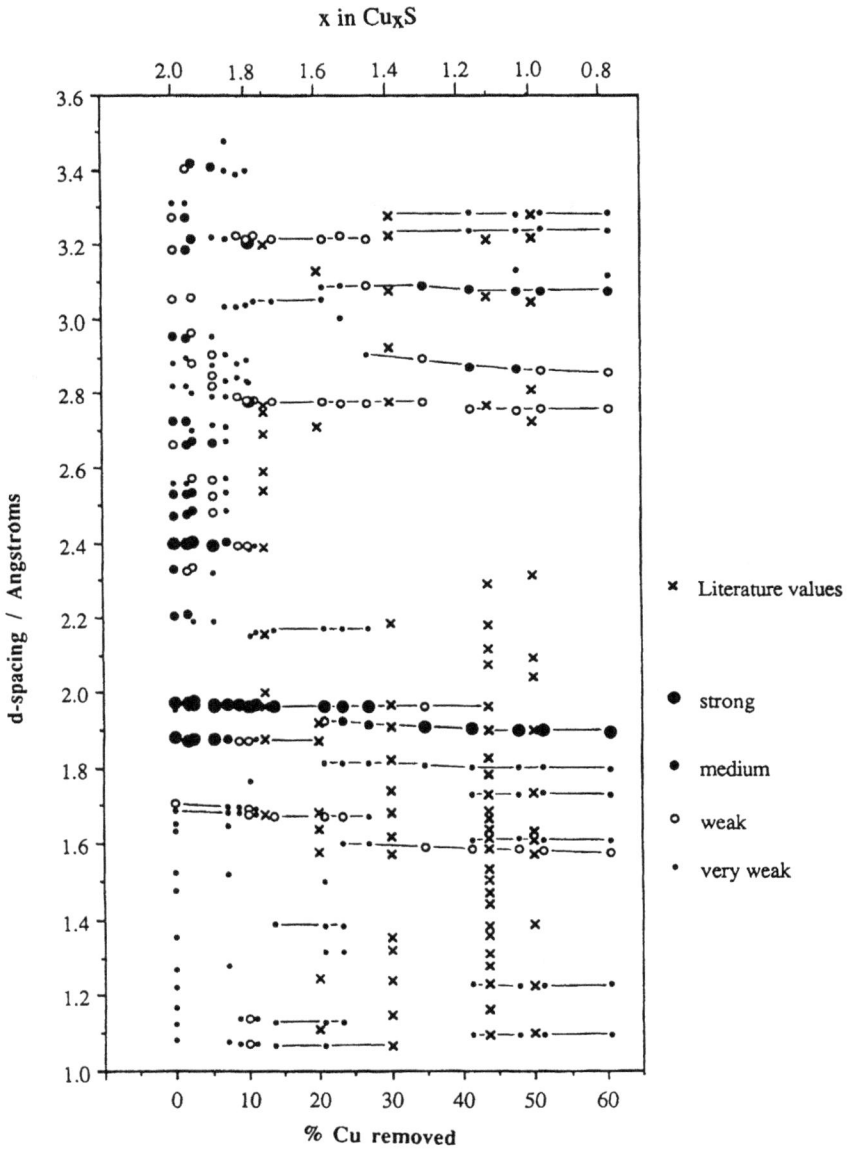

Fig. 5.11.   X-ray diffraction data for chalcocite and products obtained from it during leaching in iron (III) chloride solutions.

When using X-ray diffraction data as an analytical tool to identify the phase or phases present in a solid sample the three strongest lines are generally used for each phase with the Powder Diffraction File Search Manual.[227] The procedure is, in principle, made easier in the present case because what is to be established is the presence or absence of each of several known phases in the partially leached chalcocite samples. The four strongest lines of each of the five phases mentioned above are given in Table 5.10 together with the d-spacings of lines having similar values in the seven samples of composition $Cu_{1.966}S$ to $Cu_{1.793}S$.

This data contains a number of features which are very unusual in the analysis of X-ray diffraction data in general. Some of these can be seen most clearly in Fig. 5.11 used with Table 5.8. For example the strongest chalcocite line, $d = 1.973$ Å decreases slightly to $d = 1.966$ Å in $Cu_{1.891}S$ and remains the strongest line in all samples to $Cu_{1.460}S$. It is present in the diffraction pattern of djurleite-III ($I = 40$); the tetragonal phase djurleite-II with $d = 1.967$ Å, $I = 30$; digenite; anilite, the strongest line of which has $d = 1.962$ Å[195] not 1.956 Å as given in Table 5.9; spionkopite with I, reduced to 12; and yarrowite with $I = 5$. It is not reported in the pattern for geerite but is present in the leached sample of overall composition $Cu_{1.586}S$ with $d = 1.962$ Å, $I = 100$.

Chalcocite has another line at $d = 1.950$ Å which becomes a line of medium intensity, $d = 1.959$ Å in $Cu_{1.966}S$ and a very strong line in $Cu_{1.949}S$ with $d = 1.965$ Å, clearly resolved from the spacing $d = 1.975$ Å in the same sample. The two very strong lines remain distinct in $Cu_{1.891}S$ but in the next sample, $Cu_{1.854}S$, only a single line is present with $d = 1.966$ Å, the strongest line. The spacing of the line with $d = 1.968$ Å is stated by Goble to be approximately equal to the radius of the sulphur atoms in the cubic close packed structure of digenite.[228] This is consistent with its presence in djurleite-II and as the sole line approximating to that value in the sample of overall composition $Cu_{1.854}S$. Presumably the two lines originating from $d = 1.973$ and 1.950 Å in chalcocite, which are present in the more copper-rich materials relate to the sulphur spacings in the hexagonal close packed structure.

The chalcocite line with $d = 1.973$ Å can obviously not be used to indicate the presence of the Cu_2S phase in any of the samples of the partially leached mineral. The same is true of the line for $d = 1.878$ Å which persists in the samples to $Cu_{1.776}S$, with decreasing intensity and is present, at slightly lower d-values, in the patterns for anilite and geerite. It has not been reported in the diffraction pattern for digenite. The chalcocite line with $d = 2.400$ Å

Table 5.10. Correlation between strongest *d*-spacings of minerals and of partially leached chalcocite.

| Mineral | line | x in Cu_xS | 1.949 | 1.891 | 1.854 | 1.823 | 1.800 | 1.793 |
	d Å	1.966 *d* Å	*d* Å	*d* Å	*d* Å	*d* Å	*d* Å	*d* Å
Chalc.	1.973 vs	1.972 s	1.975 vs	1.966 vs	1.966 (100)	1.966 (100)	1.962 (100)	1.963 vs
	1.878 vs	+1.881 1.872 vs 1.872 vs	1.877 vs	1.873 vs	1.874 (70)	1.872 (20)	1.872 (20)	1.872 vw
	2.400 vs	2.398 vs	2.388 m	2.395 s 2.388 m	2.403 (40)	2.396 (20)	2.396 (20)	2.383 ew
	2.473 m	2.475 m	2.484 m	2.481 w	2.483 (5)			—
Djurl.	1.870 (100)	+1.881 1.872 vs	1.877 vs	1.873 vs	1.874 (70)	1.872 (20)	1.872 (20)	1.872 ew
	1.959 (95)	1.959 m	1.965 vs	1.966 vs	1.966 (100)	1.966 (100)	1.962 (100)	1.963 vs
	2.389 (85)	2.380 ew	2.388 m	2.388 m	2.403 (40)	2.396 (20)	2.396 (20)	2.383 ew
	1.966 (40)	1.972 s	1.965 vs	1.966 vs	1.966 (100)	1.966 (100)	1.962 (100)	1.963 (100)
Tetrag.	2.74 (100)	—	2.741 w	2.745 ew				
	2.302 (80)	—	2.307 w					
	1.883 (35)	+1.881 1.872 vs	1.885 w	—	—	—	—	
	1.994 (40)	—	1.992 ew	1.988 ew				
Dig.	1.967 (100)	1.973 s	1.965 vs	1.966 vs	1.966 (100)	1.966 (100)	1.962 (100)	1.963 vs
	2.779 (45)						2.782 (20)	2.778 s
	3.21 (35)	—	3.214 m	3.219 ew	3.217 (5)	3.223 (20)	3.217 (20)	3.204 s
Dig.	3.042 (24)*				3.031 (5)	3.031 (10)	3.040 (10)	—

*Data from D. Smith *et al.* JCPDS Powder Diffraction File 26-476.
+ This line forms a broad line with *d* = 1.872 Å.

persists in the samples to $Cu_{1.776}S$ and is present in the diffraction pattern of anilite. The line $d = 2.473$ Å of chalcocite cannot be distinguished here from $d = 2.474$ Å of djurleite-III but can be confused with no other mineral phase. As shown in Table 5.10, this line is present at low intensity is $Cu_{1.854}S$ but was not detected in $Cu_{1.823}S$.

Two lines, with $d = 2.942$ Å and $d = 2.207$ Å, were stated by Cook[193,220] to be diagnositc for chalcocite over djurleite. These lines were given by Potter[195] as $d = 2.9516$ Å and $d = 2.2097$ Å and in the material used to prepare the partially leached samples were measured as $d = 2.954$ Å and 2.208 Å. They were present as extremely weak lines in $Cu_{1.891}S$ and not detected in $Cu_{1.854}S$. It can be seen in Fig. 5.11 that several other lines present in chalcocite are shown by $Cu_{1.891}S$ but not by $Cu_{1.854}S$, while others persist in the latter material. It appears, therefore, that the chalcocite-III phase has almost if not entirely disappeared from the partially leached chalcocite residue by the time the overall composition has reached $Cu_{1.854}S$.

Cook[193] stated that the $d = 3.39$ Å line of djurleite-III is diagnostic for that phase over chalcocite. The value is given by Potter[195] as $d = 3.385$ Å, $I = 3$, the closest lines of chalcocite having $d = 3.373$ and 3.558 Å. Two lines appear in the leached sample of composition $Cu_{1.966}S$ which were not present in the material from which it was produced, their d-values being 3.401 Å and 3.363 Å. The latter is not included in Fig. 5.11 and it is not clear whether this reflection persists in some form in samples from which more copper has been removed. The weak line with $d = 3.401$ Å is presumably $d = 3.407$ Å of medium intensity in $Cu_{1.891}S$ which is not present in $Cu_{1.854}S$. This line could be attributed to djurleite-III.

Of the strong lines of djurleite-III given in Table 5.10 those with $d = 1.870$, 1.959 and 1.966 Å persist in leached samples having copper contents too small to permit the lines to be accepted as indicating the presence of that phase. That with $d = 2.389$ Å is not present in the pattern of the chalcocite used, it increases to medium intensity in $Cu_{1.949}S$ and $Cu_{1.891}S$ and is not present in the pattern of $Cu_{1.854}S$. The possibility of confusion with the $d = 2.400$ Å very strong line of chalcocite is ruled out by the clear resolution of the lines with $d = 2.401$ Å and 2.388 Å in the sample of overall composition $Cu_{1.949}S$.

Of the four strongest lines of the tetragonal djurleite-II phase the three with $d = 2.74$, 2.302 and 1.944 Å appear in $Cu_{1.949}S$ and in two cases in $Cu_{1.891}S$. It is concluded that the evidence for the formation of both djurleite-III and djurleite-II in the early stages of leaching of chalcocite is strong. As the composition of the solid approaches $Cu_{1.8}S$ it can be seen from Fig. 5.11 how the cubic digenite structure appears.

$Cu_{1.80}S-Cu_{0.79}S$.   The partially leached chalcocite samples with compositions in this range all show far fewer reflections than do spionkopite, yarrowite or covellite. In the case of geerite there is little correlation between the published d-spacings for that phase and the values observed with the sample of overall composition $Cu_{1.586}S$ or any sample. This is to be expected since geerite occurs as an epitactic replacement on sphalerite so that it has a cubic structure and its cell dimensions are related to those of the ZnS lattice. However, almost all of the reflections observed from the partially leached chalcocite samples are present in the patterns of one or more of the other three minerals of similar composition. With very few exceptions the d-spacings of the leached samples either have constant values or change slowly and regularly as copper is progressively removed from the solid, and some lines of the three minerals fit into such a sequence, or lie very close to one (Fig. 5.11). In a few cases a lattice spacing changes as copper is removed after remaining constant over a significant range of composition, for example the line $d = 3.087$ Å which first appears in $Cu_{1.586}S$.

The X-ray data indicate that when chalcocite is leached under acidic oxidising conditions the lattice contracts and changes from a hexagonal to a cubic-close packed structure as the composition of the solid approaches that of digenite. Removal of more copper causes further contraction of the lattice and the leached solids show the same basic structural characteristics as spionkopite, yarrowite and covellite. However they do not show most of the weaker reflections of spionkopite or yarrowite, perhaps because the structures have not had time to become fully ordered.

Support for the close relationship between the structures of the leached samples and these two minerals is provided by the behaviour of the strongest line as copper is leached from the digenite structure. In this mineral the value is $d = 1.967$ Å and all of the leached samples to $Cu_{1.460}S$ have the line, with $d = 1.962$ Å, $I = 100$ in that material. In anilite the strongest line is $d = 1.956$ Å[222] or 1.962 Å.[195] The leached chalcocite sample of overall composition $Cu_{1.460}S$ also has a line $d = 1.915$ Å, $I = 60$ which first appeared in $Cu_{1.586}S$ as $d = 1.921$ Å, $I = 20$, see Fig. 5.11. In the sample $Cu_{1.306}S$ the equivalent lines are $d = 1.960$ Å, $I = 30$, and $d = 1.907$ Å, the strongest line. The strongest line of spionkopite is $d = 1.910$ Å and this mineral also shows $d = 1.965$ Å, $I = 12$. Thus a change in the structure occurs during leaching of the residual solid from chalcocite and digenite when its composition contains less copper than $Cu_{1.460}S$ but more than $Cu_{1.306}S$, to give a spacing identical with that of spionkopite, $Cu_{1.40}S$. Yarrowite has its strongest line at $d = 1.899$ Å and a line $d = 1.961$ Å, $I = 5$.

The structure of the residue from leaching chalcocite of composition close to $Cu_{1.0}S$ is not that of synthetic covellite of composition $Cu_{1.000}S$. The strongest line of the residue has $d = 1.897$ Å whereas that of covellite is $d = 2.813$ Å, a spacing not shown by any leach residue of similar composition.

Whitehead and Goble[229] have suggested that the solid phases formed by leaching digenite in acidic iron (III) sulphate solutions retain the cubic close packed structure of that mineral and anilite. They prepared synthetic chalcocite and digenite by heating copper and sulphur powders at unspecified temperatures and quenched the sintered product, which was crushed gently between the fingers and sieved to give a 150–250 μm product. Under the microscope polished sections of the powder showed that the sized grains were aggregates of many smaller particles, which were generally smaller than 50 μm. Such porous sintered material is regarded by many workers as unsuitable for leaching studies because of its undefined surface area and because the solution within the pores and in contact with the reacting solid is of unknown composition because diffusion of reactants and soluble products to and from the internal surface is slow. It has been shown[230] that the kinetics of leaching porous synthetic chalcopyrite in acidic iron (III) sulphate solutions were different from the behaviour of natural mineral which behaves similarly to synthetic fully dense chalcopyrite.[231]

Leaching of the sintered sulphide[229] grains was carried out inside the vertical sample container of a diffractometer, stuck onto a thin mylar film with an adhesive. No agitation was used and the temperature was about 23°C. During the period of leaching the diffractometer was oscillated over a 6 degree diffraction angle to monitor structural changes taking place. The measurements recorded diffraction peaks in the range $d = 1.87$ Å to $d = 1.97$ Å; the line studied was the spacing $d = 1.967$ Å of digenite which, as stated above, Goble suggested corresponds to the sulphur radius in the cubic close packed structure.[228] In a separate series of experiments mineral grains were leached under similar conditions to those used in the diffractometer, washed and analysed using an electron microprobe, to obtain kinetic data. In 10^{-1} and 10^{-2} M

Table 5.11. Progressive shifts in the $d = 1.967$ Å line of digenite during leaching.[229]

Digenite	Anilite		Phase		Geerite
$Cu_{1.8}S$	$Cu_{1.75}S$	$Cu_{1.72}S$	$Cu_{1.67}S$	$Cu_{1.625}S$	$Cu_{1.6}S$
		I	II	III	IV
1.967	1.962	1.957	1.945	1.933	1.918
Spionkopite		Yarrowite			Covellite
$Cu_{1.4}S$		$Cu_{1.125}S$			$Cu_{1.0}S$
Va	Vb	VIa	VIb	VIC	VII
1.910	1.907	1.903	1.900	1.898	1.895

Fe^{III} sulphate solutions both synthetic chalcocite and digenite reacted until approximately 40% of the copper had been leached from the solid, after which reaction ceased. Thus stage 1 of the reaction was studied.

The results reported for the changes in the values of the $d = 1.967$ Å line as copper was leached from the digenite are given in Table 5.11. Phases reported as being formed during this process were identified by Roman numerals and in some cases a shift in the value of d within a single phase was indicated by a letter. Approximate compositions of phases I to III, which are not associated with a particular mineral, were estimated from histograms of compositions of grains measured by electron microprobe which were included in the paper. The results differ in a number of respects from these given in Table 5.8 and Fig. 5.11 for that range of d = spacings. The value $d = 1.957$ Å for phase I, $Cu_{1.72}S$, compares well with $d = 1.962$ Å for $Cu_{1.728}S$ in Table 5.8. However $d = 1.945$ Å for phase II, $Cu_{1.67}S$ is much lower than is indicated in Fig. 5.11 and the values for phases III and IV suggest that the measured values have moved to the lower of the two sets of d-values shown in the range 1.9 to 2.0 Å in Fig. 5.11. The value $d = 1.918$ Å for phase IV corresponds to $d = 1.921$ Å, $I = 20$ in the leached chalcocite phase of composition $Cu_{1.586}S$, which has $d = 1.962$ Å as its strongest line. The later samples of compositions $Cu_{1.532}S$ and $Cu_{1.480}S$ have as their strongest line $d = 1.960$ Å and $d = 1.962$ Å. They also have lines $d = 1.921$ Å, $I = 40$ and $d = 1.915$ Å, $I = 60$ respectively which are fairly close to the values of Whiteside and Goble, as can be seen on Fig. 5.11. However it is not until the leached chalcocite sample of composition $Cu_{1.306}S$ that the spacing $d = 1.907$ Å becomes the strongest line, with $d = 1.960$ Å having $I = 30$. Since Whiteside and Goble do not report the presence for reflections in the region of $d = 1.96$ Å for their phases II to V, and argue that the spacings they report for their leached solids are derived from $d = 1.967$ Å of digenite by contraction of the lattice as copper is removed, the two sets of results conflict.

Electrochemical Studies

The anodic dissolution of chalcocite has been studied by many workers, using both sulphate and chloride systems. For electrorefining of copper matte the former was the obvious choice but as interest in the possible use of chloride systems for hydrometallurgical processes developed in the 1960's and 1970's, electrochemical investigations of several kinds were undertaken in the Cu–Cl system. A major point of interest was the stabilisation of Cu^I by complexing with Cl$^-$ and consequent possibility of producing copper metal by a one-electron reduction reaction.

A potentially commerical process for the production of cathode copper by electrolysis of copper matte was patented in 1877[232] and much work has been carried out on electroleaching simple copper matte since then.[233,234] The sequence of reactions and phase changes taking place were not completely determined by use of electrochemical methods although use of particulate solid in the leaching step and study of the products by X-ray powder diffraction resulted in a major advance.[218] In addition there were conflicting views as to whether the diffusion process of interest took place within the solid or solution phase.[235] Much of the data published has been analysed again in terms of the electrochemical methods discussed in Sec. 4.2.5.[236] A model which explains the results involves chemical interaction between the solid and the solution, and diffusion of copper in the lattice. The diffusion coefficients found depend somewhat on temperature and the composition of the solution but are in the range 1.8×10^{-12} to 2×10^{-11} m^2 s^{-1}, in agreement with the value for the diffusion of copper in Cu$_2$S at 30°C, 1.7×10^{-12} m^2 s^{-1}.[237]

Anodic dissolution of chalcocite in chloride solutions was studied using voltammetry and a rotating ring disk electrode.[238] This was a rotating disc electrode within a Teflon sleeve with a platinum ring around the reacting face so that the solution containing the soluble products of the reaction flowed across the ring. This is particularly useful for determining unstable species such as Cu^I produced by the reaction, since the ring electrode can detect dissolved electroactive species as soon as they are produced at the reacting disc. The radius of the disc used was 0.254 cm; the gap between disc and ring was 0.121 cm and the thickness of the ring was 0.035 cm. Voltammograms were obtained for disc potentials between 0.0 and 1.4 V (versus SCE) and in other experiments a constant potential in the range 0.1 to 1.0 V (versus SCE) was applied to the disc and the current was measured at 10 minutes intervals. At the same time ring current measurements were made at 0.0 V (versus SCE) to measure Cu^{II} and at 0.8 V (versus SCE) to measure Cu^{I}. A separate disc of chalcocite was used in each experiment.

Different behaviour occurred at low, intermediate and high potentials in 1 M hydrochcloric acid. At low values the current density at the disc increased slowly with applied potential and the soluble copper was largely Cu^{I}. At intermediate values current density increased more rapidly with applied potential and both Cu^{I} and Cu^{II} were produced. At the high potential end of the range current density increased rapidly and the main soluble product was Cu^{II}. Final values give results for discs which have been significantly depleted in copper. At constant chloride concentration a change of pH had no detectable effect on the results. At constant pH addition of potassium chloride increased the $Cu^{I} : Cu^{II}$ ratio at disc potentials of 0.2 to 0.4 V (versus SCE). At the higher chloride concentration the reaction rate, and so the current density, was higher. Sulphur was not detected after reaction at 0.0 to 0.2 V and no soluble sulphur species was formed at any potential. Electron microprobe analyses were carried out to determine the compositions of solid products of the experiments and it was found that at a constant applied potential the surface material reached a fixed composition within 60 minutes, indicating that the value of the applied potential determines the composition of final solid product.

It has been shown[239,240] that copper sulphide electrodes behave reversibly with respect to copper ions in aqueous copper (II) sulphate and copper (I) and copper (II) chloride solutions, the second electrode being copper metal. Thus the potential of the cell, E, is given by

$$E = \frac{RT}{zF} \ln a_{Cu} \qquad (5.23)$$

where a_{Cu} is the activity of copper in the sulphide electrode phase, the activity of the metal in the copper electrode being unity. The electrolyte containing the electrodes must be capable of conducting the ions of the metal in each valency. Thus provided no oxidation-reduction of the sulphur species (S^{2-} in the solid sulphide) occurs measurements of the potential provide activity data for copper in the solid sulphide of the composition of the electrode. This in turn can be used to determine the free energies of formation of the sulphides. A major difficulty in using this electrochemical method is in determining the identity of the phases present in the electrodes.

The experimental method has been used[196] to measure the values of $\Delta G°$ of formation of covellite; $Cu_{1.1}S$; $Cu_{1.4}S$; anilite, low digenite ($Cu_{1.765}S$); djurleite ($Cu_{1.934}S$); djurleite ($Cu_{1.965}S$); low chalcocite and high chalcocite over appropriate ranges of temperature between $0°C$ and $250°C$. The two blaubleibender covellites, low digenite, tetragonal phase and solid of the djurleite range of compositions were found to be metastable. Optical microscopy and X-ray powder diffraction were used to identify phases present, diffraction measurements of high precision and high sensitivity, to measure weak reflections, being made on synthetic sulphides to supplement published data.[195]

The procedure necessary to obtain the emf data is to measure the value of E at a given temperature as a function of the value of x in Cu_xS. One method of doing this is to add or substract metal from the sulphide electrode by coulombic titration; that is by passing a known quantity of electric current between the metal and sulphide electrodes. A coulometer can be used to measure the amount of current passed and the amount of copper transferred can easily be measured to $\pm 6.5 \times 10^{-8}$ g and, with some difficulty, to $\pm 6.5 \times 10^{-9}$ g. Thus in principle it is possible to add copper to a covellite electrode until it becomes a chalcocite electrode, if the metal diffuses to the required positions in the crystal lattice as the process continues. In practice this is not possible because the solid disintegrates as the volume changes become too large to be accommodated. The second method, which was used by Potter, is to prepare many electrodes of different compositions and add or remove small quantities of copper by coulombic titration, at each temperature in the range to be studied for the particular phase or phases. The success of this procedure in providing values of cell potentials which gave apparently good thermodynamic data shows that in the Cu–S system transformations between phases can occur in both directions as the composition of the solid is changed by addition or removal of copper from the lattice.

It was pointed out above that use of Eq. (5.22) to calculate the activity of copper in the suphide phase is valid provided no oxidation-reduction reaction involving S^{2-} in the solid phase occurs. In covellite two thirds of the sulphur atoms are bound together as S_2 groups like those in pyrite, i.e. as S_2^{2-}. Goble[210] considers that the formula of spionkopite can be represented in terms of conventional valencies as $Cu_{15}^{I}Cu_{4.5}^{II}(S_2^{2-})_2(S^{2-})_{10}$ and yarrowite as $Cu_{20}^{I}Cu_{7}^{II}(S_2^{2-})_7(S^{2-})_{10}$. In the general conclusions to the discussion above of the individual copper suphide phases it was pointed out that the conventional valency formula $Cu_4^{I}Cu_2^{II}(S_2^{2-})_2(S^{2-})_2$ for covellite is not applicable and this is of course also true for the other two sulphides. However as these phases change into one another and eventually form solids containing no S_2^{2-} groups the reaction

$$S_2^{2-} + 2e = 2S^{2-}$$

must occur. Potter does not state whether or not this redox reaction was taken into account in measurements involving the blaubleibender covellites and covellite itself. Since he provides no experimental data or statistical analysis of such data it is not possible to determine this.

Whiteside and Goble[229] stated that the phases they produced by iron (III) sulphate leaching which showed the X-ray diffraction patterns resembling those of spionkopite and yarrowite (based on measurements of a single line) did not possess the optical properties of the natural minerals when polished sections through grains were examined. There is general agreement that the X-ray diffraction pattern of the phase of composition $Cu_{1.00}S$ formed by leaching solids containing more copper than this is not that of the natural mineral covellite. Whiteside and Goble argue that the phases produced by leaching digenite in iron (III) sulphate solutions retain the cubic close packed structure of digenite. The X-ray diffraction data in Table 5.8 is not inconsistent with this, since far fewer lines are shown for samples with compositions approximating to those of spionkopite and yarrowite than are present in the published data for those phases. The weight of evidence suggests that the phases produced by leaching copper from digenite or from sulphides containing less copper produces phases with a continuous range of compositions to $Cu_{1.00}S$ (or one containing a little less copper) which subsequently alter to form the more stable (even though in some cases thermodynamically metastable) solids which occur as natural minerals. Some additional support for this can be seen in Table 5.8 where the d-spacings for a number of the lines of the samples of overall compositions $Cu_{1.460}S$, $Cu_{1.306}S$, $Cu_{1.176}S$ and $Cu_{1.042}$ show a linear decrease as copper is removed. This suggest that the samples are not a mixture of two phases of constant composition such as were synthesised by Potter or measured by Goble and Smith.[241] The number of samples is small however.

5.3.4.   *Oxidative Leaching of Bornite in Acidic Solutions*

Bornite, Cu_5FeS_4, is a minor copper mineral in many ores and if these or concentrates produced from them are to be treated in a hydrometallurgical process its behaviour during the leaching step is of some importance. Such a process would, in the forseeable future, use an acidic liquor and oxidising conditions. A great deal of laboratory work has been carried out on bornite using both acidic and alkaline solutions as leaching agents however, largely because of the complexity of its behaviour and the controversial nature of the results published.

Modern work on bornite leaching began with that of Sullivan[242] who used crushed natural mineral and iron (III) sulphate solutions. His results indicated that: (i) particle size did not affect the dissolution rate; (ii) the rate of dissolution was practically independent of the concentration of sulphuric acid and (iii) of iron (III) provided enough solution was present; (iv) the rate of dissolution increased as the temperature was raised; and (v) air and water alone had little action on bornite. He reported that when the reaction started copper was preferentially leached from the mineral and the iron and sulphur content were not attacked appreciably. He used the old (incorrect, see Sec. 1.2) assumption that a mineral of composition Cu_5FeS_4 could be written and considered to behave as $2Cu_2S \cdot CuS \cdot FeS$ and that the following reactions took place:

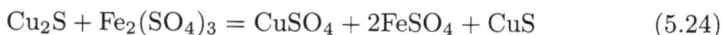

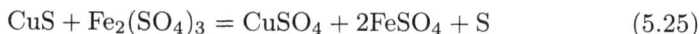

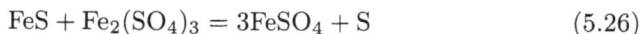

$$Cu_2S + Fe_2(SO_4)_3 = CuSO_4 + 2FeSO_4 + CuS \qquad (5.24)$$

$$CuS + Fe_2(SO_4)_3 = CuSO_4 + 2FeSO_4 + S \qquad (5.25)$$

$$FeS + Fe_2(SO_4)_3 = 3FeSO_4 + S \qquad (5.26)$$

giving the overall reaction as

$$Cu_5FeS_4 + 6Fe_2(SO_4)_3 = 5CuSO_4 + 13FeSO_4 + 4S. \qquad (5.27)$$

The stoichiometry of Eq. (5.27) is correct but any mechanism based on Eqs. (5.24) to (5.26) is not.

Specimen grade mineral bornite was leached in water under oxygen at elevated temperatures[243] and it was found that if sufficient sulphur was added in the form of H_2SO_4 or FeS_2 to provide the quantity of sulphate required for complete dissolution of the metals, recovery of the copper at 200°C with p_{O_2} 115 lb in^{-2} (0.8 MPa) was 99% complete in 30 minutes. It was reported that incompletely leached particles in residues had altered to chalcopyrite and covellite. When heated at 200°C under nitrogen (p_{N_2} 110 lb in^{-2}; 0.76 MPa) for 2 hours no copper dissolved but the mineral had altered to form chalcocite at the outside of the particles whereas the centres of the particles consisted predominantly of bornite with a small amount of chalcopyrite.

Using the rotating disc technique with natural bornite in $Fe_2(SO_4)_3$ solutions, the rate of dissolution of copper was found to increase with $[Fe^{III}]$ in the concentration range 9–36 g $L^{-1}Fe^{III}$ and to increase slightly with increasing rate of rotation.[244] Linear kinetics were observed with apparent activation energy 23 ± 5.6 kJ mol^{-1}. The most important observation in this work was that an outer layer of a product formed within a short time of the commencement of the reaction, which was identified by X-ray powder diffraction as chalcopyrite. In a later paper[245] it was stated that this outer layer of solid was not chalcopyrite but a previously unreported material.

Sintered discs of synthetic bornite were used in acidifed iron (III) sulphate solutions at temperatures between 5°C and 94°C.[246] The mineral was converted completely to "nonstoichiometric" bornite at temperatures below 40°C whereas reaction at higher temperatures produced two new solid phases. A section cut perpendicular to the leached surface of the disc showed a dark blue layer of nonstoichiometric bornite, adjacent to the unreached mineral, the copper content of which decreased progressively towards the leached surface, and an outer layer which was identified by X-ray powder diffraction as a mixture of chalcopyrite and sulphur. The only reaction at temperatures below 25°C was

$$Cu_5FeS_4 = xFe_2(SO_4)_3 = Cu_{5-x}FeS_4 + xCuSO_4 + 2xFeSO_4 . \tag{5.28}$$

The maximum value of x in the nonstoichiometric bornite was estimated as 1.2. The lattice spacings of the solid decreased with increasing distance from the interface with the unreacted bornite. Electron microprobe analyses showed that across both the bornite and the nonstoichiometric bornite the S : Fe ratio remained at 4:1 and that the Cu : S and Cu : Fe ratios decreased steadily.

At temperatures above 40°C, the following additional reactions were said to occur

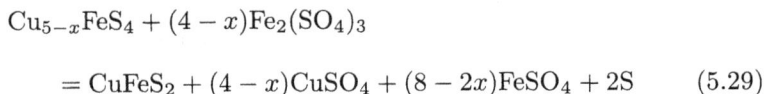

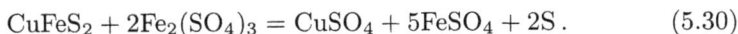

$$Cu_{5-x}FeS_4 + (4-x)Fe_2(SO_4)_3$$

$$= CuFeS_2 + (4-x)CuSO_4 + (8-2x)FeSO_4 + 2S \tag{5.29}$$

$$CuFeS_2 + 2Fe_2(SO_4)_3 = CuSO_4 + 5FeSO_4 + 2S . \tag{5.30}$$

The fact that sulphur appeared only in leaching experiments which had continued for a considerable time was explained by stating that reactions (5.28) and (5.29) are very much faster than (5.30), which is certainly a very slow reaction when natural or synthetic chalcopyrite mineral is used (Sec. 5.3.5). A discrepancy was found in the amount of elemental sulphur actually produced during the reaction and the amount produced by Reaction (5.29) based on the amount of $CuFeS_2$ in residual solids. This was explained by assuming that the chalcopyrite was sulphur-rich and that some sulphur was oxidised to sulphate and could not be measured because of the high background concentration of this.

In view of the results of King,[219] obtained when studying the two-stage leaching reaction of chalcocite in acidic solutions of $FeCl_3$, the reaction of bornite in iron (III) sulphate solution was reexamined because structural considerations led to doubts as to the mechanism by which bornite could change to chalcopyrite by a leaching reaction at low temperatures.[247] Fully dense bornite was synthesised from the stoichiometric quantities of sulphur, copper rods, and iron sponge reduced in H_2 at 500°C and cooled in that gas shortly before weighing, all of Specpure quality.[248] After reaction at 600°C was apparently complete (a few days) the solid in the silica reaction tube was maintained at 900°C for 10 days, during which time crystals of bornite grew on the surface of the solid mass. Some of the best formed of those produced in later preparations were used in other studies. To allow for phase transformations the temperature of the bornite was lowered at the rate of 0.5°C min^{-1} to 700°C, kept constant for 10 hours, lowered to 200°C, kept constant for 6 hours, after which the furnace was turned off.

The results of kinetic studies using in general 1 g of +100–150 μm bornite in 200 ml of solution 0.1 M in H_2SO_4 containing $Fe_2(SO_4)_3$, with a stirring speed 950 rpm, confirmed in general the earlier work. Bornite is not attacked by sulphuric acid using such conditions at 30°C. With 0.005 and 0.01 M Fe^{III} reaction of bornite ceased when all iron had been reduced to Fe^{II} but under air very slow dissolution of copper continued due to reoxidation of some Fe^{II}. Increasing [Fe^{III}] above 0.03 M did not increase the rate of dissolution of copper significantly. At 90°C dissolution of copper from bornite occurred in two stages, an initial fast reaction being followed by a much slower leaching process. At this temperature the rate of dissolution during the first stage was identical using 0.065 M and 0.100 M Fe^{III}. The rate of the second stage depended on [Fe^{III}] up to 0.065 M but was independent of it above that value. The effect of temperature using 0.1 M Fe^{III} is shown in Fig. 5.12. At 40°C and at lower temperatures the reaction became very slow when about 27% of the copper had been dissolved from the solid. Dutrizac *et al.*[246] gave the transition as

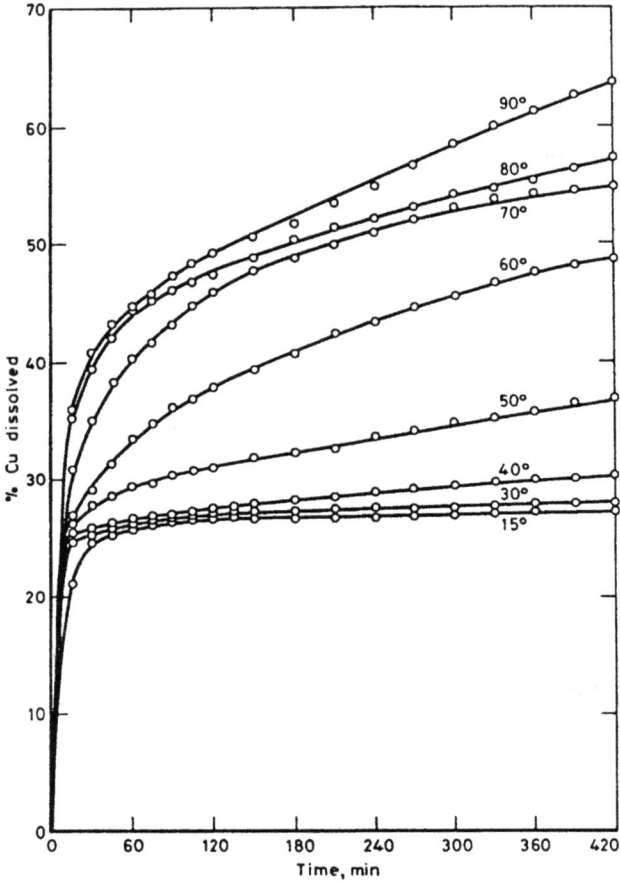

Fig. 5.12.   Leaching of bornite in iron (III) sulphate solutions: effect of temperature.

occurring at about 25% copper loss. At 70°C, 80°C and 90°C, the rate curves
show an initial fast reaction lasting until about 40% of the copper has been
dissolved and a second stage which is much more sensitive to temperature than
the first. During this second stage some flotation of the particles occurred as
hydrophobic sulphur was formed as a product of the leaching. This caused
some experimental error because of a lack of reproducibility in the effective
degree of dispersion of the solid in the leach solution, sulphur tending to accumulate at the air–water interface. During the first part of the reaction at low

or high temperatures no iron was detected as being dissolved from the bornite and no elemental sulphur was formed. The critical increment of energy for the initial stage of the leaching reaction was calculated from the amounts of copper dissolved after 3 minutes of reaction at 50°C, 60°C, 70°C, 80°C and 90°C, giving $E_A = 8.8 \pm 0.4$ kJ mol^{-1}, indicating a diffusion controlled reaction. Attempts to determine the value for the second stage of the reaction were unsuccessful, probably because of experimental factors such as the flotation of particles.

Examination of partially leached bornite by optical microscopy, electron microprobe and X-ray powder diffraction. Partially leached residues were mounted in resin and ground and polished to reveal sections through them. Their overall compositions were determined in the same way as those of chalcocite (previous section).

When 7.9% of the copper had been removed from bornite two zones could be distinguished in the particles at a magnification of about 400×. At the centre was solid which looked like unreacted bornite, and surrounding it was a layer of material in which pores and cracks had begun to appear, caused by the contraction of the unit cell during removal of copper by diffusion. The aspect of this material under polarized light was similar to that reported for natural anomalous (nonstoichiometric) bornite.[249] It gave a large number of X-ray diffraction lines,[248] including all but one of those reported for x-bornite,[250] some of which were not present in the diffraction patterns of bornite.

Many of the X-ray diffraction lines shown by this material disappeared as more copper was removed from the solid. Several of the lines which then remained showed a systematic shift caused by shrinkage of the lattice as the additional copper was removed, which ceased after 40% of the copper present in bornite had been lost. Beyond this the lines remained constant within experimental error, including any uncertainty due to the proximity of lines due to sulphur.

Particles from which 14.6% of the copper had been dissolved out consisted entirely of nonstoichiometric bornite, the porosity extending throughout the whole of the solid. The edges remained very well defined which suggests that up to this stage in the leaching reaction removal of copper had occurred by diffusion of the metal through the solid and transport across the solid–liquid interface rather than by attack on the solid surfaces. When 37% of the copper had been lost the residual solid showed some pronounced pits along the edges of the polished particles. These increased in extent as more copper was removed until, with 64% of the copper gone, the edges were skeletal. The residues from the later stages of leaching tend to disintegrate on polishing, leaving only some central regions of solid surrounded by holes in the resin mount.

The attack on particle surfaces when more than about 40% of the copper had been removed from the solid, together with the appearance of elemental sulphur in the grains of residues, suggested that in the second stage of the reaction the solid phase formed during the first stage reacted as a whole. This is in agreement with the fact that during the second stage iron is dissolved from the solid, whereas during the first stage it is not.

Quantitative analyses were carried out using the electron microprobe and iron and copper metal standards, together with pyrite for sulphur. Results for particles of residues from which 37.5% or more of the copper had been leached are given in Table 5.12. Scans showed that the particles were homogeneous except for areas containing elemental sulphur, which were avoided for quantitative measurements. Two compositions were reported for particles in the residue from which 64.1% of the copper had been removed; (i) was obtained using particles about 200 μm and (ii) using particles around 50 μm in diameter. The analyses suggest that the product of the first part of the leaching reaction has a composition close to Cu_3FeS_4 and that this composition remains constant as the second stage proceeds.

The analytical data in Table 5.12 indicate clearly that the product of the first stage of the leaching reaction does not have the composition of chalcopyrite. However the formation of chalcopyrite by this reaction had been consistently reported, with one retraction and statement that the product was a previously unreported material.[245] The values of all the d-spacings for the

Table 5.12.

% Cu dissolved	Cu, wt %	Fe, wt %	S, wt %	Total, %
37.5	51.5	13.8	33.0	98.3
64.1 (i)	49.9	13.3	31.3	94.5
64.1 (ii)	50.5	14.4	33.7	98.6
80	48.6	16.7	35.1	100.4
Calculated compositions of some copper-iron sulphides				
$CuFeS_2$	34.64	30.42	34.94	100.00
Cu_3FeS_4	50.87	14.90	34.23	100.00
Cu_5FeS_6	56.14	9.87	33.99	100.00

Table 5.13. *d*-spacings of bornite from with 49.1% of the copper content has been leached, and for α-chalcopyrite (JCPDS 37–472).

Bornite, 49.1% Cu removed								
*d*A	3.215	3.018	2.622	1.855	1.583	1.315	1.207	1.076	1.011
I	w	vs	w	s	m	w	w	w	vw

α-chalcopyrite								
*d*A	3.039	2.645	2.606	1.870	1.8561	1.5926	1.5757	1.3223	1.3027
I	100	3	3	16	25	12	6	2	1
*d*A	1.3027	1.2126	1.2049	1.0770	1.0128				
I	1	2	3	5	5				

leach residue from which 49.1% of the copper had been leached are given in Table 5.13 together with published data for chalcopyrite.[251] Every value measured in the pattern of the residue has a line in the chalcopyrite pattern with a *d*-spacing of very similar magnitude. Only two lines of significant intensity are not shown by the residue, those with $d = 1.869$ Å and $d = 1.5753$ Å. Three weak lines of chalcopyrite are matched closely by lines in the residue. Comparison with the matching of data for leached chalcocite in Table 5.8 shows that it is to be expected that the data for the bornite residue would be accepted as firm evidence for chalcopyrite.

Since it had long been accepted that chalcopyrite was formed when bornite was leached under acidic oxidising conditions work was carried out to clarify the position.[252] The behaviour of the solid residues formed during leaching of (i) a bornite concentrate produced by froth flotation; (ii) discs of synthetic bornite produced by sintering the powder; and (iii) massive natural bornite, was reexamined using acidified iron (III) chloride and sulphate solutions over the temperature range 10°C to 90°C. A very detailed study confirmed the formation of a phase of composition Cu_3FeS_4 by further removal of copper from the nonstoichiometric bornite produced during the first stage of leaching of bornite. The formation of minor quantities of chalcopyrite, chalcocite and possibly digenite was also reported.

The mechanism of formation of the Cu_3FeS_4 phase from bornite by leaching reactions. When the mechanism outlined below was first proposed the Cu_3FeS_4 phase was identified as the mineral idaite on the basis of its optical reflectivity dispersion profile.[248] The validity of idaite is now highly questionable and the confusion caused by the naming of a mineral without providing a reference sample, an adequate chemical analysis of it, or definitive measurement of its physical properties is briefly discussed below.

There are three forms of bornite: a cubic high temperature form stable at temperatures above $228 \pm 5°C$, a rhombohedral metastable modification which appears below that temperature and a low temperature form generally referred to as the tetragonal form, which is produced on cooling the metastable phase to room temperature. In 1964, Morimoto determined the crystal structures of the high temperature and metastable phases.[253,254] The structure of the high temperature form is essentially the antifluorite structure, only slightly more complicated. The sulphur atoms occupying the nodes of the face-centred cubic lattice ($a = 5.50$ Å) are cubic close packed. Each tetrahedron of sulphur atoms on average contains 3/4 of a metal atom. This fractional atom is itself distributed statistically over 24 equivalent sites inside the sulphur tetrahedron. In the whole cell six metal atoms are statistically distributed over 24×8 sites.

The metastable form is transitional between the low and high temperature forms. It shows two characteristics: (i) the existence of a cubic subcell with $a = 5.47$ Å and (ii) the fact that only reflections with indices $(4m \pm 1, 4n \pm 1, 1)$ can appear. This forms a special extinction rule. Attempts to find a single crystal structure which would account for the cubic reflections were unsuccessful, and these were attributed finally to the twinning of a large number of small domains in eight different orientations. This structure of the rhombohedral form can be derived from that of the high temperature form by considering the latter along the body diagonal (111) of the cubic cell. All the sulphur atoms remain in place, retaining the cubic close packing. Of the four partial metal atoms in the sulphur tetrahedra in the high temperature form, two do not change at all. One tetrahedron becomes vacant, and the metal 3/4 atom which occupied it in the high temperature form becomes redistributed among the other three sites. The corresponding three sulphur tetrahedra now contain one full metal atom each. The last metal atom M_I, moves slightly in its tetrahedron to compensate for the vacant tetrahedron. The statistical distribution of 3/4 of the metal atom among 24 possible sites inside each sulphur tetrahedron in the high temperature form changes to the statistical distribution of one metal atom among four possible sites in the metastable form.

This structure proposed for the metastable form is a layer structure parallel to (0001). There are two kinds of sulphur layers, S_I and S_{II} and three kinds of metal layers, M_I, M_{II} and M_{III}. The S_I layers are sandwiched between the M_{II} and M_{III} layers and the S_{II} layers have only the M_I layers, on one side.

The X-ray data for the low temperature form of bornite were said to give tetragonal symmetry with cell dimensions $a = 10.94$ Å, $c = 21.880$ Å. The strong and medium reflections generally have similar intensitites in the metastable and low temperature form which indicates that the two crystal structures have a similar basic structure. However the reflections of the low temperature form do not obey the special extinction rule. The basic structure which was proposed consisted of a stack of layers of sulphur and of metal atoms, each layer having one kind of metal. The sequence was $S_IM_{III}S_{II}M_IM_{II}S_IM_{III}$ $S_{II}M_IM_{II}$ ... ; the question was which metals existed in each kind of M position. Manning[255] considered the apparent 5- and 7-coordination of the sulphur atoms in the structure proposed by Morimoto and suggested a model for the low temperature form of bornite with a sphalerite-type skeleton containing layers of ionically bound interstitial Cu_I atoms. The remaining iron and copper atoms would then be distributed among the M_I and M_{III} sites. He suggested that on the basis of this model the structural formula of bornite could be written as $[Cu_3FeS_4]^{2-}$ $2Cu^+$. When a ball model of this hypothetical structure of the low temperature form of bornite was constructed the ionically bound copper layers, parallel to $(111)_{rh}$, were readily observed and it was clear that there would be no hindrance by other atoms in the structure to the free movement of these copper atoms within their layers.[248] Complete removal of these atoms from the structure results in the formation of a chalcopyrite-like unit cell with the formula Cu_3FeS_4, as proposed by Manning. This cell is derived by replacing four $(0, 1/2, 3/4)$ iron atoms of the chalcopyrite unit cell by copper atoms.

It was suggested[248] on the basis of these arguments that during the leaching of bornite under oxidising conditions in acidic solutions the interstitial ionically bound Cu^I ions are removed from the bornite lattice by diffusion in the plane of their layers and transfer to the solution phase at the solid–liquid interface. Initially this gives rise to nonstoichiometric bornite and the changes in lattice parameters as a function of the amount of copper removed can be determined. When the $2Cu^+$ in Manning's suggested structural formula have been removed, 40% of the total copper content of bornite, this reaction ceases. If the temperature and chemical conditions are suitable the Cu_3FeS_4 phase will react and the structure will be destroyed as this occurs. Under severe conditions and with large pieces of bornite decomposition of nonstoichiometrc bornite and the Cu_3FeS_4 phase will take place at the surface of the solid while copper is still diffusing through the lattice from the interior.

Table 5.14. Changes in lattice parameters as copper is leached from bornite.

% Cu removed	d_{440}	$a, \text{Å}$	$d_{44.16}$	$c, \text{Å}$
0.0	1.937	10.94	1.119	21.88
7.9	1.922	10.87	1.112	21.81
14.6	1.916	10.84	1.108	21.73
27.7	1.870	10.58	1.086	21.17
28.4	1.870	10.58	1.080	21.17
30.8	1.863	10.54	1.078	21.15
37.5	1.860	10.52	1.075	21.08
43.7	1.860	10.52	1.070	20.93
49.1	1.860	10.52	1.070	20.93
57.9	1.860	10.52	1.070	20.93
64.1	1.860	10.52	1.070	20.93
80.0	1.860	10.52	1.070	20.93
96.0	1.854	10.49	1.070	20.96

Variation of lattice parameters as a function of copper removed was measured[248] using the (440) and (44.16) reflections. These were selected because (i) the intensities permitted measurements to be made on all leach residues and (ii) the same results are obtained from the indexing corresponding to the chalcopyrite-like structure.[220,228] An average of readings taken at both edges of the lines was used in measuring the angular positions of these reflections. For the tetragonal system the general formula $1/d^2 = [(h^2+k^2)/a^2] + l^2/c^2$ applies. The values obtained for the a and c parameters are shown in Table 5.14.

A considerable contraction took place in the unit cell during the first part of the leaching reaction, particularly while between 14% and 27% of the copper was removed. There was a further decrease between 28% and 40% copper removal, but no further change occurred until almost all of the copper had been removed. It is interesting to note that the rate curves in Fig. 5.12 show that the rate of leaching of copper drops appreciably after 28% of the copper has been removed. The second section of the first part of the reaction curves for temperatures below 40°C extends from this point to where about 40% of the copper has been leached.

Taking the values $a = 10.52$ Å and $c = 20.93$ Å for leach residues with constant lattice dimensions and dividing by 2 gives the unit cell size for Cu_3FeS_4 as $a = 5.26$ Å and $c = 10.46$ Å, close to that of chalcopyrite, $a = 5.24$ Å, $c = 10.30$ Å. The estimated decrease in unit cell volume, based on these measurements, on changing from Cu_5FeS_4 to Cu_3FeS_4 is 10%. From measurements of the difference in the relative densities of bornite and Cu_3FeS_4 the change is 9%. Dutrizac *et al.*[252] compared their X-ray diffraction data and calculated values of a and c for the Cu_3FeS_4 phase with those of Ugarte and Burkin[248] and concluded that there is generally excellent agreement between the two sets of data.

Electrochemical Studies

In Sec. 4.2.5, the most important electrochemical techniques for following leaching reactions which involve oxidation or reduction were outlined and examples of the experimental curves which can be obtained by oxidation of bornite were given in Figs. 4.9 to 4.12. It was pointed out that quantitative treatment of the data requires information which cannot be obtained by electrochemical methods. It has been obtained in the work discussed above, in this section.

The first electrochemical study of the leaching behaviour of bornite which could lead to quantitative results used chronopotentiometry in acidic sulphate solutions.[256,257] The first and second stages were described as being controlled by diffusion in the solid and the electrochemical measurement were analysed using the equations for parallel reactions. The diffusion coefficient and concentration of the electroactive species, Cu^+, were apparently assumed to be the same for both stages of the reaction. In view of the overwhelming evidence that the reactions are consecutive, and the different structural characteristics of bornite and the Cu_3FeS_4 phase the analysis is not accepted here.

Chronopotentiometry, chronoamperometry, and linear and cyclic voltammetry were used to test the mechanism which had been proposed for the formation of the Cu_3FeS_4 phase from bornite during oxidative leaching in acidic solutions.[258] This postulated that 40% of the copper atoms in the bornite lattice occurred in layers and that these could diffuse within the solid without hindrance from other atoms in the structure, leading to a rapid first stage reaction. Thus the diffusion coefficient of copper should be much greater along these planes than in any other direction. The electrochemical characteristics of the reaction using fully dense synthetic polycrystalline bornite and a sample of natural mineral were established, and diffusion coefficients of copper were measured using different orientations of the crystal lattice of single crystals of bornite.

The model of the crystal structure of bornite was changed in view of the demonstration[259] that the true symmetry of the mineral is orthorhombic and it has a pseudo-tetragonal cell with $a = 10.95$ Å, $b = 21.862$ Å and $c = 10.95$ Å, and a cell content of 16 Cu_5FeS_4. The sulphur atoms are arranged in cubic close packing and ordering of the metal atoms in the tetrahedral or triangular interstices between the sulphur atoms results in two structural units: the antifluorite type and sphalerite type cubes respectively. Koto and Morimoto calculated the atomic coordinates and metal–sulphur distances in bornite, and proposed possible sites for the iron atoms. A ball model of the structure was built after calculating the atomic coordinates of all the atoms within the unit cell, using these data in conjunction with the equations for the equivalent atomic positions for the orthorhombic cell, space group *Pbca*.[260] As in the previously proposed structure copper atoms lie in two types of position in the lattice. One type lies only within the sphalerite-type cubes and these atoms are approximately parallel to one of the (111) planes of the sulphur skeleton and hence at 70.53°C to be other three. Removal of these copper atoms does not result in the formation of a chalcopyrite-like unit cell so that the structure remaining must rearrange somewhat to produce this.

Using polycrystalline synthetic bornite in 1 M sulphuric acid and constant current chronopotentiometry with different current densities in the range 100–770 Am^{-2} curves such as that shown in Fig. 4.9 were obtained. The transition times were estimated by a standard graphical method and used in the Sand Eq. (4.20). In order to calculate the diffusion coefficient of i, the electroactive species, Cu^+, during stage 1 of the reaction the number of electrons transferred per molecule of i oxidised is required. Since the reaction in stage 1 is

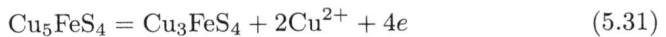

$$Cu_5FeS_4 = Cu_3FeS_4 + 2Cu^{2+} + 4e \qquad (5.31)$$

then $n = 2$. The initial concentration of i is also required, corresponding to the amount of copper available in bornite for the first stage reaction. This is given by

$$C_{Cu}^* = 0.4\frac{0.6333\rho}{63.546}$$

where 0.4 is the fraction of the copper in bornite which is available for reaction in the first stage; 0.6333 is the weight fraction of bornite which is copper; 63.546 is the "molecular" (atomic) weight of Cu; ρ is the mass density of bornite, 5.08 g cm^{-3}. The diffusion coefficient for copper ions in synthetic polycrystalline bornite was found to be 22.27×10^{-9} cm^2 s^{-1} and in the natural mineral sample 24.35×10^{-9} cm^2 s^{-1}.

Chronoamperometry was used in polarisation experiments with natural mineral bornite, Fig. 4.10. All of the data required by the Cottrell Eq. (4.21) are available and the diffusion coefficient for copper was measured as 39.14×10^{-9} cm^2 s^{-1}.

The cyclic voltammogram shown in Fig. 4.12 has a first oxidation peak at 0.5111 V versus SHE and a second oxidation peak at 1.41 V versus SHE. Using the value -314 kJ mol^{-1} for the free energy of formation of the Cu$_3$FeS$_4$ phase[260] the value of E° for Eq. (5.31) is 0.529 V versus SHE which agrees reasonably well with the experimental value. However the standard potential for the second stage was calculated to be 0.222 V versus SHE, very much lower than the experimental value. This is probably because the products of the second stage of the reaction are Cu^{2+}, Fe^{2+} and S$^\circ$ so that the full error in the value of ΔG° for Cu$_3$FeS$_4$ is revealed whereas in a reaction involving two metal sulphides only part of the free energy is involved.

Using single cystals of synthetic bornite the predominantly copper plane was indentified by Laue studies (using "white" X-rays directed along a principal axis of the stationary crystal). Each crystal was cut so as to produce a face at an angle α to that plane. A schematic diagram of a crystal is shown in Fig. 5.13. As the copper plane is rotated about the central point α is the

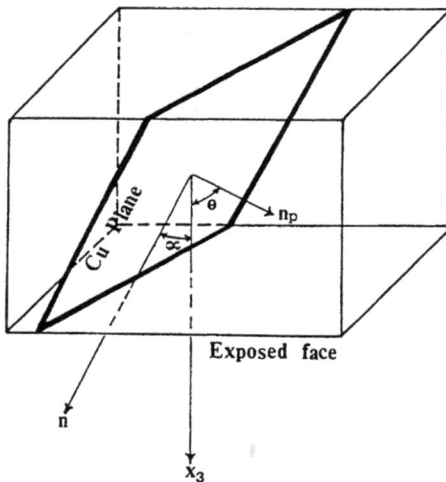

Fig. 5.13. Schematic diagram of a crystal of bornite showing the predominantly copper plane making an angle $\alpha = 90 - \theta$ with the normal of the exposed face.

angle made by the flux, J_{Cu}^+, and gradient concentration vectors with the reference axis X_3, which is perpendicular to the face of the crystal to be used as the working electrode. Since Θ is known from the Laue studies and $\alpha = 90 - \Theta$ it is possible to estimate the difference in the rate of diffusion of Cu^+ to the electrode surface at the angle α relative to that in a polycrystalline bornite electrode, for which the predominantly copper plane is situated statistically in the middle, i.e. $\Theta = 45°$. For all other orientations the calculated (theoretical) values of the ratio of the atomic flux with respect to that for $\Theta = \alpha = 45°$, R_T is given by

$$R_T = J_\Theta/J_{45} = \cos^2 \alpha/\cos^2 \Theta$$

The Sand Eq. (4.20) states that product of the current density and the square root of the transition time for the first stage, $jt_1^{1/2}$, is proportional to the square root of the diffusion coefficient. Then if it is assumed that $D_{Cu}^+ = D_{Cu^+}^* + \cos^2 \alpha$, where $D_{Cu^+}^*$ is the diffusion coefficient in the predominantly copper plane, then for purposes of comparison

$$jt_\theta^{1/2}/jt_{45}^{1/2} = \cos \alpha/\cos 45 .$$

Thus in chronopotentiometric measurements the experimental ratios $R_E^{1/2}$ should be compared with $R_T^{1/2}$. In the case of linear voltammetry comparison with the polycrystalline case was carried out by considering that the peak current densities are proportional to $D^{1/2}$.

Calculated values of $R_T^{1/2}$ for $\Theta = 0$, 45, 54.73 and 90 degrees are given in Table 5.15 together with transition times using current density 170 Am^{-2} in some chronopotentiometric experiments. Except in the case of $\Theta = 0°$, there is reasonable agreement between the calculated and experimentally detemined values of $R_T^{1/2}$ and $R_E^{1/2}$. However with $\Theta = 0°$, the predominantly copper plane is parallel with the face of the working electrode and it would be expected that no first stage of the leaching reaction would be observed. The values of the transition times found when this orientation was used appear to indicated that the first stage of the leaching reaction of bornite occurs as if another predominantly copper plane were at $\Theta = 54.73°$. Examination of the ball model of the unit cell indicated that there is a plane of such atoms making an angle of about 55°C with the exposed face. Calculations showed that in the orthorhombic lattice of bornite there is a plane of type (111) wich makes an angle of approximately 54.83° with a (121) type plane.[260] The experiments using linear voltammetry confirmed the results obtained using chronopotentiometry.

Table 5.15. Summary of experimental ratios, $\cos\alpha/\cos 45$, obtained from chronopotentiometry in 1 M sulphuric acid.

θ	α	$j(\mathrm{Am^{-2}})$	t min	$R_{\mathrm{E}}^{1/2}$
Low-resistance single crystals				
0	90	170	23.1	1.11
0	90	170	19.95	1.03
45	45	170	18.66	1.00
54.73	35.27	170	23.10	1.11
90	0	170	38.69	1.44
High-resistance single crystals				
45	45	179.74	9.3	1.00
54.73	35.27	170	14.25	1.17
70.53	19.47	170	16.35	1.25
50.53	19.47	170	15.90	1.24
70.53	19.47	170	16.95	1.28
90	0	170	28.5	1.66
Theoretical ratios				$R_{\mathrm{T}}^{1/2}$
0	90			0
45	45			1.00
54.73	35.27			1.155
70.53	19.47			1.333
90	0			1.414

The mineral idaite and the identification of solid phases.   Idate was first described from the Ida mine, Khan, South West Africa (Nambia) by Frenzel[261,262] during studies of natural bornite in the process of decomposition, but because of the small amount present in samples he was not able to analyse it. He considered that the optical properties and general structure were similar to a phase synthesised by Merwin and Lombard[250] which had a similar X-ray powder diffraction pattern to that of the new mineral and the composition Cu_5FeS_6. Frenzel decscribed idaite as a supergene sulphide which is formed by the alteration of bornite by oxidation. The mineral has been observed from very many localities and Ramdohr[4] states that in each case the idaite is apparently the first oxidation product of bornite.

Sillitoe and Clark[249] carried out three analyses of lamellar idaite grains using the electron microprobe and found them to contain on average about 3 wt % less copper and 6% more iron that is required by the formula Cu_5FeS_6 proposed by Frenzel. The analyses of the three gains correspond to the formulae

$Cu_{3.1}FeS_{3.8}$; $Cu_{3.0}FeS_{3.9}$; and $Cu_{2.8}FeS_{3.6}$. The composition determined for optically normal bornite was Cu_5FeS_4. Levy[263] analysed five reasonably pure specimens of idaite from two localities, using the microprobe, and concluded that idaite occurring in association with bornite and presumably of supergene origin has the composition Cu_3FeS_4. He also showed that the spectral reflectivity dispersion curves of natural idaite differ markedly from that given by the original material synthesised by Merwin and Lombard.[250] Two areas of polygranular idaite aggregates gave dispersion curves parallel to those found for pure idaite by Levy[263] but lying at slightly lower values.[249] The curve obtained using Cu_3FeS_4 phase prepared by leaching synthetic bornite on the same microphotometer as was used by Sillitoe and Clark very closely resembles those for natural idaite and is very different from curves for chalcopyrite and bornite.[248] No reliable X-ray powder diffraction data has been obtained for mineral idaite owing to the difficulty of obtaining a sufficient quantity of material of the required purity.[249,263] However Levy has pointed out that the powder data given by Frenzel[203,262] bear close comparison with the tetragonal pattern of mawsonite, an intermediate member of the idaite (Cu_3FeS_4)-stannite (Cu_2SnFeS_4) series.

Because of this uncertainty over the X-ray powder data on the original material named idaite and the fact that none of this material is available the validity of idaite is in doubt. A further complication is that the composition Cu_5FeS_6 has been eatablished as the mineral nukundamite,[264,265] which was obtained from Nukundamu, Fiji.[266] As part of a critical comparison of this with a specimen of idaite from the Ida mine reflectance values were determined on areas of this assemblage which were as free from chalcopyrite as possible. They bore some resemblance to those reported by Levy[263] and Sillitoe and Clark[249] for idaites with composition close to Cu_3FeS_4 and were different from nukundamite. This can be taken as establishing that there are two different minerals and the latter has been properly validated. In the idaite from Khan a phase which occurs as spindles was identified as chalcopyrite. A phase occurring as lamellae was almost impossible to resolve optically but X-ray powder diffraction patterns were obtained. These were said[264] to show the lines of chalcopyrite and some additional lines which could be attributed to covellite if that material did not have normal optical properties of the natural mineral. The assemblage appeared to have formed from the breakdown of bornite and to be itself undergoing further alteration. This would involve dissolution of copper which could reprecipitate as a sulphide phase if the oxidation potential fell to a sufficiently low value. The difficulty encountered in the interpretation of X-ray data on assemblages of this kind lies in the fact that the Cu_3FeS_4 phase prepared in the laboratory by leaching bornite has a diffraction pattern very closely resembling that of chalcopyrite.

Although X-ray powder diffraction has proved to be an extremely powerful method of identifying solid phases it has some limitations, as can be seen from the discussions in this chapter on the phase changes which occur during the leaching of chalcocite and bornite under acidic oxidising conditions. One of these is that in a sample containing two or more solid phases it can be difficult to identify them, particularly when a large number of lines are present. Another is that the individual lines each provide a *d*-spacing but no other information about the substances present. It has now been demonstrated that the use of infrared spectra together with X-ray powder diffraction measurements provides an extremely powerful method of identification.[267] One reason for this is that infrared absorption peaks arise from individual vibrations, the frequencies of which are characteristic of the atoms involved. Thus a sample of a precipitate obtained from a leach solution containing iron and arsenic in sulphuric acid by heating at 150°C gave a broad peak in the 2500–4000 cm^{-1} range with a sharp spike at about 3510 cm^{-1}. This is characteristic of scorodite, $FeAsO_4 \cdot 2H_2O$, the peak being due to water of crystallisation. The sample also showed peaks at around 1200 and 1080 cm^{-1} which are due to the v_3 and v_1 vibration modes of SO_4^{2-} and indicate that it contained some sulphate, probably in the scorodite lattice.[268] Infrared spectral data bases for minerals are now being developed.

5.3.5. *Oxidative Leaching of α-Chalcopyrite and Mooihoekite (γ-Chalcopyrite) in Acidic Solutions*

Introduction

Chalcopyrite is the most important copper mineral which is used for production of the metal. Most is smelted, but oxidised ore and low grade material and waste rock from the mines are being used increasingly for treatment by hydrometallurgical routes. Because of the steadily decreasing grade of copper ores being mined a considerable amount of work was carried out during the 1960's and 1970's developing new hydrometallurgical processes which could be used to treat copper sulphide concentrates, which did not necessarily have to reach the specification required for smelting. None of them have been used on a full scale commercial plant although several have been tested on a pilot scale and costed. These processes consist of two main sections; the first to transfer the copper content of the minerals containing it into an aqueous solution phase; the second to recover the copper from the solution and produce a saleable material from it, preferably metal of high quality. The processes were developed before solvent extraction of copper from leach liquors was accepted technology for use in large scale commercial plant and so have not been used. However some of the methods used to decompose the minerals may well be used commercially in future. They illustrate modern trends in the chemical processing of minerals; sophisticated developments of "leaching" processes. Copper would at present be recovered from sulphate process liquors by the solvent extraction-electrowinning (SXEW) process, or from a chloride solution.

The most important reaction being used at present to leach copper from chalcopyrite is oxidation by Fe^{III} sulphate in solutions at pH 2 to 3. Such solutions are continuously formed during heap- and dump-leaching of the low-grade ore and waste rock referred to above. The sulphides present are attacked by bacteria which occur in the orebody and use the oxidation of S^{2-} to SO_4^{2-} and Fe^{II} to Fe^{III} as their sources of energy. Hydrolysis of the Fe^{III} maintains the pH of the solution within approximately this range by balancing the precipitation and redissolution of basic sulphates. The leaching reaction is very slow but capital and operating costs are very low. The copper in the liquor which is collected in ponds at the base of the heaps is usually recovered at present by cementation.

Chalcopyrite reacts much more rapidly in iron Fe^{III} chloride solutions than in sulphate and this system, with continuous removal of the sulphur which forms on the surface of the particles, was suggested[269] for industrial use. Later work[270] confirmed that the process is feasible. The kinetics of the reaction were investigated and the parabolic rate curves obtained were attributed to progressive thickening of a film of sulphur on the particle surfaces.

It was found that heating chalcopyrite in the absence of air at 550°C considerably increased the rate at which it reacted with dilute nitric acid.[271] Similarly, heating the mineral at 825°C for 2 hours in the absence of oxygen caused loss of about 6% of its sulphur content and considerably increased the rate at which it reacted in sulphuric acid under oxygen pressure.[272] After the treatment at 825°C the X-ray diffraction lines corresponding to $d = 1.57$ and 1.59 Å had merged, as had lines $d = 1.85$ Å. This was interpreted as indicating that some disorder in the lattice had occurred and it was suggested that further work was required to determine whether the sulphur lost on heating came from the chalcopyrite itself or from some impurity present.

Synthetic particulate chalcopyrite was sintered to form rotating discs which were used to study the kinetics of the reaction with acidic Fe^{III} sulphate solutions.[230] After partial reaction the solid product was elemental sulphur and unreacted chalpopyrite. Neither the acid concentration nor the speed of rotation of the disc had any effect on the rate of the reaction, and increasing the concentration of Fe^{III} above 10^{-2} M also had no effect. The apparent activation energy of the reaction was 71 ± 12 kJ mol^{-1}. The synthetic chalcopyrite reacted much more rapidly than did natural mineral, a fact attributed to the larger surface area of the sintered material or to effects due to impurities in the natural mineral. The sintering temperature must also have contributed to the increased reactivity.

An early study of chalcopyrite[273] showed that there are three forms; α-chalcopyrite, the natural low temperature phase of composition $CuFeS_2$; the β-form of composition approximately $CuFeS_{1.82}$; and the γ-form $CuFeS_{2-x}$ where x had a large range of values. Thus the β- and γ-phases were said to be sulphur-deficient forms of α-chalcopyrite. When this was heated under N_2 it changed to a high temperature phase at 550°C. The sulphur content of the solid decreased as the temperature was raised above that value until at about 720°C only the β-form was present and no more sulphur was driven off at higher temperatures. The effects of heating chalcopyrite on some of its d-spacings[272] had clearly been due to the phase change which had occurred and had made the residual solid much more reactive on leaching than the original mineral. This behaviour was investigated.[231]

Leaching of Mooihoekite, $Cu_9Fe_9S_{16}$ (γ-Chalcopyrite)

Although γ-chalcopyrite was reported[273] to be a high temperature phase unstable at room temperature it was subsequently shown[274] that the synthetic material of composition $CuFeS_{1.82}$ is a low temperature phase and is identical with the mineral mooihoekite. A second new mineral of composition $Cu_4Fe_5S_8$, named haycockite, was found in the same small samples of massive sulphide collected at Mooihoek Farm, Transvaal. Thus there are now four minerals closely resembling one another in the chalcopyrite group

Chalcopyrite $CuFeS_2$;   Mooihoekite $Cu_9Fe_9S_{16}$;

Talnakhite $Cu_9Fe_8S_{16}$;   and Haycockite $Cu_4Fe_5S_8$.

$Cu_9Fe_8S_{16}$ is equivalent to the β-phase of Hiller and Probsthain.[273] Thus the X-ray diffraction data in Ref. 231 for β-chalcopyrite are here described as being the data for γ-chalcopyrite (mooihoekite). It is emphasised[274,275] that the X-ray diffraction powder patterns of mooihoekite and haycockite cannot be distinguished from one another or from those of chalcopyrite and talnakhite except under suitable conditions, but key reflections for identification are tabulated and differences in relative intensities of some lines are given.[274]

When α-chalcopyrite and mooihoekite were synthesised in the manner described in Sec. 5.3.4 for bornite the products after cooling from 900°C were mixtures of chalcopyrite, bornite and pyrite.[231] When the mixtures were lowered slowly (2 cm h^{-1}) thorough a hot zone at 990–1000°C in a vertical tube furnace melting occurred and fully dense solids of the two phases were obtained. The mooihoekite had the composition $CuFeS_{1.83}$ and the α-chalcopyrite $Cu_{1.12} \cdot Fe_{1.09}S_2$, well within the range of composition metal to sulphur ratio 1 to about 1.17, and copper to iron ratio 0.5 to 1.5, which produces tetragonal low temperature chalcopyrite.[276] The melting point of mooihoekite was determined under a pressure of 5 MPa. Melting began at 891°C and was complete at 931°C. On cooling solidification began at 926°C and was complete at 901°C.

The general characteristics of the leaching reaction between mooihoekite and 0.03 M $Fe_2(SO_4)_3$ solution with pH 1.0 are shown in Fig. 5.14. It is a three stage process. Stage 1 takes place from the start of the reaction until about 20% of the copper has been lost by the solid; stage 2 from this point

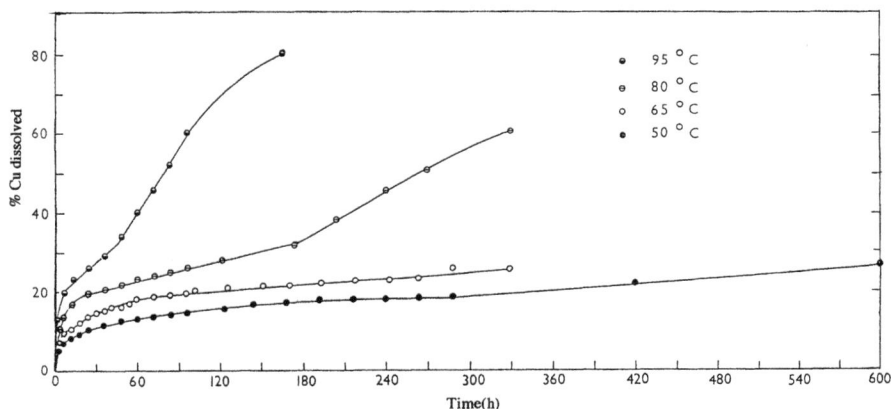

Fig. 5.14. Leaching of mooihoekite in 0.03 M iron (III) sulphate solution: effect of temperature.

until 30 to 35% has been lost; and stage 3 during the remainder of the reaction. Using hydrogen peroxide as the oxidant it was shown that some iron passes into solution during stage 1 as well as some sulphur as sulphate. Solid samples of residues from experiments using Fe^{III} as oxidant showed no evidence of the presence of elemental sulphur, using the optical microscope, until more than 14.5% of the copper had been removed, when traces of it could be seen. Attempts to measure the apparent activation energy during stage 1 were unsuccessful. Use of Fe^{III} concentrations greater than 0.03 M had no significant effect on the rate of leaching.

During stage 2 of the reaction the rate curves are approximately linear, and the rates at 50°C and 65°C are very similar whereas those at 80°C and 95°C are much faster. Accordingly the Arrhenius plot using rates for these four temperatures shows two lines, suggesting that there are two competing reactions with apparent activation energies about 6 and 80 kJ mol^{-1} in the lower and higher temperature ranges respectively. This corresponds to a heterogeneous transport controlled reaction at low temperatures and a homogeneous chemically controlled reaction above 65°C.

Stage 3 of the reaction could be studied conveniently only at the higher temperatures since only 26% of the copper in the solid was leached out after 25 days at 50°C. The start of stage 3 was indicated by an abrupt increase in leaching rate when between 30% and 35% of the copper had been lost by the solid. The rate was then almost constant until, after about 60% copper dissolution it gradually fell. This was due at least in part to the formation of coatings of amorphous sulphur on the surfaces of particles. During stage 3 the solution became turbid due to the presence of sulphur and a jarosite-like phase in suspension. Flakes of goethite covered the walls of the reaction vessel above the level of the solution, formed by reaction between water flowing from the condenser, and leaching solution or jarosite transferred to the walls by splashing of the liquid.

Leach residues from runs in which mooihoekite was leached in acidic $Fe_2(SO_4)_3$ solutions under different conditions were examined by X-ray powder diffraction over the range 6.50–98.74% of the copper removed. Co K_α radiation and a Debye–Scherrer camera were used, line intensities being measured with a microdensitometer plus integrator. This was not sufficiently sensitive to detect weak lines. The diffraction data obtained using the synthetic mooihoekite are given in Table 5.16(a) together with the published data. Examples of the way

Table 5.16(a). X-ray powder diffraction data for mooihoekite, Cabri and Hall (274, JCPDS Powder Diffraction File 25–286) and R.C.H. Ferreira (PhD Thesis, University of London (1972)).

25–286 d Å	7.49 (20)	5.30 (5)	4.78 (2)	4.35 (5)	3.77 (30)	3.35 (10)	3.07 (100)
Ferreira d Å	7.54	5.25	–	–	3.77 (8)	3.35 (2)	3.063 (100)

25–286 d Å	2.84 (20)	2.69 (10)	2.64 (30)	2.49 (10)	2.360 (2)	2.310 (5)	2.260 (10)
Ferreira d Å	2.84 (2)	–	2.65 (20)	2.50	2.37	2.26	–

25–286 d Å	2.160 (20)	2.090 (10)	1.889 (80)	1.871 (40)	1.832 (5)	1.615 (30)	1.597 (60)
Ferreira d Å	2.164 (2)	2.073	–	1.876 (94)	–	1.612	1.589 (68)

25–286 d Å	1.533 (20)	1.346 (20)	1.323 (40)	1.223 (50)	1.214 (20)	1.196 (5)	1.182 (10)
Ferreira d Å	1.533 (3)	1.340	1.325 (29)	–	1.216 (45)	–	1.186 (5)

25–286 d Å	1.091 (30)	1.083 (60)	1.033 (10)	1.025 (20)	1.018 (40)	0.9430 (30)
Ferreira d Å	1.089	1.082 (81)	–	1.020 (59)	–	0.9350 (30)

Table 5.16(b). X-ray diffraction lines of mooihoekite which shift and split as the mineral is leached; their spacings when 31.58 percent of its copper has been leached out (Ferreira); and the equivalent spacings of α-chalcopyrite (274, JCPDS File 25–291).

Mooihoekite			Leach residue	α-Chalcopyrite	
JCPDS 25–286 hkl	d Å	Ferreira d Å	Ferreira d Å	hkl	JCPDS 37–472 d Å
221	3.07 (100)	3.063 (100)	3.04	112	3.039 (100)
			3.025 (100)		
400	2.64 (30)	2.651 (20)	2.634	200	2.645 (3)
			2.609 (27)	004	2.606 (3)
440	1.871 (40)	1.876 (94)	1.863 (180)	220	1.870 (16)
			1.855	204	1.8561 (25)
621	1.597 (60)	1.598 (68)	1.589 (154)	312	1.5926 (12)
			1.576	116	1.5757 (6)
800	1.323 (40)	1.325 (29)	1.320 (61)	400	1.3223 (2)
			1.306	008	1.3027 (1)
661	1.214 (20)	1.216 (45)	1.211 (80)	332	1.2126 (2)
			1.205	316	1.2049 (3)
				JCPDS 35–752	
842	1.083 (60)	1.082 (81)	1.077 (157)	424	1.0770 (5)
			1.070		
1021	1.018 (40)	1.020 (59)	1.018 (117)	512	1.0173 (4)
			1.014	336	1.0128 (5)
			1.006		

in which some of the more important lattice spacings changed during leaching
are shown in Figs. 15(a), (b) and (c).

After 6.5% copper loss there was a relatively smooth variation in *d*-spacings
until about 17.5% copper loss, corresponding to a range of nonstoichiometry
in the phase and contraction of the unit cell. Intensity measurements on the
lines showed a disordered variation, possibly associated with a broadening of
some lines, but the others remained sharp. This broadening corresponds to
a deviation from Bragg's law, the reflection appearing over a range of angles.
The effect cannot be caused by small crystallite size since this would result in
all lines being broadened, and in this case it appears to be due to structural
faults. When these simulate a certain structure, those reflections which are
common to the basic and simulated structures are sharpened, whereas those
which are not common are broadened. During the first stage of leaching most
of the very weak superstructure lines disappeared. These had been revealed
by the use of cobalt K_α radiation.

It can be seen from the behaviour of the (661), (842) and (800) reflections
that the lines split to form doublets when 6.5% of the copper present in the
mooihoekite had been leached out. The (440) and (221) reflections split when

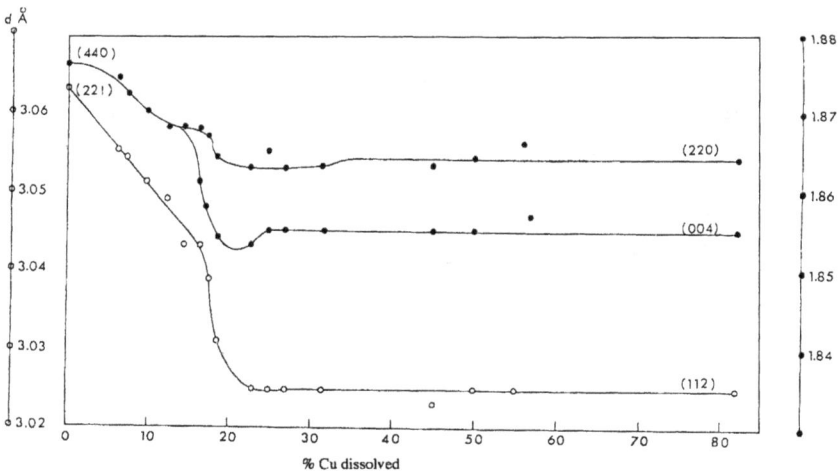

(a)

Fig. 5.15.   (a), (b) and (c) Changes in important lattice spacings of mooihoekite during
leaching as a function of the amount of copper removed from the solid.

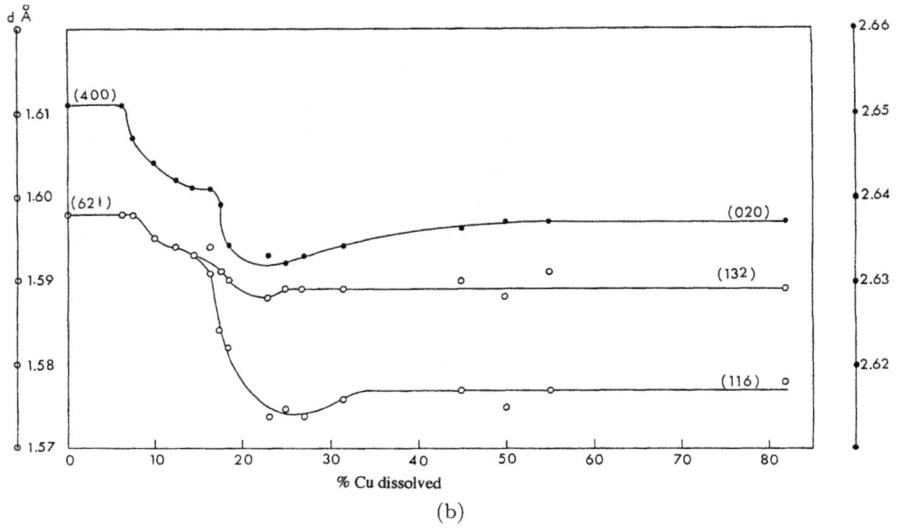

(b)

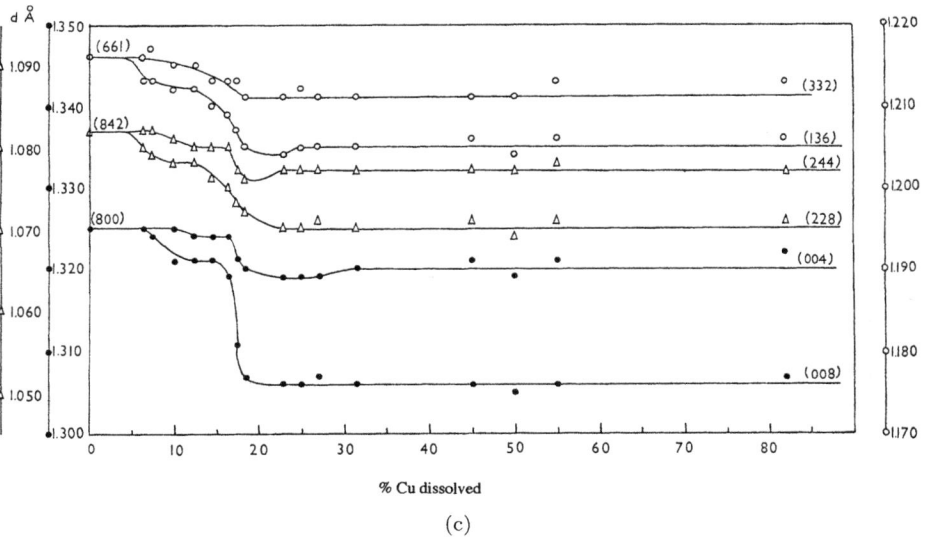

(c)

Fig. 5.15 (*Continued*)

about 17.5% of the copper had been lost and at this point an abrupt collapse of the structure occurred. In most cases the d-spacings fell sharply to smaller values than were measured using residues from which more copper had been leached. The structure only "recovered" from this collapse after about 31.5% of the copper had been lost. This range from about 17.5 to 31.5% corresponds to the second stage of the leaching reaction, Fig. 5.14.

During stage 3, the structure of the solid did not change and the X-ray diffraction pattern corresponds closely to that of α-chalcopyrite, Table 5.16(b). The lines are referred to here as though they are derived from that phase. However the intensities of lines which were measurable in the pattern from the leach residues are all much higher relative to the reference reflection (hkl 112) than in the pattern from α-chalcopyrite.

If the line showing how d for the (221) line changed as 0 to 15% of the copper was removed from the solid is extrapolated it cuts the constant value for the (112) line of α-chalcopyrite at about 35% copper removed. This extrapolation corresponds to the change in the lattice dimension which would occur in the absence of the collapse of the structure, Fig. 5.15(a).

Electron microprobe analyses of leach residues were carried out on particles which had been mounted and polished. Using normal electron beam energies, as the beam swept once across a particle cracks formed. During a second pass the solid broke up and apparently shrank. This behaviour was probably due to two facts: (i) in the early stages of leaching the crystal structure was severely strained and so collapsed when the solid had to dissipate a considerable amount of energy, and (ii) the solid from the later stages was very porous and so crumbled readily under similar conditions. Point counting in fixed positions with a very low beam energy had been used successfully to analyse partially leached bornite samples which decomposed using normal beam energy and was moderately successful with the residues from mooihoekite leaching. Particles of the unleached mineral behaved normally under the electron beam at normal energy and it was shown that the sample was homogeneous. Consistent and reproducible results were obtained only with this and with a sample from which 6.5% of the copper had been removed. The results obtained from some of the partially leached samples which had been analysed using both the electron microprobe and the atomic absorption method are given in Table 5.17. The composition of the solid which had the d-spacings of α-chalcopyrite had the composition $CuFe_{1.2}S_{2.22}$.

Table 5.17. Analysis of mooihoekite leach residues.

% Cu removed	Electron microprobe analyses				Atomic ratios Cu = 1					
	Cu wt %	Fe wt %	S wt %	Total %	Microprobe analyses			Chemical analyses		
					Fe	S	Metal : S	Fe	S	Metal : S
0	35.3	31.4	32.5	99.2	1.01	1.83	2 : 1.82	1.00	1.83	2 : 1.83
6.5	34.1	32.1	33.4	99.6	1.07	1.94	2 : 1.87	1.07	1.95	2 : 1.88
10.0	33.3	34.2	34.1	101.6	1.17	2.03	2 : 1.87	1.13	2.04	2 : 1.92
18.5	30.6	32.9	34.7	98.2	1.22	2.25	2 : 2.03	1.14	2.18	2 : 2.04
31.5	30.9	33.1	33.8	97.8	1.22	2.17	2 : 1.96	1.19	2.28	2 : 2.08
50.0	30.2	33.9	34.0	98.1	1.28	2.23	2 : 1.96	1.21	2.25	2 : 2.04
82.0	31.2	35.0	34.5	100.7	1.28	2.19	2 : 1.92	1.20	2.24	2 : 2.04

The chemical analyses of the solid samples removed during leaching runs indicated that both copper and iron were leached from mooihoekite during the reaction with Fe^{III} sulphate solution, but at different rates during the early stages of the process. Confirmation of this behaviour was obtained using hydrogen peroxide of initial concentration 0.45 M as the oxidant in HCl solution at pH 1.0. 0.200 g of 76–53 μm mineral was leached in 200 ml of the solution at 65°C. Samples of clear solution were taken at intervals; copper and iron were determined in the solutions by atomic absorption analysis and sulphate by nephelometry. Since the analyses could be performed rapidly large numbers of samples could be taken. Each was of accurately known volume so that allowance could be made for the volume removed when the compositions of the solids remaining in the liquors were calculated from the total amounts of the three elements which had been transferred to the leach liquor. A few samples of solid were also taken during the run and analysed. The results are shown in Figs. 5.16(a), (b) and (c).

At the start of the reaction 6 to 7% of the copper in the mineral was leached out with no loss of iron by the solid, Fig. 5.16(a). This is the range in which no change takes place in the cell dimensions. After this both copper and iron appeared in the solution but at different rates until after 17.5% of the copper had been leached out, when the rate of dissolution of iron relative to that of copper increased somewhat. The crystal structure collapsed when 17.5% of its copper content had been lost. At this point the ratio of copper plus iron to sulphur reached the value 2:2, having risen from 2:1.83. Thus this is the composition range over which the γ-chalcopyrite phase is stable towards removal of copper by leaching.

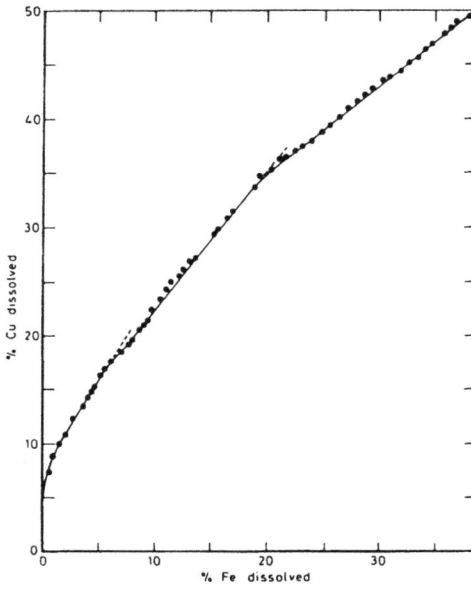

(a)

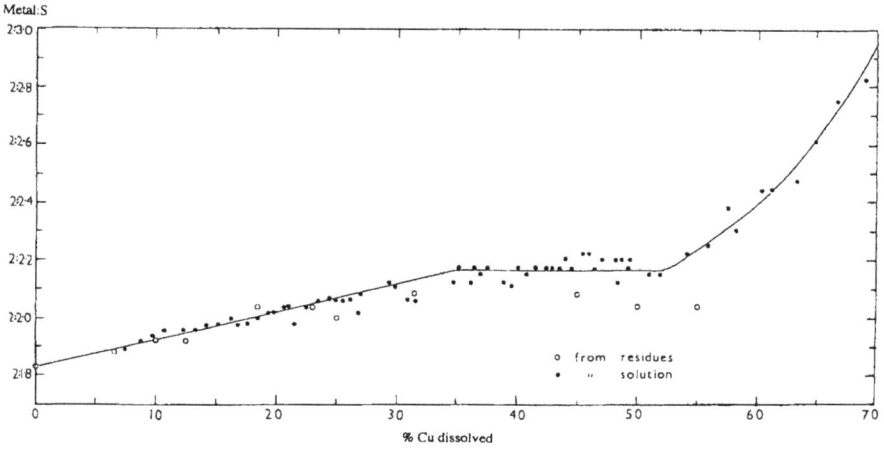

(b)

Fig. 5.16.   (a) and (c) Changes in ratios of the metals present in mooihoekite as leaching proceeds. (b) Changes in the metal : sulphur ratio.

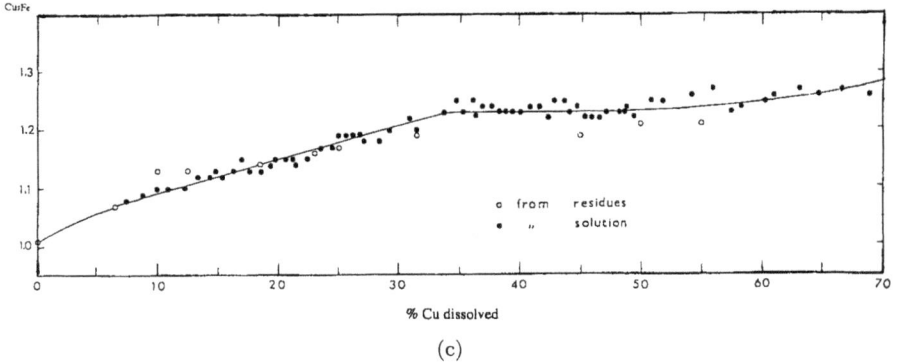

(c)

Fig. 5.16. (*Continued*)

The rate of removal of iron relative to copper increased again when 34% of the copper had been lost and the solid reached a composition between $CuFe_{1.2}S_{2.3}$ and $CuFe_{1.2}S_{2.4}$. The composition of the solid as determined by analysis of the solution for copper, iron and sulphate remained constant beyond this point until about 50% copper loss had occurred. An abrupt increase in its metal : sulphur ratio appeared to take place in this region, Fig. 5.16(b). However it was noted above in the description of the general characteristics of the reaction that the liquid then had in suspension a small amount of sulphur and of a white solid which contained appreciable amounts of iron. Thus when less than 50% the copper present in the mooihoekite remained in the residual solid, analysis of the solution for dissolved copper, iron and sulphate did not permit the chemical composition of the leach residue to be calculated. Up to that point, the amount of elemental sulphur formed must have been small.

The results show that when mooihoekite is leached with iron (III) sulphate solution acidified with sulphuric acid, or with hydrogen peroxide solution acidified with hydrochloric acid the composition of the solid changes progressively and eventually produces a material of composition approximating to $CuFe_{1.2}S_{2.3}$ having lattice parameters very close to those of α-chalcopyrite. This composition can be explained in terms of conventional valencies if it is assumed that sulphur remains present as S^{2-} only, none as S_2^{2-} groups. The metals can be present as Cu^{I} or Cu^{II} and Fe^{II} or Fe^{III}. If all copper is present as Cu^{I} in γ-chalcopyrite then for charge balance the formula is $Cu^{I}Fe_{0.34}^{II}Fe_{0.66}^{III}S_{1.83}^{2-}$. Thus 34% of the iron is present as Fe^{II}; during the first and second stages of the leaching reaction about 34% of the copper was leached out, the solid achieved the α-chalcopyrite structure and its composition remained constant during subsequent reaction. The oxidising agent in the solution accepted the electrons from the Fe^{II} as it formed Fe^{III} to maintain charge balance.

However, while the 34% of the copper present was being removed during stages 1 and 2 of the reaction, it was found experimentally that approximately 19% of the iron and 15% of the sulphur present in the solid was also leached from it. This leads to a final theoretical composition for the solid remaining of $Cu^I Fe^{III}_{1.277} S^{2-}_{2.341}$, in close agreement with that found experimentally, Fig. 5.16(c). Further oxidation of the solid could in principle take place by oxidation of Cu^I in the solid to Cu^{II} coupled with leaching out of Fe^{III}, but this does not happen under the conditions used. Instead slow attack on the solid continues with S^{2-} being oxidised to elemental sulphur. During the final stages in the leaching of *chalcocite* some oxidation of S^{2-} occurs with the formation of the S^{2-}_2 groups required to form the covellite lattice structure, and possibly the structures of yarrowite and spionkopite[229] but this does not occur apparently during the leaching of mooihoekite. In view of the instability of the Cu^+ ion in sulphate solution and in dilute chloride solution, in which chloro-complexes do not form, oxidation occurs so that Cu^{II} is the soluble product of the reaction.

Although the d-spacings of the X-ray diffraction pattern of the product are very similar to those of α-chalcopyrite their relative intensities are not, Table 5.16(b). The position in space of the X-ray beams diffracted by crystals are determined by the basic pattern of atoms in the unit cell. The intensities of the diffracted beams depend on the arrangement and positions of the atoms in the unit cell. Thus as mooihoekite is leached its unit cell changes progressively to resemble that of α-chalcopyrite more closely. This does not produce the unit cell structure of either talnakhite or haycockite as it occurs. The diffusion of cations in copper-containing sulphides is considered in Sec. 5.3.6.

Leaching of Natural α-Chalcopyrite, $CuFeS_2$

Leaching using oxygen as oxidant. Warren[211] studied the leaching of a chalcopyrite concentrate, which contained about 20% pyrite, in sulphuric acid using oxygen under pressure, about 150 lb in^{-2} (1 MPa), as the oxidant at temperatures between 120°C and 180°C. Elemental sulphur was formed together with some sulphate; copper dissolved as did all or part of the iron, the extent of hydrolysis and precipitation of the iron depending on the conditions.

When the pH was 1 and p_{O_2} 100 lb in^{-2} (0.7 MPa) hydrogen sulphide was produced. An extensive study of the pressure leaching of chalcopyrite with oxygen was carried out using the shrinking core model to describe the kinetics of the reaction, Sec. 4.3.2.[120]

During a temporary shortage of sulphur during the 1960's, plans were made to produce it as a by-product in other processes. One proposal[277] was to use the oxidation of chalcopyrite concentrates by oxygen in sulphuric acid to give sulphur and a copper (II) sulphate solution. In order to recover copper metal from this by electrowinning it was necessary that the concentration of iron should not exceed 5 g L^{-1} Fe whereas the copper content should be as high as possible. This was achieved by using a large stoichiometric excess (about 50%) of concentrate over the amount of sulphuric acid available for dissolution of copper, which caused the pH of the solution to rise to a level at which hydrolysis of the iron occurred, causing precipitation of what was described as iron (III) oxide, but probably contained basic sulphates. The concentrate was ground to 99.5% finer than 44 μm to obtain acceptably fast reaction at 115°C with p_{O_2} 500 lb in^{-2} (3.45 MPa), a retention time of 2.5 hours in the reaction vessel then being sufficient. The resulting slurry consisted of a solution containing about 80 g L^{-1} Cu, 5 g L^{-1} Fe, together with elemental sulphur, residual unreacted chalcopyrite, "iron oxides", and unreacted gangue (waste impurities) which had been present in the concentrates. The slurry was heated to about 138°C to melt the sulphur, which then wetted the particles of sulphide minerals. These agglomerated so that while being allowed to cool the sulphur and sulphides could be pelletised, pellets 1–3 mm in diameter being convenient. The oxides and nonsulphidic gangue could be readily separated from the solution by clarification or filtration and discarded to waste. The elemental sulphur could be separated from the sulphide minerals in the pellets by hot filtration, distillation, extraction by organic solvents, or some other convenient method. The unreacted chalcopyrite could be recycled for further leaching, but if the concentrate contained pyrite this would not have reacted during the leaching period and in order to prevent it building up in the circuit it would be necessary to remove it by froth flotation before recycling the chalcopyrite.

Rejection of iron during the leaching step is desirable or necessary in many hydrometallurgical processes and where possible is achieved by causing hydrolysis of Fe^{III} to occur at a pH value at which the elements which are to be recovered are not precipitated. In the case of metal sulphides oxidative leaching can either form elemental sulphur, sulphate, or both, when O$_2$ or Fe^{III} is used as the oxidant in strongly acidic solutions (pH < 3).

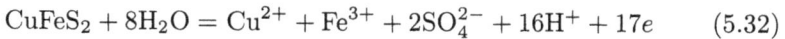

$$CuFeS_2 + 8H_2O = Cu^{2+} + Fe^{3+} + 2SO_4^{2-} + 16H^+ + 17e \qquad (5.32)$$

$$CuFeS_2 = Cu^{2+} + Fe^{2+} + 2S^\circ + 4e. \qquad (5.33)$$

Reaction (5.32) occurs under strongly oxidising conditions, very often at high temperatures, (5.33) under milder conditions. In the latter case the Fe^{II} may be oxidised to Fe^{III} by a separate reaction unconnected with the decomposition of the chalcopyrite. Elemental sulphur has only a limited range of stability in aqueous systems if equilibrium is achieved, Sec. 3.4.4. and Fig. 3.2, and the extent of that region is particularly dependent on the activity of $H_2S(aq)$, and HS^- derived from it. At constant activities of species the extent of the area of sulphur stability is smaller the higher the temperature, Sec. 3.6.2. Thus use of a high temperature when leaching sulphide minerals favours the formation of SO_4^{2-} or HSO_4^- by increasing the range of E and pH values at which S° is thermodynamically unstable, and also by increasing the rate at which it is oxidised if it forms as a metastable product of a reaction. This can occur because the activity of the solid chalcopyrite is 1 at the surface of the mineral and it is acting as the reducing agent in the chemical reaction which is taking place. This may result in the production of a higher thermodynamic activity of $H_2S(aq)$ or HS^- in the reaction zone than in the bulk of the solution.

The sequence of reactions which occur during th oxidative leaching of chalcopyrite in sulphuric acid using oxygen, under conditions causing hydrolytic precipitation of iron was studied by Le Houillier and Ghali[278] who used the following stoichiometric equations as the basis of the interpretation of their experimental results:

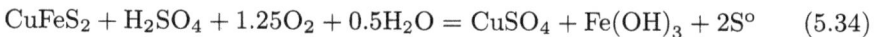

$$CuFeS_2 + H_2SO_4 + 1.25O_2 + 0.5H_2O = CuSO_4 + Fe(OH)_3 + 2S^\circ \qquad (5.34)$$

$$S^\circ + 1.5O_2 + H_2O = H_2SO_4. \qquad (5.35)$$

Whether the acid produced according to (5.35) is formed directly from S^{2-} in chalcopyrite or from the S° formed by reaction (5.34) is of no consequence for the stoichiometry, although the distinction may be important for the kinetics of the reaction. In practice the conditions used in the proposed process outlined above resulted in a reaction which after 2.5 hours could be represented by (5.34) with about 15% of the sulphur remaining in elemental form.

The conditions selected for use in the experimental work were as follows. Oxygen pressure was not taken as being a variable and was kept constant at 70 lb in^{-2} (0.48 MPa) throughout each experiment. Initial sulphuric acid concentrations were in the range 0.1–1.8 M and temperature 50–110°C, with most experiments at 90°C.

Two size fractions were taken from the sample of chalcopyrite concentrate and used in the experiments. Their particle sizes and compositions were:

$$-200 + 270 \text{ mesh } (53\text{–}74 \ \mu\text{m}); \text{ Cu, 28.4; Fe, 31.6\%}$$

$$-400 \text{ mesh (smaller than 5 } \mu\text{m}); \text{ Cu, 24.1; Fe 36.9\%}$$

$$\text{CuFeS}_2 \text{ requires; Cu, 34.625; Fe, 30.430; S, 34.945\%}.$$

Analytical data for the concentrate as received were:

$$\text{Cu, 24.6; Fe 34.7; S, 35.4\%; total, 94.7\%}.$$

Other materials listed as present in the concentrate included:

$$\text{Zn, 1.03; SiO}_2, 1.22\%. \quad \text{Total listed; 5.22\%}.$$

Both size fractions had a low copper and high iron content relative to chalcopyrite. The analytical data for the as-received concentrate almost quantitatively account for the whole of the material so that it must be concluded that both fractions used for leaching experiments contained iron sulphides in addition to chalcopyrite.

Each experiment was carried out using 50 ml of a sulphuric acid solution and 10 g of chalcopyrite in a glass reactor with a magnetically driven stirrer, held in a thermostat bath. No samples were taken. At the end of the leaching period the slurry was filtered and the solution analysed. The solid was analysed by chemical methods. In addition in some cases the solids were examined by (i) X-ray powder diffraction; (ii) scanning electron microscopy. The specific surface areas of the weighed residues from some experiments were determined.

The relative stoichiometric quantities of mineral and H_2SO_4 present initially in the reaction vessel are important for the interpretation of the experimental results. 10 g of the 53–74 μm chalcopyrite contains 0.0450 g atom Cu^{II} and 0.0566 g atom Fe^{II}. In many experiments 0.9 M H_2SO_4 was used. 50 ml of this contains 0.0450 g mol H_2SO_4. Suppose that 35% of the Cu and Fe present in the 10 g of mineral has dissolved. Then 0.0158 g atom of Cu^{II} and 0.0198 g atom of Fe^{III} are present in the solution, the iron having been

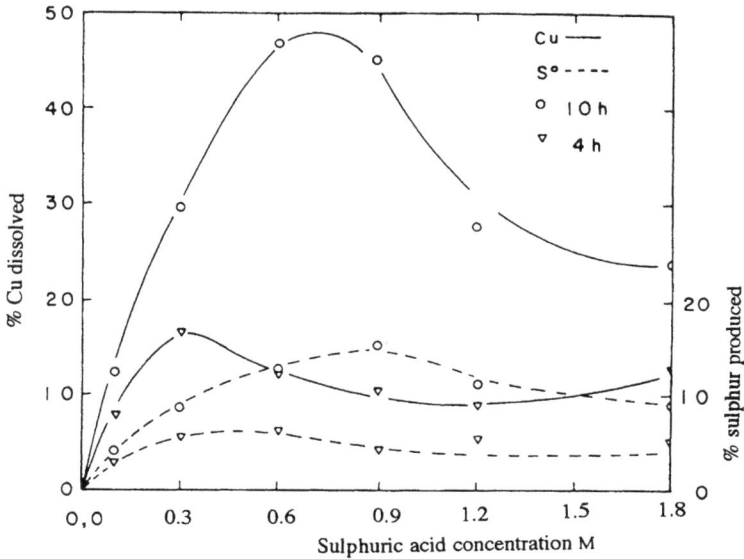

Fig. 5.17. Effect of acid concentration on the amount of copper dissolved and sulphur produced after leaching for 4 hours and 10 hours at 90°C. Particle size 63 μm.

oxidised according to Reaction (5.32). The total quantity of H_2SO_4 required to form soluble sulphates of these metals is 0.0455 g mole, which is more than was present initially. Thus when leaching has procceded to the extent of 35% metal dissolution the only free acid remaining in solution is that produced by oxidation of S^{2-}, directly or by Reaction (5.35).

The experimental results of the investigation are given in Figs. 5.17 to 5.24. The points of interest are as follows.

Acid concentration. Figure 5.17. The amounts of copper in solution and of elemental sulphur in the residues after leaching for 4 hours and for 10 hours at 90°C increase to maximum values as the acid concentration is increased and fall again at higher concentrations.

Acid concentration. Figure 5.18. The amount of iron in solution increased with time initially but in 0.3 and 0.9 M acid it reached a maximum value and then fell. With 0.3 M acid the pH of the solution rose sharply as reaction commenced since a substantial fraction of the acid present was

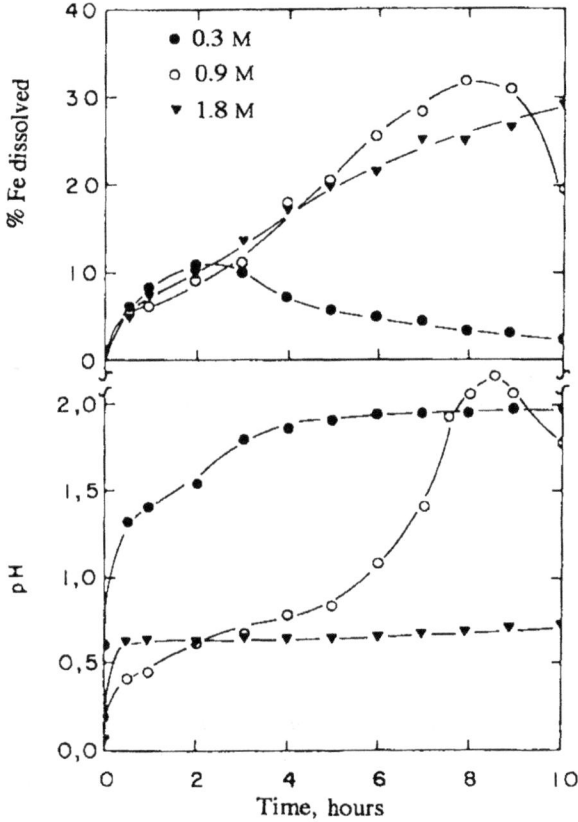

Fig. 5.18. Correlation between initial acid concentration and pH change as iron dissolves and is subsequently hydrolysed as the pH rises above 2.0.

consumed rapidly. It then remained steady at just below 2 as iron precipitated out of the solution. In 0.9 M acid the increase in iron concentration in solution was accompanied by a rise in pH as acid was consumed, until the pH rose to slightly above 2.0, when about 30% of the iron present in the chalcopyrite had dissolved. It can be seen from Fig. 5.19 that at that point, after about 8 hours leaching at 90°C, approximately 35% of the copper present had also dissolved. This corresponds to almost complete comsumption of the H_2SO_4 initially present, as was calculated above. The amount of iron in solution then decreased and the pH fell to below 2.0 as hydrolysis occurred, see Secs. 3.3.9

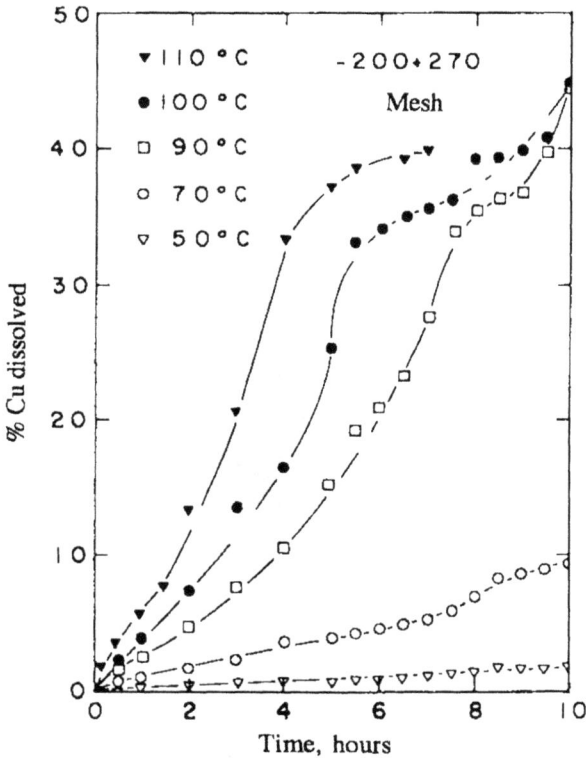

Fig. 5.19.   Data for measurement of apparent activation energy, see text.

and 3.8.6. In 1.8 M sulphuric acid dissolution of iron was continuing steadily after 10 hours and the pH of the solution remained low because the proportion of the acid present initially which had been neutralised was only about 30%.

Temperature.   Figures 5.19 and 5.20. The data for copper dissolution, Fig. 5.19, were obtained in order to measure the apparent activation energy of the leaching reaction. This was achieved by using a mathematical model described below, which was applicable to the first 2 to 6 hours of the leaching period. The initial rate of iron dissolution, Fig. 5.20, was faster than that of copper, which may be due to rapid reaction of pyrrhotite if the excess iron content of the sample leached was due to the presence of that mineral. At 110°C, the concentration of iron fell sharply after 4 hours of leaching, when about 30%

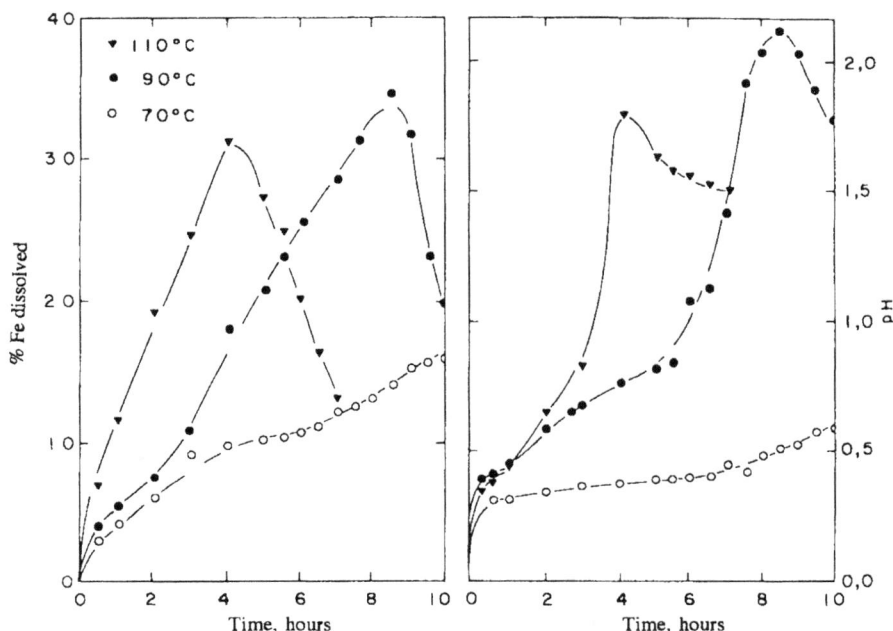

Fig. 5.20.   Data for measurement of apparent activation energy, see text.

dissolution had occurred, the pH of the solution falling simultaneously from about 1.7 to 1.5. At 90°C, dissolution of iron was slower than at 110°C but the maximum concentration reached was about 35% dissolution after 8 hours, and the pH about 2.1. The subsequent rapid decrease in concentration was accompanied by a fall in pH. Data for both Fe and Cu are given for temperatures 70°C, 90°C and 110°C and it should be noted that the time at which the maximum Fe concentration was reached at 90°C and 110°C coincided with the start of a shoulder in the rate curves for copper at the same temperatures. No such maximum and shoulder was reached at 70°C in 10 hours because the pH had not risen to a value at which hydrolysis of iron occurs.

Particle size.   Figures 5.21 and 5.22. The shoulder in the rate curve for Cu dissolution at 90°C from 53–54 μm chalcopyrite in 0.9 M H_2SO_4 occurred when approximately 35% of its copper had been dissolved. The minus 5 μm material also showed a shoulder at which little or no copper dissolution

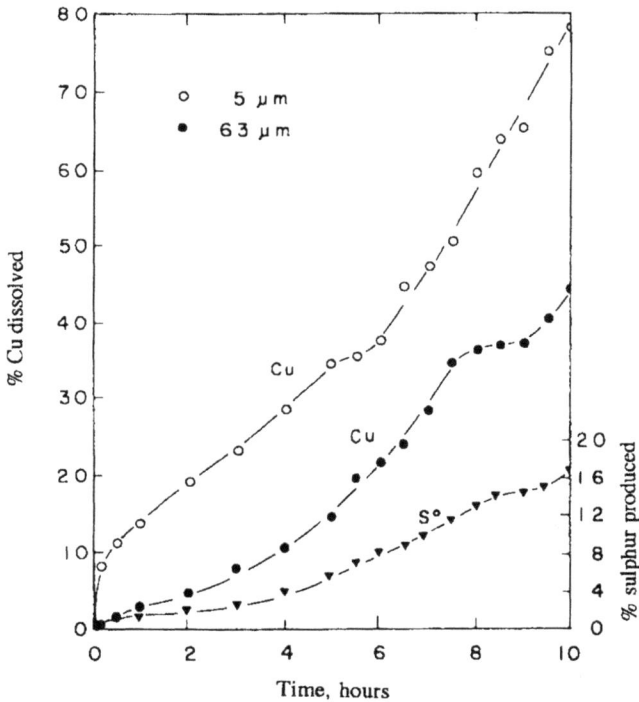

Fig. 5.21. Rates of dissolution of copper and iron, formation of sulphur, and change of pH in 0.9 M sulphuric acid at 90°C.

occurred for about 1 hour, again when about 35% copper dissolution had occurred. The rate of formation of elemental sulphur from the 53–74 μm solid also appeared to decrease significantly during the period when copper dissolution was slow. The rate curves for iron dissolution and pH change in the solutions for the two size fractions of the chalcopyrite are shown in Fig. 5.22. The maximum concentrations of iron in both cases occur at the maximum values of pH in the solutions and also during the period when dissolution of copper has almost ceased.

Surface areas of chalcopyrite particles and partially leached residues. The samples of solid residues were washed with CS_2 to remove the orthorhombic sulphur produced by the reaction, before surface areas were measured. The results obtained for the 53–74 μm and minus 5 μm solids,

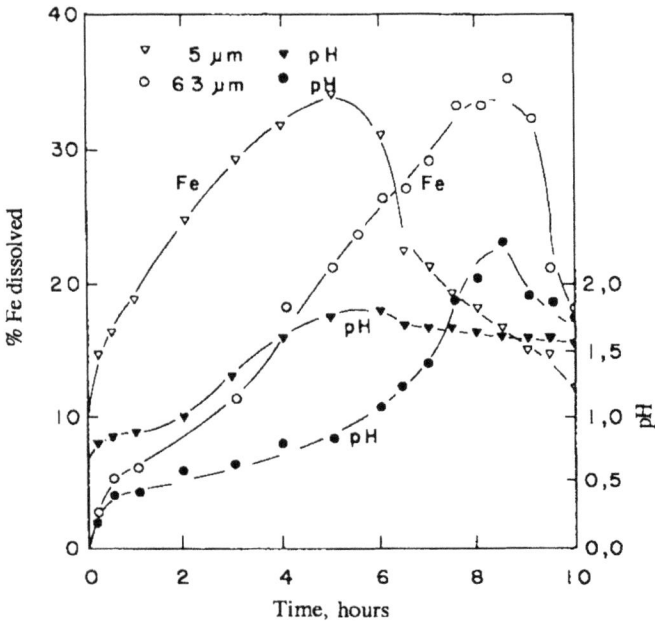

Fig. 5.22.   Rates of dissolution of copper and iron, formation of sulphur, and change of pH in 0.9 M sulphuric acid at 90°C.

and partially leached residues from them are given in Table 5.18. The specific surface area and total mass of each initial size fraction and leach residue are given so that the total area of surface of each leach residue derived from 10 g of the chalcopyrite can be seen.

After leaching the 53–74 μm mineral for 1 hour, 2.7% of the copper content and 6.2% of the iron had been leached from it. The residual solids had only 26% of the surface area of the original material. The explanation offered for this was that the original particles had very rough and irregular surfaces and that these irregularities would be removed during a short initial period of leaching. The minus 5 μm mineral behaved similarly but to a lesser extent. Subsequently the total surface areas of the two materials remained practically constant as leaching continued for some time, so that the specific surface areas of the residues increased as leaching continued. During the period between 4 and 5 hours leaching of the minus 5 μm mineral the total surface area more than doubled, although the total mass of solid fell from 8.49 to 8.00 g. This coincided

Table 5.18. Surface areas of chalcopyrite fractions leached and of partially leached residues.

Leaching conditions 0.9 M H_2SO_4; p_{O_2} 0.48 MPa; 90°C.
Leached solids washed with CS_2 to remove sulphur produced.

Leaching time hours	Specific surface $m^2 g^{-1}$	Weight of residue g	Total surface m^2
53–74 μm chalcopyrite			
0	0.50	10.00	5.0
1	0.14	9.35	1.3
3	0.15	8.86	1.3
5	0.16	8.04	1.3
6.5	0.23	7.24	1.7
minus 5 μm chalcopyrite			
0	5.18	10.00	51.8
1	2.55	8.56	21.8
3	2.75	8.04	22.1
4	2.48	8.49	22.1
5	5.77	8.00	46.2

with the beginning of the rapid decrease in the concentration of iron in the solution. The increase in total surface area of the 53–74 μm sample between 5 and 6.5 hours leaching was much less than for the smaller particle size material, but occurred just before the iron concentration reached its maximum value. The authors state "The appearance of iron sulphate or another compound in the residue contributed without doubt to the increase in the specific surface area".

Microscopic examination of surfaces of leach residues. Surfaces were examined using a scanning electron microscope. After 4 hours leaching in 0.9 M H_2SO_4 at 90°C, the surfaces were covered with a porous deposit of sulphur which does not, at a magnification of 1700× appear in the published photomicrographs to be fibrous. After removal of the sulphur by washing with CS_2 the surfaces of the particles can be seen to be heavily pitted and severely attacked. The particles are said to be much more rounded after 8 hours leaching, with most surfaces covered with deposits but with some regions free from reaction products. After washing with CS_2 the pitting could be seen to be far more pronounced than after 4 hours. Many small particles can be seen in photomicrographs (600×), associated with the residual mineral. Even with the sulphur present this appears to be heavily pitted and after washing with CS_2 it is obvious that the residual mineral particles are skeletal. Using X-ray powder difraction, no sulphate or hydroxide of Fe^{III} was detected in residues after leaching 53–74 μm chalcopyrite in 0.9 M H_2SO_4 at 90°C for 4 hours. However after 7 or 8 hours $Fe_2(SO_4)_3 \cdot 10\ H_2O$, or $\cdot 9H_2O$, and $Fe(OH)_3$ were found to be present. This indicates that the $Fe(OH)_3$ was crystalline. See Sec. 3.8.6.

Thus as leaching takes place the sulphur formed by the reaction coats the particles, but since the rate of the reaction increases during this first few hours, as can be seen in Fig. 5.19, the coating is obviously porous. The residual mineral is heavily pitted and its appearance is compared to that of metals being severely corroded, with strongly anodic and cathodic areas. Precipitation of basic iron compounds, accompanied by rising acid concentration caused by the hydrolysis and attributed also to oxidation of elemental sulphur at this stage of the reaction, creates more dense coatings on the mineral surfaces and decreases the rate of reaction. The photomicrographs show that the particle size of the precipitate is very small and if some of it is crystalline $Fe(OH)_3$, as is stated, it could be expected to block the pores in the coating of sulphur.

The rate curves for copper dissolution shown in Fig. 5.19 are parabolic for the early stage of the reaction and a mathematical model was proposed to describe them. If the reaction at the surface remains unchanged then $y = kt^{1/2}$ where y is the percent Cu dissolved, k is the rate constant and t the time the reaction has been in progress. If the increase in rate as the reaction proceeds is due to the formation of pits and peaks on the particle surfaces, which are clearly visible in the photomicrographs, then the rate of leaching at a given time is a function of the amount of copper dissolved, since this contributes to the increase of surface. Then

$$dy/dt = f(y, t^{-1/2}).$$

Suppose that

$$dy/dt = f(y) \cdot f(t^{-1/2}) \tag{5.36}$$

then

$$dy/dt = K'y(t^{-1/2})$$

where K' is a proportionality constant. On integrating

$$y = y_o \exp(Kt) \tag{5.37}$$

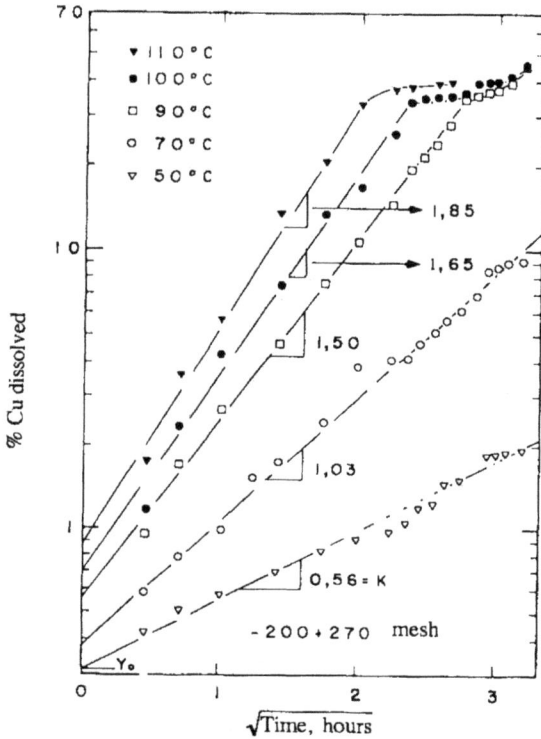

Fig. 5.23. Percentage of dissolved copper as a function of $t^{1/2}$ according to Eq. (5.36).

where y_o is an integration constant and K is a constant for given conditions. It will be a function of the partice size, p_{O_2}, initial pH of the solution and temperature of the reaction. Thus plotting log y against $t^{1/2}$ gives a straight line for a given temperature with intercept y_0 and slope K. The data in Fig. 5.19 provide Fig. 5.23, the slopes of the lines being

Temperature, °C	50	70	90	100	110
Slope, K	0.56	1.03	1.50	1.65	1.85.

Values of y_o are not given in the paper. In order to calculate the apparent activation energy Eq. (5.38) is derived from (5.37).

$$y(t) = 1/2y_oKt^{-1/2}\exp(Kt^{-1/2}). \qquad (5.38)$$

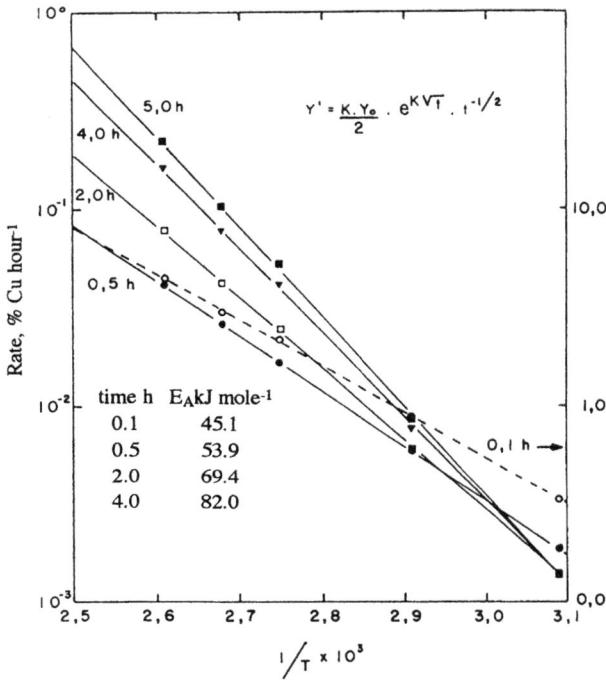

Fig. 5.24.   Lines providing the apparent activation energies.

Since the rate curve is parabolic the rate at any given temperature changes as the reaction proceeds and the rate at a particular time is given by the slope of the curve at that time. This is usually estimated, with doubtful accuracy, from the extents of reactions at two closely spaced times. These rates are used in the Arrhenius equation to obtain apparent activation energies at different stages (times) of reaction. In the procedure used here values of $\ln y'(t)$ are calculated from (5.38) for a number of times of reaction and plotted against $1000/T$, Fig. 5.24; the units of rate are % Cu dissolved h^{-1}. The values of E_A at different reaction times t are:

$$E_A, \text{kJ mol}^{-1} \quad 45.2 \quad 54.0 \quad 69.5 \quad 82.4 \quad 88.7$$

$$\text{Time } t, \text{ hours} \quad 0.1 \quad 0.5 \quad 2.0 \quad 4.0 \quad 5.0.$$

The value of 45.2 kJ mol^{-1} is in good agreement with the value of 46 kJ mol^{-1} reported in the literature, calculated using the equation $y = kt^{1/2}$. This is not applicable however at longer reaction times. Other values of E_A reported in the literature generally lie between about 38 and 134 kJ mol^{-1} and this range now appears to be due to the use of kinetic data obtained under different chemical conditions in the system.

Equation (5.38) can be broken down into a series

$$y = y_o \left(1 + Kt^{1/2} + \frac{K^2 t}{2!} \cdot \frac{K^3 t^{3/2}}{3!} \cdots \right). \tag{5.39}$$

It is said that the terms in t and $t^{3/2}$ can be neglected at temperatures below 70°C and for leaching times of about 1 hour because the value of K is fractional. These terms may also be neglected for higher temperatures when small percentages of copper have dissolved. With higher temperatures and leaching times of 1 hour or longer the term in t should be used, giving logarithmic kinetics. The other terms in the series become important for increasing temperatures and leaching times.

Kinetics of Leaching Using Iron (III) Sulphate and Chloride as Oxidants

Hot concentrated sulphuric acid decomposes chalcopyrite readily, forming only the metal sulphates and sulphur, together with some SO_2 produced because the H_2SO_4 acts as the oxidant required for the reaction to occur.[279] Acidified solutions of chlorine decompose chalcopyrite with formation of sulphur and some sulphate.[280] Because of the practical difficulties which were inherent in the use of Cl_2 in leaching processes before the development of new technology for nickel production the reaction was not used commerically. Aqueous iron (III) chloride solutions were suggested as an alternative.[269] These are extremely corrosive towards metals commonly used in industrial process equipment, and iron (III) sulphate in sulphuric acid solution was examined as an alternative,[281] although the reaction is much slower than with chloride solutions. The problems associated with the use of chlorine and chlorides in leaching processes have now been overcome and several chloride routes for the treatment of chalcopyrite concentrates have been proposed. However, use of iron (III) sulphate is still attractive because the anion is derived from the mineral itself and is usually more easy to dispose of as an effluent from a plant than is chloride.

A large number of research studies on the leaching of chalcopyrite using iron (III) as the oxidant were carried out during the 1970's and early 1980's, mainly in the USA and Canada, as part of the programme to develop alternative routes to smelting for producing copper from sulphide ores. The amount of published information which resulted was very considerable and was summarised in several reviews, which are referred to here as sources of detailed information. There is no evidence which suggests that natural chalcopyrite mineral changes its composition during leaching reactions, forming new copper–iron sulphide phases. The solids which are formed when that mineral is heated so that it loses part of its sulphur content without oxidation taking place, leach much more rapidly than the original mineral, as was discussed above. Some of these materials, for example mooihoekite (γ-chalcopyrite), do change their composition during leaching reactions. For these reasons experimental work carried out using synthetic or sintered "α-chalcopyrite" is not considered here. If the material behaved differently from the natural mineral, which was certainly the case in some published work, confusion would arise.

When considering results of experimental work those obtained using particulate mineral should be distinguished from those where a flat of a piece of α-chalcopyrite was used. This latter experimental arrangement is frequently employed when electrochemical methods are to be used, the specimen being mounted in resin and the exposed face ground flat and, usually, polished. As reaction proceeds particles of mineral become smaller, the surface area of the flat face remains constant or increases as it becomes roughened. Also the sulphur formed by the reaction adheres to the surface to a greater or lesser extent. As is the case in many leaching reactions, the tenacity with which it does so depends on the leaching conditions and the physical nature of the mineral surface with which it is in contact. Particles may become completely surrounded by a shell of this sulphur which is then highly tenacious. It is, however, porous and the rate of reaction of the residual mineral particles is determined by the rate of diffusion of reactants and soluble products through the layer of sulphur, and the surface area of the residual mineral. The meanings to be attributed to the terms linear and parabolic kinetics referring to the leaching reactions of α-chalcopyrite need careful consideration in some cases. See Sec. 4.3.2.

The general aspects of the behaviour of α-chalcopyrite when leached with iron (III) sulphate and chloride solutions were clearly shown by Jones and Peters.[282] They used an exceptionally pure specimen of chalcopyrite approximately 10 cm in diameter. Individual small pieces were cut from it and crushed as required for particulate leaching experiments. Wet grinding was avoided. Cubes of side approximately 1 cm were cut for use as massive samples for leaching and electrochemical measurements, mounted in epoxy resin so that one face was in contact with the stirred solution with which it was reacting.

A typical rate curve for a flat surface leached in 1 M $Fe_2(SO_4)_3$ + 0.2 M H_2SO_4 at 90°C shows a slow, increasing rate for about 2 days followed by linear copper dissolution during a total leaching period of 55 days. The published graph indicates that 3×10^{-3} g atom cm^{-2} of Cu dissolved during a 40 day period of constant reaction rate.

In the chloride system, using 0.1 M $FeCl_3$ + 0.2 M HCl at 90°C the method of preparing the flat surface affected the rate of reaction during the initial period. About twice as much copper dissolved during the first 80 hours from a saw-cut surface as from one which had been polished. The rate of dissolution from the former then decreased rapidly and became equal to that from the surface which had originally been polished, both then having linear kinetics, that is a uniform rate constant. The data plotted for a total leaching period of 330 hours show that during the final 250 hours, when the leaching rate was constant, 1.83×10^{-3} g atom cm^{-2} of Cu dissolved. Thus the rates in 1.0 M Fe^{III} sulphate and 0.1 M Fe^{III} chloride were 0.075×10^{-3} and 0.178×10^{-3} g atom cm^{-2} of Cu per day respectively.

The rate curve obtained when particulate chalcopyrite (smaller than 37 μm) was leached in 0.1 M $Fe_2(SO_4)_3$ + 0.18 M H_2SO_4 at 90°C was not linear; the rate slowly decreased until the reaction ceased when about 70% of the copper had been lost by the solid. When $[1 - (1 - \alpha)^{1/3}]$ was plotted against time, α being the fraction of the mineral already reacted, a straight line was obtained up to about 40% copper dissolution. This is Eq. (4.29) derived in Sec. 4.3.2 for the shrinking particle model, which corrects measured reaction rates for decreasing surface area (particle size) as reaction proceeds, if certain assumptions are made. The alternative, parabolic plot of rate versus $t^{1/2}$ showed approximate linearity up to 70% copper dissolution, when reaction ceased. In separate experiments it was shown that the rate and total extent of reaction when leaching under these conditions were not increased if the particle size of the solid was smaller than the 150–300 μm fraction.

This was not the case in the chloride system. The rate curves obtained when particulate chalcopyrite size fractions smaller than 37 μm; 150–300 μm; and 1000–1400 μm were leached in 0.1 M $FeCl_3$ at 90°C gave excellent straight lines when $[1 - (1 - \alpha)^{1/3}]$ was plotted against time, the rate of leaching being considerably faster the smaller the particles. This is described by the authors as showing that linear kinetics are observed in the chloride system up to 90%

copper dissolution compared with linear kinetics in the sulphate system up to 40% reaction. It could also be said that the experimental data are consistent with the postulate that over the course of the leaching reaction for which the data used apply, the rate of dissolution of copper is determined by a single rate constant k, having a different value in each system, and the surface area of the unreacted mineral at time t. The data might, of course, fit equally well other models which have not been tested with it.

Addition of Fe^{II} chloride to a solution of $FeCl_3$ had no effect on the rate at which particulate chalcopyrite was leached. However when $FeSO_4$ was added to an Fe^{III} sulphate solution the leaching rate was greatly reduced. For example with other conditions the same, copper extraction in 0.1 M $Fe_2(SO_4)_3$ was 20% after 70 hours with no $FeSO_4$ present in the solution initially, but only 4% if the solution was initially 0.1 M in $FeSO_4$. If this same solution was diluted with 0.1 M $Fe_2(SO_4)_3$ solution until it was only 0.01 M in $FeSO_4$, the rate of leaching was identical with that in the solution containing no $FeSO_4$. Decreasing the Fe^{III} to Fe^{II} ratio lowers the oxidation potential of the solution and hence the driving force of the oxidation reaction. In commerical plants leaching uranium oxide ores the Fe^{III} to Fe^{II} ratio is maintained at a required value in the sulphate leaching solution in order to achieve an adequate rate of oxidation of the U^{IV} in the ore to U^{VI}, which passes into solution.

Since it had been suggested that the rate of leaching of chalcopyrite in Fe^{III} sulphate solutions was significantly affected by sulphur produced on the solid surfaces, polished specimens were leached and examined using a scanning electron micrscope. After treatment in 1 M $FeCl_3$ at 90°C for 11 hours and for 5 days the surfaces of the two specimens were covered with sulphur which at magnifications 100× and 400× appeared as porous solid and at 1000× and 2000× was clearly a network of fibres. After leaching in 1 M $Fe_2(SO_4)_3$ at 90°C a specimen showed some areas with a coating of sulphur after 71 hours and 305 hours, but with polishing marks still present on the mineral surface. Other areas of the same leached specimen showed that attack had produced fissures which contained elemental sulphur. These areas indicated that reaction had been occurring, with sulphur not adhering to the surface but retained in fissures produced by the reaction.

Very closely sized natural α-chalcopyrite, 47–64 μm; 12–16 μm and 2–5 μm, gave parabolic rate curves when leached in acidic Fe^{II} sulphate solutions.[283] The elemental sulphur formed appears on a photomicrograph (3700×) to completely cover partially leached particles. The shrinking core model, see

Sec. 4.3.2, considers spherical particles at the surface of which a reaction rate is controlled by transport through a solid reaction product at the outside of the residual mineral particles. The relationship between the fraction already reacted, α, and time for a batch system is

$$F(\alpha) = 1 - (2\alpha/3) - (1 - \alpha)^{2/3} = k_p t \tag{5.40}$$

where k_p is the parabolic rate constant, which includes many factors. The rate data gave good straight lines when used with this equation and it was concluded therefore, in agreement with other workers, that the rate controlling step was transport through the sulphur produced by the reaction. The values of k_p, given by the slopes of the lines, should be inversely proportional to the square of the initial diameter of the particles, and the three values obtained give an excellent straight line. Samples of the 4 μm and 12 μm particle fractions were leached in 0.25 M $Fe_2(SO_4)_3 + 1.0$ M H_2SO_4 and in 0.03 M $Fe_2(SO_4)_3 + 0.01$ M H_2SO_4 at 60°C, 75°C and 90°C and the values of k_p were normalised with respect to particle size. The value of the apparent activation energy given by the single straight line obtained using the results was 83.7 kJ mol^{-1}. The fact that the rate data obtained using the different concentrations of Fe^{III} gave a single line showed that the rate was independent of $[Fe^{III}]$ in the range 0.03–0.25 M.

Early work on the leaching of copper sulphide ores was reviewed by several authors[284–287] and in 1981 Dutrizac[288] published a literature survey which had been carried out to identify those factors in the leaching of α-chalcopyrite in acidic Fe^{III} sulphate and chloride solutions which appeared to be well established, and to identify areas requiring further work. The following factors appeared to be well established for the leaching of the natural mineral.

(i) The rate of solution in either chloride or sulphate solutions is independent of stirring speed provided the particles are kept in suspension. (This is the case because the rate of reaction is very slow so that it is independent of the rate of transport of Fe^{III} to the interface between the bulk of the solution and the surface of the solid, whether this is chalcopyrite or, when present, porous reaction product, as long as the particles are in suspension.)

(ii) The rate is independent of acid concentration provided iron is prevented from precipitating from the solution.

(iii) Rates are faster in the chloride system; $E_A \sim 42$ kJ mol^{-1}.

(iv) E_A in the sulphate system is ~ 75 kJ mol^{-1}.

(v) Rate is proportional to the surface area of chalcopyrite used in both systems.

(vi) In the sulphate system increasing the concentration of sulphate, particularly of $FeSO_4$, decreases the leaching rate considerably.

(vii) The presence of copper in the sulphate system does not increase the leaching rate.

(viii) Addition of chloride to a sulphate solution slightly increases the leaching rate at elevated temperatures.

Additional experimental work published in this paper showed that:

(ix) In the sulphate system rate is proportional to $[Fe^{III}]^{0.12}$. That is, the rate is almost independent of the concentration of Fe^{III}, over the range 0.01 to 2.0 M.

(x) In chloride solutions the dependence of rate on $[Fe^{III}]$ is greater and appears to be independent of temperature between 45°C and 100°C.

Dutrizac observed that samples of chalcopyrite from different localities had been reported as having widely differing reactivity when leached. Under bacterial leaching conditions the range of leaching rates reported was so wide that it had been suggested that more than one type of the mineral existed, the difference being related to differences in the distribution of oxidation states between the copper and iron. He therefore carried out leaching studies on chalcopyrite concentrates collected from 11 different geographical localities, using both sulphate and chloride solutions.[289] They all reacted similarly and, under identical sets of conditions, within 50% of the same rates. As a first approximation the rate was independent of

(i) electrical conductivity;

(ii) whether the sample was an *n*- or *p*-type semiconductor;

(iii) density of crystal dislocations; and

(iv) the presence of minor amounts of other sulphides or silver minerals.

It was concluded that the greatly differing rates of leaching of different samples of the mineral reported in the literature were due to problems associated with obtaining well defined surface areas of the materials leached. Production of damaged surfaces during comminution must also be a factor.

The production of dislocations in the crystal lattices of minerals has been suggested as a means of increasing reactivity of the solids under mild conditions, such as leaching for example. Problems arise in proving that any change observed after treatment is due to dislocations rather than to damage to the surfaces of the particles of mineral, caused by producing them. This was avoided by shock loading of polycrystalline natural, relatively pure chalcopyrite,[290] which provided a means of obtaining regular increments of crystal dislocation, independent of any mechanical damage which would be caused by grinding. The effect of the dislocations on the rate of leaching of the mineral in acidic dichromate solutions was small, but detected with certainty for the largest mineral grains (210–350 μm) at the lowest temperature used, 50°C. The reaction rate constants under these conditions increased for one sample of chalcopyrite from 1.44×10^{-3} min^{-1} for the material which had not been shock loaded to 1.62×10^{-3} min^{-1} after shock loading to 1.2 GPa. For another sample from a different locality the increase was from 1.44×10^{-3} to 2.73×10^{-3} min^{-1} after shock loading to 18 GPa. The dislocation densities increased by a factor of approximately 10^3 and 10^4 respectively in the two cases.

In a detailed study[291] of the leaching of 37–44 μm particulate α-chalcopyrite in FeCl$_3$ solutions it was found that 3.0% of the copper present in the solid dissolved in a solution of composition 1 M HCl + 3 M NaCl containing no oxidant. This was attributed to the wet grinding which had been carried out and allowance was made for the presence of this excess copper in solutions produced during leaching experiments. The "standard" conditions used in these were 1 M HCl + 3 M NaCl + 0.1 M FeCl$_3$ at 96°C, the concentration of one constituent or the temperature being varied as required in a set of experiments. The following results were obtained.

(i) Plots of $[1-(1-\alpha)^{1/3}]$ against time were linear for FeCl$_3$ concentrations 0.5, 1.0, 2.0, 3.0, 4.0 and 5.0 M. The slopes of the lines were used as the experimental rate constants.

(ii) The order of reaction with respect to [FeCl$_3$] is 0.5 over the concentration range 0.02–0.50 M.

(iii) The rate of reaction is inversely proportional to the initial particle size, r_o μm, deduced from a log r_o – log experimental rate constant plot.

(iv) The rate increases with [NaCl] at concentrations below 1 M but is independent of it at higher values.

(v) The apparent activation energy using the standard concentrations of reagents is 83 kJ mol^{-1}. With 0.2 M HCl present $E_A = 62$ kJ mol^{-1}.

No eletrochemical experiments were reported in the paper but the general behaviour of chalcopyrite leaching in the iron (III) sulphate and chloride systems was discussed and it was pointed out that it was consistent with the reactions being electrochemical in nature. See, however, Sec. 5.2.3.

In view of the evidence that the behaviour of chalcopyrite during at least the first stages of its leaching reactions with iron (III) sulphate and chloride solutions could be affected by the method of treatment of the mineral surfaces during sample preparation,[282,289,291] an investigation was carried out using museum grade chalcopyrite crystals in which particular attention was paid to the nature of changes taking place on the leached surface.[292] This was followed by other papers in which the kinetics of the reactions in the two systems were reexamined using conventional chemical and electrochemical techniques.[293,294]

Two types of specimen were used for examination of the leached surfaces of the mineral. The first was the flat face of a piece of the mineral mounted in epoxy resin, the area of the exposed surface having been measured when kinetic measurements were to be made. This surface was ground wet, finishing with fine emery. After leaching, the crystal was split using a knife to give a cross section through the specimen and any layer of solid reaction product adhering to it. The second type of mineral sample used was particulate solid broken from the original sample and leached without being ground wet. In both cases the leached materials were examined using a scanning electron microscope and by electron microprobe analysis.

When a mounted ground face of chalcopyrite was leached in 1 M Fe$_2$(SO$_4$)$_3$ + 0.2 M H$_2$SO$_4$ at 70°C the leaching rate steadily decreased for about 100 hours, after which the rate increased and became linear for about 150 hours, when the experiment was terminated. The published graph indicates that during this linear 150 hours period approximately 0.23×10^{-5} g atom of Cu dissolved per day. Jones and Peter[282] using a flat face of mineral in a solution of the same composition, but at 90°C, obtained a rate of 7.5×10^{-5} g atom cm^{-2} of Cu per day. This is considerably faster, allowing for the higher temperature.

The solid products on the ground face of the mineral, obtained by Majima *et al.*[292,293] by leaching at 90°C in the Fe^{III} sulphate solutions were examined after different periods of treatment. Initially a layer of sulphur formed on the surface of the mineral. It was a fairly compact aggregate of very thin flakes of the element which tended to peel off from the solid surface and could be seen on the surface of the liquid during the leaching period. After a longer time essentially all of the sulphur was lost leaving a roughened surface on the minerals to which sulphur produced by further reaction did not adhere well. The rate curves then became linear. Jones and Peters reported that the sulphur produced on leaching the polished face of their chalcopyrite in the $Fe_3(SO_4)_3$ solution did not adhere to a significant extent, only small amounts being present in fissures, and only some areas having a coating. The fact that the ground face of the chalcopyrite surface became coated with sulphur at the beginning of the reaction period whereas the polished surface had a much smaller area covered may be one reason for the difference in the shapes of the rate curves and rates of leaching during that period. Munoz *et al.*[283] found that particulate chalcopyrite became completely coated by sulphur produced on leaching in Fe^{III} sulphate solutions, in spite of the agitation of the liquid.

Leaching in 1 M $FeCl_3 + 0.2$ M HCl at 90°C caused the appearance after 24 hours of plate-like structures on the chalcopyrite surface.[292] Elemental sulphur had been produced at the edge of each plate with unreacted mineral forming the central area. After leaching for 168 hours the formation of sulphur had continued at the edges and reaction had also taken place at sites on the previously uncreacted areas of the plates, with the formation of pits. After 240 hours a porous layer of sulphur had formed over the whole surface of the mineral and after 336 hours the coating appeared to consist of a network of fibres similar to that in the photomicrographs of Jones and Peters showing sulphur produced by leaching in ferric chloride solutions.[282] The development of the plate-like structures is characteristic of etching procedures carried out on mineral and metal polished sections. They show that chemical attack first occurs along grain boundaries and other discontinuities in the crystal and lattice structure, being followed later by reaction at other sites. With particulate minerals this frequently results in the formation of deeply pitted surfaces. These are seen in the published photomicrographs of sections through the chalcopyrite specimens leached in the $FeCl_3$ solution for 240 hours and 336 hours, appearing as very much roughened surfaces, resulting in an increased surface area of the exposed face of the specimen during leaching.

It was concluded by the authors that the period of 100 hours after commencing the leaching of the chalcopyrite specimen corresponded to that during which a layer of sulphur built up on the mineral surface, slowing the reaction and leading to a parabolic rate curve. The sulphur then broke away from the surface, the leaching rate increased substantially, and became linear. It was found that pieces of chalcopyrite from the same deposit, used as mounted discs, showed a wide range of leaching rates. When a leaching experiment was repeated using the same conditions and the same disc after repolishing on the same automatic machine, the leaching rate was different from that found previously. Thus in order to study the effects of changing the chemical conditions on the leaching rate in sulphate solutions a different procedure was adopted. The polished disc was leached at 70°C in 1 M $Fe_2(SO_4)_3$ + 0.2 M H_2SO_4 solution until the constant leaching rate condition was reached. The disc was removed, rinsed with water, then with the solution in which the next experiment was to be carried out, and replaced in the reaction vessel. The fresh solution was added and the experiment commenced when convenient. It was emphasised[293] that this procedure can only be used when no reaction products remain on the surface to affect the rate of the subsequent reation, which must show linear kinetics, that is have a uniform reaction rate constant.

The results obtained in iron (III) sulphate media were as follows:[293]

(i) Using a single disc in three solutions 0.2 M in H_2SO_4 and having different concentrations of $Fe_2(SO_4)_3$ the apparent activation energies over the temperature range 50–90°C were, Fe^{III} 0.01 M, 87.7, Fe^{III} 0.1 M, 83.2; Fe^{III} 1.0 M, 76.8 kJ mol^{-1}. This suggests that the reaction between Fe^{III} and the mineral is chemically controlled when the rate is not lowered by the presence of a sulphur layer. The values are similar to that obtained from earlier published data by Dutrizac, $\sim$ 75 kJ mol^{-1}.[288]

(ii) At 70°C with 0.2 M H_2SO_4 the rate was first-order with respect to the concentration of $Fe_2(SO_4)_3$ between 10^{-3} M and about 10^{-1} M but at higher concentrations the rate increased only slightly. It was suggested that this could be explained in terms of the speciation of Fe^{III} in the solution as the sulphate concentration increased. The species present were considered to be Fe^{3+}, $FeHSO_4^{2+}$ and $FeSO_4^+$, and calculations showed that between [Fe total] 10^{-3} and 1.0 M, the molar fraction of $[FeSO_4^+]$ increased linearly, containing between about 90% and 99.9% of the iron. Fe^{3+} and $FeHSO_4^{2+}$ are present in almost equal quantities and their mole fractions increase almost linearly between 10^{-3} and 10^{-1} M iron, remaining almost constant at higher total iron concentrations. Thus the concentrations of these two species correspond to the effect of iron (III) sulphate on the leaching rate and it was suggested that this indicates that they play an important role in the oxidation of chalcopyrite. It had previously been reported that the rate of sphalerite (ZnS) dissolution in acidic iron (III) sulphate solutions was proportional to the sum of the concentrations of the Fe^{3+} and $FeHSO_4^{2+}$ species present.[295] It was later[307] stated that if the leaching reaction of chalcopyrite was allowed to continue until the sulphur layer had peeled off and the reaction rate curve had become linear, the order of reaction with respect to iron (III) sulphate concentration became 0.5 over the concentration range 0.01 to 0.1 M at 70°C. Increasing the concentration above this value had little effect on the leaching rate.

(iii) Iron (II) sulphate was added to a solution 0.2 M in H_2SO_4 and 0.1 M in $Fe_2(SO_4)_3$ to give solutions 0.1, 0.5 and 1.0 M in Fe^{II}. With $[Fe^{II}] = 1$ M the leaching rate at 70°C fell to about 30% of that in the solution containing no Fe^{II}. The results are similar to those reported previously.[282,288]

Using the same experimental procedure for the treatment of the chalcopyrite specimens, leaching studies were carried out using $FeCl_3$ as the oxidant.[294]

(i) At 70°C with 0.2 M HCl the rate was 0.5 order with respect to the concentration of $FeCl_3$ between 10^{-2} M and 1.0 M. The six rate curces were each linear within experimental error. Two chalcopyrite specimens showed the same order of reaction although the reaction rates were different.

(ii) Iron (II) chloride was added to 0.2 M HCl + 0.1 M $FeCl_3$ to produce solutions 0.5 and 1.0 M in $FeCl_2$. Since the leaching rate was affected by the total chloride concentration the value of this was maintained at 2.3 M by adding NaCl. On leaching at 70°C, with N_2 used to prevent oxidation of Fe^{II} by air, the leaching rate with the Fe^{II} present was identical with that without Fe^{II}. This agrees with results published previously.[282]

(iii) The effect of adding NaCl to 0.1 M $FeCl_3$ on the leaching rate of chalcopyrite was examined. It was not stated that HCl was present in the solution. The leaching rate was doubled in the presence of 2 M NaCl. Earlier work[291] indicated that increasing the concentration of NaCl up to 1 M increases the leaching rate, which is not affected by raising [NaCl] further. No data are given in Ref. 294 for values of [NaCl] lower than 2 M.

(iv) The apparent activation energy for the leaching of chalcopyrite in 0.2 M HCl + 0.1 M FeCl$_3$ was found to be 69.0 kJ mol^{-1} in the temperature range 52 to 85°C.

Several studies of oxidative leaching of chalcopyrite in acidic solutions have been carried out using electrochemical techniques. These include the use of copper (II) chloride as the oxidant in moderately concentrated chloride solutions. The thermodynamic basis of this reaction differs from that using Fe^{III} and the results of the electrochemical investigations are discussed after the basic chemistry of the use of the Cu–Cl system has been described.

Thermodynamics of leaching chalcopyrite using copper (II) chloride solution as oxidant. In solutions containing sufficiently high concentrations of chloride the copper (I) state is thermodynamically stable and the Cu^{II}–Cu^I couple can be used as an oxidising system (Sec. 3.11.1). Its use for leaching copper sulphide ores was suggested at the end of the 19th Century[296,297] but the corrosion problems associated with the use of solutions containing chloride, particularly under oxidising conditions, prevented application of the reaction at that time. Several processes in which it is used in the leaching of chalcopyrite concentrates were developed in the 1970's and the chemical basis of the "Cyprus process" is discussed here.

Leaching of chalcopyrite using copper (II) chloride as the oxidant differs from the reactions using iron (III) sulphate or chloride. Whereas the Fe^{III}–Fe^{II} couple provides an adequate thermodynamic driving force as long as the metal ion activities and pH are suitable, in the copper chloride system the thermodynamics of the complete reactions have to be considered in order to achieve suitable conditions for leaching to proceed. Wilson and Fisher[298] analysed the situation using the thermodynamic data given in Table 5.19. It differs from that used in this book, substantially in the cases of iron species.

Table 5.19.

Species	Cu$^+$	Cu^{2+}	Fe^{2+}	Fe^{3+}	Cl$^-$
ΔG° kJ	50.20	64.96	−84.91	−10.58	−131.14
Species	CuCl$_3^{2-}$	CuCl$^+$	FeCl$_2$(aq)	FeCl$_2^+$	CuFeS$_2$
ΔG° kJ	−373.46	−66.80	−349.45	−277.00	−242.70

They consider that the reation between chalcopyrite and Fe^{III} in moderately concentrated chloride solutions is probably

$$CuFeS_2 + 4FeCl_2^+ + 3Cl^- = CuCl^+ + 5FeCl_2(aq) + 2S \qquad (5.41)$$

for which ΔG° at 25°C is −69.93 kJ. That for the reaction involving copper (II) chloride is given as

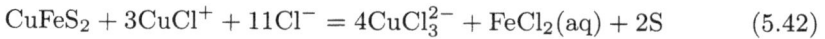

$$CuFeS_2 + 3CuCl^+ + 11Cl^- = 4CuCl_3^{2-} + FeCl_2(aq) + 2S \qquad (5.42)$$

for which ΔG° at 25°C is +42.35 kJ. Thus this reaction does not take place unless the activity terms in the free energy equation for the reaction are favourable. The justification for selecting the species shown is discussed below. Three $CuCl^+$ are involved since the Cu^{II} in the chacopyrite is also reduced to Cu^I in the reaction. The change in free energy in a reaction

$$A + B = C + D$$

which is not at the equilibrium point can be expressed in terms of the equilibrium constant and the activities of the reactants and products

$$-\Delta G = RT \ln K - RT \ln\{C\}\{D\}/\{A\}\{B\}.$$

This equation expresses in terms of free energies the same relationship as that in Eq. (3.3) for half-cell reactions, in which electron transfer occurs

$$E = E^\circ - \frac{RT}{zF} \ln \frac{\{\text{Reduced State}\}}{\{\text{Oxidised state}\}} \qquad (3.3)$$

since $\Delta G^\circ = -RT \ln K = -zE^\circ F$, and also for full cell redox reactions. In some cases it is convenient to work in terms of potentials, in others the use of free energies is to be preferred, even when oxidation and reduction are involved.

Wilson and Fisher analysed the leaching reaction in term of the free energy change ΔG_r using a series of specified value of activities.

$$\Delta G_r = \Delta G^\circ + RT \ln \frac{\{CuCl_3^{2-}\}^4\{FeCl_2(aq)\}}{\{CuCl^+\}^3\{Cl^-\}^{11}}. \qquad (5.43)$$

If the leaching reaction between chalcopyrite and copper (II) chloride takes place as suggested in (5.42) then the thermodynamic driving force is given by (5.43).

Table 5.20.

Change in ΔG_r as chalcopyrite reacts with 1.5 M CuCl$_2$ solution, 6 M in Cl$^-$.

[CuCl$^+$], M	1.5	1.425	1.125	0.938	0.563	0.375
[CuCl$_3^{2-}$] M	10^{-3}	0.10	0.5	0.75	1.25	1.50
[FeCl^2(aq)] M	10^{-3}	0.025	0.125	0.188	0.313	0.375
ΔG_r kJ	-95.2	-41.1	-23.41	-13.47	-2.90	$+2.37$

The relative importance of the terms in the equation for ΔG_r can be seen in the following examples.

	Term	ΔG°	CuCl$_3^{2-}$	FeCl$_2$	CuCl$^+$	Cl$^-$	
[CuCl$_3^{2-}$]10^{-3} M	$\Delta G_r =$	$+42.35$	-68.51	-17.13	-3.016	-48.87	$= -95.2$ kJ
[CuCl$_3^{2-}$]0.5 M	$\Delta G_r =$	$+42.35$	-6.87	-9.15	-0.88	-48.87	$= -23.42$ kJ
[CuCl$_3^{2-}$]1.0 M	$\Delta G_r =$	$+42.35$	-0	-3.44	$+2.14$	-48.87	$= -7.82$ kJ
[CuCl$_3^{2-}$]1.25 M	$\Delta G_r =$	$+42.35$	$+2.21$	-2.87	$+4.27$	-48.87	$= -2.91$ kJ
[CuCl$_3^{2-}$]1.5 M	$\Delta G_r =$	$+42.35$	$+4.02$	-2.43	$+7.30$	-48.87	$= +2.37$ kJ

Typical initial conditions for the reaction were taken to be:

[CuCl$^+$], 1.5 M; [CuCl$_3^{2-}$], 0.001 M; [FeCl$_2$(aq)], 0.0001 M; [Cl$^-$], 6.0 M.

Activities are assumed to equal concentrations. Rewriting (5.43) gives

$$\Delta G_r = +42.35 + 22.84 \log\{CuCl_3^{2-}\} + 5.709 \log\{FeCl_2(aq)\}$$
$$- 17.13 \log\{CuCl^+\} - 62.80 \log\{Cl^-\}.$$

On this assumption, Eq. (5.42) indicates that three quarters of the copper in the solution, present as CuCl$_3^{2-}$ is derived from the CuCl$^+$ intially present, so that in calculations if [CuCl$_3^{2-}$] is defined [CuCl$^+$] in the same solution is known. Also [FeCl$_2$(aq)] = 0.25 [CuCl$_3^{2-}$].

The change in ΔG_r as the reaction between 1.5 M copper (II) chloride solution 6 M in Cl$^-$, and excess chalcopyrite proceeds is shown in Table 5.20.

The consumption of 11 g atoms of chloride ion and the use of a high total chloride concentration leads to the term for $\{Cl^-\}$ in the equation for ΔG_r being of sufficient magnitude to compensate for the positive value of ΔG° for the reaction, Table 5.20. The major reason for the rapid decrease in the thermodynamic driving force in the reaction is the increasing concentration of the product of the reation, CuCl$_3^{2-}$. This is four times that of the other metal-containing product, FeCl$_2$(aq), because it is formed by reduction of the oxidant as well as by oxidation of the chalcopyrite. The redox couple is given as

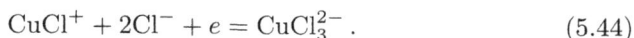

$$CuCl^+ + 2Cl^- + e = CuCl_3^{2-}. \tag{5.44}$$

The selection by Wilson and Fisher of the species containing copper and iron shown in reactions (5.41) and (5.42) requires examination. The Fe^{II} species shown in Fig. 3.12 as being predominant with chloride concentrations above 1 M is $FeCl_3(aq)$ and this is supported by the fact that this species can be extracted from aqueous solutions of such compositions by solvent extraction using a number of nonionic organic solvents. $FeCl_2^+$ is not an important species at any chloride concentration. For Fe^{II} there is no apparent alternative to $FeCl_2(aq)$ at the chloride concentrations being considered.

The speciation of Cu^I in concentrated chloride solutions was discussed in Sec. 3.11.1 where it was stated that Fritz[83] had used the methods of Pitzer to reexamine published data on the stabilities of the copper(I)-chloro-complexes, based on the solubility of CuCl in choride solutions, and had calculated stability constants for them. A predominance area diagram is shown in Fig. 3.11. Fritz published further work on the system[299,300] and later[301] presented the data given in Table 5.21 for the equilibrium constants and changes in thermodynamic properties for the formation of $CuCl_2^-$ and $CuCl_3^{2-}$ from the reaction

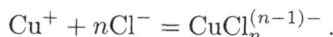

$$Cu^+ + nCl^- = CuCl_n^{(n-1)-}.$$

The equilibrium constants for the chloro-complexes of Cu^{II} were determined[302] using extraction of the anionic species by tri-n-octylamine. The distribution data are shown in Fig. 3.13, the predominant species in 6 M chloride solutions being $CuCl_4^{2-}$. Thus the species involved in (5.41) are considered to be

Table 5.21. Thermodynamic data for the formation of $CuCl_2^-$ and $CuCl_3^{2-}$ from Cu^+.

Species	Temp °C	K	ΔH° kJ mol^{-1}	ΔG° kJ mol^{-1}	ΔS° J mol^{-1} K^{-1}
$CuCl_2^-$	15	3.1×10^5	-11.97	-30.29	63.6
	25	2.6×10^5	-11.21	-30.92	66.1
	35	2.3×10^6	-10.46	-31.63	68.6
$CuCl_3^{2-}$	15	7.9×10^4	-26.36	-27.03	2.5
	25	5.6×10^4	-24.60	-27.11	8.4
	35	4.0×10^4	-24.18	-27.15	9.6

$FeCl_3(aq)$, $FeCl_2(aq)$ and $CuCl_4^{2-}$, in (5.42) $CuCl_4^{2-}$, $CuCl_3^{2-}$ and $FeCl_2(aq)$ leading to the following equation for the reaction

$$CuFeS_2 + 3CuCl_4^{2-} + 2Cl^- = 4CuCl_3^{2-} + FeCl_2(aq) + 2S. \qquad (5.45)$$

See also Ref. 302.

In the "Cyprus process"[303,304] chalcopyrite concentrates are leached in two stages. In the first a solution containing both $CuCl_2$ and $FeCl_3$ produces a very concentrated solution of CuCl from which the solid is crystallised out and reduced in H_2 to produce the metal and HCl gas. The HCl is fed into the mother liquor from which part of its iron content is precipitated by oxidation and controlled hydrolysis. The slurry is used as the leach liquor in the second leaching stage, which is essentially an iron (III) chloride leach.

Kinetics of Leaching Chalcopyrite Using Copper (II) Chloride as Oxidant

The kinetics of the leaching reaction between chalcopyrite and copper (II) chloride solutions have been investigated by three groups. The general behaviour was found to be similar to that reported previously for the reactions using iron (III) sulphate and chloride.

(i) The chalcopyrite particles in a concentrate became covered with a porous layer of sulphur. Using particulate solid of particle size 74–105 μm with a stirring speed in the range which gave leaching rates independent of this factor, the rate curves at 90°C in 0.79 M Cu^{II} + 6.35 M Cl^- were linear up to 28% iron removal.[298] The reaction was carried out under N_2 and it was assumed that all iron in solution was derived from chalcopyrite since any pyrite in the concentrate does not react under these conditions.

Using particulate solid and solutions containing $[Cu^{II}] = [Cu^I] = 1$ M and $[NaCl] = 4$ M at several temperatures, the rate curves up to about 35 to 60% iron removal were parabolic[305] and fitted Eq. (4.35) for the shrinking core model, in agreement with results using iron (III) sulphate, see above. Parabolic rate constants and root mean square deviations in α are tabulated in the paper.

Using polished faces of mounted discs it was found in Ref. 306 that as when iron (III) sulphate and chloride were used, linear rate curves and reproducible rates from an individual disc were obtained if it was not repolished between experiments. After about 10^{-4}g atom of iron cm^{-2} had been leached from the surface the leaching rate gradually increased with time due to roughening of the surface, as with FeCl$_3$ as the oxidant. This explanation was tested using a disc wich was giving such an increased rate constant and leaching it again in a fresh, newly prepared leaching solution. The leaching rate was much greater than it had been in the linear stage of its first treatment and was almost the same as in the final stage before leaching was stopped. Kinetic experiments reported by these authors used discs from which less than 10^{-6} g atom of Fe cm^{-2} had been leached, which were not repolished between experiments, and gave linear rate curves.

(ii) For particulate solid giving linear rate curves, the rate was proportional to the surface area.[298] With particulate solid and parabolic kinetics k_p was inversely proportional to the square of the initial particle size for particles larger than 50 μm. Smaller particles tended to float at the air–water interface, lowering the rate at which they reacted.[305]

(iii) The effect of temperature on the reaction rate was unusual with particulate chalcopyrite. Wilson and Fisher[298] reported that with 74–105 μm solid in a solution 0.79 M in Cu^{II} and 6.35 M in Cl$^-$ no reaction was detected after leaching for 4 hours at 50°C. Reaction occurred at 80°C but the temperature at which it started was not determined. For the temperature range 80–97.5°C E_A was 134.7 kJ mol^{-1}. Bonan *et al.*[305] detected some reaction at 75°C and 80°C and the parabolic rate constants for 4 M NaCl and $[Cu^{II}]/[Cu^I] = 1$ gave E_A over this temperature range as about 330 kJ mol^{-1}. The three values of k_p measured at 85°C, 95°C and 103.9°C gave a good straight line and $E_A = 71$ kJ mol^{-1}. The value varied with the $[Cu^{II}][Cu^I]$ ratio, falling to about 59 kJ mol^{-1} when it was increased to 2.5, again in 4 M NaCl. With a chacopyrite disc in 0.1 M CuCl$_2$ + 0.2 M HCl the linear rate curves gave a single straight line Arrhenius plot for the rates at 60°C, 70°C, 80°C and 90°C, and $E_A = 81.5$ kJ mol^{-1}.[306]

(iv) Wilson and Fisher[298] showed that under the experimental conditions they used there was no significant difference in the reaction rate constant when $[Cu^{II}]$ was changed from 0.79 to 1.46 M and $[Cl^-]$ from 2.82 to 6.21, at 90°C. They concluded therefore that neither of these reactant species plays· a direct part in the rate controlling step of the dissolution process. Hirato *et al.*[306] using a chalcopyrite disc showed that at 70°C in 0.2 M HCl with [CuCl$_2$]0.001–1.0 M the order of reaction with respect to CuCl$_2$ is 0.5. The effect of the reaction product Cu^I on the dissolution rate at 70°C was also determined. Since the leaching rate had been found to depend on the chloride concentration to some extent sodium chloride was added in such quantity that changing the concentration of CuCl did not significantly affect the value of [Cl$^-$]. The solution used was 0.1 M CuCl$_2$; 0.2 M HCl; 3 M NaCl with [CuCl] 0.005 M to 0.1 M. The leaching rate of chalcopyrite under the experimental conditions used was inversely proportional to [CuCl]$^{0.5}$. The effect of [Cl$^-$] was determined at 70°C using solutions 0.1 M CuCl$_2$ + 0.2 M HCl containing NaCl in the range 0.0–4.0 M. The leaching rate increased up to 2 M NaCl, slightly increased between 2 and 3 M, with no further increase to 4 M NaCl.

(v) Bonan *et al.*[305] were interested in the effects of the redox potential of the solution on the rate of leaching of chalcopyrite and measured this continuously during reactions, using a platinum electrode protected from contact with solid suspended in the solution, and a reference electrode separated from the leaching solution by a salt bridge. Their leaching experiments were designed to maintain the potential and the pH of the solution approximately constant during an experiment. Thus a small quantity, about 5 g of chalcopyrite concentrate, and a large amount of copper (II) and copper (I) chlorides (from 0.1 to 0.5 M solutions) were introduced initially into a 1 litre leaching tank together with, generally, HCl to give a 1 M and NaCl to give a 4 M solution. Air was excluded by N$_2$.

Chalcopyrite reacted faster at 95°C with [NaCl] = 4 M as the ratio Cu^{II}/Cu^I increased from 1 to 7.5 and also, with a constant Cu^{II}/Cu^I ratio of one, as the chloride ion concentration was increased, using either NaCl or CaCl$_2$. Two experiments were carried out in which E(versus SHE) was set at practically the same value by using different values of [NaCl] and Cu^{II}/Cu^I. Thus with Cu^{II}/Cu^I = 3.2 and [NaCl] = 2.5 M, E = 585 mV; and with, Cu^{II}/Cu^I = 1.0 and [NaCl] = 4.0 M, E = 586 mV; the rates of dissolution of Fe were practically identical for 150 minutes, corresponding to about 45% metal extraction, and then diverged slightly as the reaction almost ceased. At a lower value of E, 557 mV with Cu^{II}/Cu^I = 1 and [NaCl] = 3 M; and 559.5 mV with Cu^{II}/Cu^I = 0.5 and [NaCl] = 4 M, the correspondence of rates was almost as good, but the reaction slowed greatly when 20% metal extraction was reached. The concentrations of Cu^{II} and Cu^{I} are not given and it is not possible therefore to analyse the rate data in term of orders of reaction.

Electrochemical studies of the leaching of α-chalcopyrite in strongly acidic solutions. Electrochemical studies were carried out by Jones and Peters,[282] and by Majima and his coworkers[292–294,306,307] the last of these papers being a summary of their work and final conclusions. They used iron (III) sulphate, $FeCl_3$, and $CuCl_2$ as oxidants and measured the mixed potentials and kinetics of the leaching reactions with no potential applied from an external source. The rates of oxidation at a chalcopyrite electrode of $FeCl_2$, $FeSO_4$, and $CuCl$ were studied and also the electrochemical leaching of the mineral with no added oxidant.

The chalcopyrite-oxidant leaching system differs in several respects from the UO_2-oxidant system, discussed in Sec. 5.2.3, as is shown by the reaction scheme proposed by Majima[307]

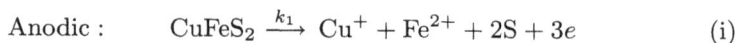

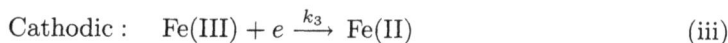

$$\text{Anodic}: \quad CuFeS_2 \xrightarrow{k_1} Cu^+ + Fe^{2+} + 2S + 3e \quad \text{(i)}$$

$$Cu(I) \xrightarrow{k_2} Cu(II) + e \quad \text{(ii)}$$

$$\text{Cathodic}: \quad Fe(III) + e \xrightarrow{k_3} Fe(II) \quad \text{(iii)}$$

or

$$Cu(II) + e \xrightarrow{k_3} Cu(I) \quad \text{(iv)}$$

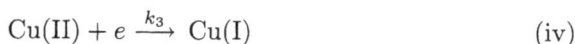

where k is the rate constant for each half-cell reaction. Oxidation of the mineral takes place in two stages, the first involving the formation of elemental sulphur and the second oxidation of Cu(I). Usually the sulphur forms a coating on the reacting solid and so affects the rate to an extent which depends on whether the layer is protective or highly porous. It was necessary to develop an experimental procedure which avoided the production of such a protective coating. The formation of a film of uranyl carbonate on the solid during the leaching of UO_2 in carbonate solution stops the reaction. Thus the application of results from the study of the electrochemical mechanism of leaching chalcopyrite would be of practical application only when the reaction rate was not controlled by the presence of a sulphur or some other coating.

The experimental investigation was not carried out in the same way as that used for uranium dioxide. No direct comparison was made of the rates of leaching at different potentials by chemical oxidants and the rates at the same values applied from an external source with no chemical oxidant present. Also, during leaching experiments using chemical oxidants the potential between the chalcopyrite and the reference electrode in the solution was not maintained at a required value while the leaching current was measured. Instead, the mixed potential resulting from the reaction between the chalcopyrite and oxidant was measured, and the rate of the reaction determined by analysing samples of the solution by the atomic absorption method. In some cases these rates were converted into current densities but the values of j are, of course, relevant to the conditions of chemical leaching and not to electrochemical leaching. Rates of electrochemical leaching of chalcopyrite in the absence of chemical oxidants were however measured as current flowing, as also were the rates of oxidation at a chalcopyrite surface of the products of the reduction of the oxidants, Fe(II) and Cu(I), in solutions having appropriate compositions.

The measured values of the mixed potentials obtained during the work with chalcopyrite were used as follows.[306,307] The anodic and cathodic current densities on the surface of the chalcopyrite are given by

$$j_A = 3Fk_1 \exp[\alpha_1 FE/RT] + Fk_2 c_{\text{red}} \exp[\alpha_2 FE/RT] \qquad \text{(v)}$$

$$j_c = -Fk_3 c_{\text{ox}} \exp[-(1-\alpha_2)FE/RT]. \qquad \text{(vi)}$$

In the steady state $j_A = j_c$ at the mixed potential E_M. Values of transfer coefficients are obtained experimentally from the Tafel slopes of the electrochemical reactions, and that for electrochemical oxidation of chalcopyrite was found to be 0.44. The value for the oxidation of Cu(I) and reduction of Cu(II) was not determined for reasons discussed below. Since for simple electrode reactions $\alpha_c + \alpha_A = 1$ and $\alpha_c \approx \alpha_A \approx 0.5$, it was assumed these were the values in the chalcopyrite system. Then

$$E_M = \frac{2.303RT}{F} \log\left(\frac{k_3 c_{\text{ox}}}{3k_1 + k_2 c_{\text{red}}}\right). \qquad \text{(vii)}$$

At $T = 343K$ (the temperature used) $2.303RT/F = 0.068$. Substitution of Eq. (vii) into the first term of (v) gives the current density j_L corresponding to the chemical leaching rate of chalcopyrite

$$j_L = 3Fk_1 \left(\frac{k_3 c_{ox}}{3k_1 + k_2 c_{red}} \right)^{0.5}. \tag{viii}$$

Experimental results. Solutions of $FeCl_3$ of concentration between 0.01 and 0.1 M containing 0.2 M HCl gave a linear relationship between the mixed potential and log $[FeCl_3]$, slope $+72$ mV decade^{-1} at 70°C. In solutions containing 0.1 M $FeCl_3$ and 0.2 M HCl the presence of $FeCl_2$ at concentrations 0.01–0.1 M had no effect on E_M.

Solutions of $CuCl_2$ of concentration between 0.01 and 0.1 M containing 0.2 M HCl gave a linear relationship between the mixed potential and log $[CuCl_2]$, slope $+66$ mV decade^{-1} at 70°C. In solutions containing 0.1 M $CuCl_2$ and 0.2 M HCl, with total chloride ion concentrations adjusted to 3 M using NaCl, the presence of CuCl at concentrations between 0.005 and 0.5 M gave a linear relationship between E_M and log [CuCl], slope -69 mV decade^{-1}.

Solutions of $Fe_2(SO_4)_3$ of concentration between 0.01 and 0.1 M containing 0.2 M H_2SO_4 gave leaching curves different in shape from those obtained using $FeCl_3$ or $CuCl_2$, which were linear. They were parabolic-like for about 100 hours and then became linear, after the protective coating of sulphur formed initially had peeled off leaving a roughened surface on the mineral. At 70°C, with linear kinetics, there was a linear relationship between the mixed potential and log $[Fe_2(SO_4)_3]$, slope $+79$ mV decade^{-1}. The presence of $FeSO_4$ in 0.1 M iron (III) sulphate and 0.2 M H_2SO_4 at 70°C lowered the mixed potential somewhat. Between 0.01 and 0.1 M $FeSO_4$ there was a total nonlinear fall of 28 mV.

Solutions in which the mixed potentials of chalcopyrite were measured were also used to study the kinetics of the leaching reactions and determine the orders of reaction of the oxidants. The leaching rate at 70°C is proportional to $[FeCl_3]^{0.5}$ and independent of added $FeCl_2$; and is proportional to $[CuCl_2]^{0.5}$ and added $[CuCl]^{-0.5}$. The apparent activation energies found were 69 kJ mol^{-1} for $FeCl_3$ and 82 kJ mol^{-1} for $CuCl_2$. When the leaching reaction with $Fe_2(SO_4)_3$ is following linear kinetics the rate is half-order in the concentration range 0.01–0.1 M at 70°C. Increasing the concentration above 0.1 M has little effect on the leaching rate. In a solution 0.1 M in Fe^{III} sulphate and 0.2 M in sulphuric acid, addition of $FeSO_4$ to give a 1 M solution of this lowered the rate of leaching to 30% of that before addition of the iron (II). The results obtained were in agreement with those of other workers.

The reactivities of the products of the reduction of the oxidants, Fe^{II} and Cu^I, at a chalcopyrite anode in solutions having compositions similar to the leach liquors at 70°C, were investigated using constant applied potentials. Current–time curves were recorded at 656 mV versus SHE in 0.2 M HCl or H_2SO_4 solution, aliquots of $FeCl_2$ or $FeSO_4$ respectively being added to produce solutions 0.1, 0.2 and 0.5 M in $FeCl_2$ and, with an additional value of 0.1 M, in $FeSO_4$. A similar experiment was carried out at 706 mV in 0.2 M HCl + 2 M NaCl at 70°C, CuCl being added to give concentrations 10^{-4}, 10^{-3} and 10^{-2} M.

Addition of $FeCl_2$ to the solution caused a few tens of μA cm^{-2} increase in the current flowing, but after a few minutes this had returned to its previous value. Thus Fe^{2+} ions in chloride solutions have no significant influence on the rate of electrochemical leaching of chalcopyrite. Their rate of oxidation on the chalcopyrite surface is slow. In the sulphate solution, addition of $FeSO_4$ caused a significantly larger rise in the current, the new value remaining constant until the next addition of Fe^{3+}, after about 40 minutes, when a further increase occurred. This suggests that Fe^{2+} may affect the rate of leaching of chalcopyrite in iron (III) sulphate solutions.

The rate of reaction of chalcopyrite in the chloride solution, and hence the current flowing, was greatly increased by addition of the CuCl. For example the rise was from about 10^{-1} mA cm^{-2} in 10^{-3} M CuCl to about 8 mA cm^{-2} in 10^{-2} M CuCl. Thus CuCl is readily oxidised at a chalcopyrite electrode. The different rates of oxidation of $FeCl_2$ and CuCl on the surface of the mineral presumably cause their different effects on the leaching rate and mixed potentials of the mineral.

Majima examined the experimental results in term of Eqs. (v) to (viii) as follows. In solutions containing $FeCl_3$, $FeCl_2$ and HCl the rate of oxidation of Fe^{II} is very slow so that this reaction need not be considered, and since k_2 is much smaller than k_1, Eq. (vii) for $T = 343$ K gives the dependency of E_M on log $[FeCl_3]$ as 68 mV decade^{-1}. Agreement with the measured value 72 mV decade^{-1} is considered to be satisfactory. Equation (viii) indicates that the leaching rate should be proportional to $[FeCl_3]^{0.5}$, as was found experimentally.

Equation (vii) also gives the dependency of E_M at 343 K on log $[CuCl_2]$ in the $CuCl_2$–CuCl–HCl system as 68 mV decade^{-1}; also the dependency on log $[CuCl]$ is given as -68 mV decade. The experimental values were 68 and -69 mV decade^{-1} respectively.

Equation (viii) is rewritten as

$$E_M = 0.136 \log j_L + \text{constant}$$

at 343 K by eliminating the concentration terms. Chemical leaching rates for a chalcopyrite specimen were measured in acidic $CuCl_2$ solutions and in acidic CuCl solutions containing 0.1 M $CuCl_2$ and 3 M NaCl, and were converted into corresponding current densities j_L. The values of log j_L for these solutions were plotted against E_M, giving straight lines of slopes 121 mV decade^{-1} and 135 mV decade^{-1} respectively, in good agreement with the theoretical value. However anodic polarisation of the same chalcopyrite specimen as was used in the chemical leaching experiments, in 0.1 M HCl gave a curved line on plotting log j_L against E. It had previously been suggested[294] that such behaviour could be caused by a surface layer of a solid having a high electrical resistance. The resistivity of the surface layer was measured and the values of the potential E of chalcopyrite in the anodic polarisation experiments were corrected. The corrected log j_L versus E plot gave a straight line with slope 130 mV decade^{-1} and its position was close to that for chemical leaching in an acidic $CuCl_2$ solution.

It was stated above that the transfer coefficient for the electrochemical oxidation of chalcopyrite was determined and found to be 0.46, but the values for the oxidation of Cu(I) and reduction of Cu(II) at a chalcopyrite surface were not obtained. The values of log j obtained as E was changed gave curves for both electrochemical reactions.[306] There were no linear portions corresponding to a Tafel region. Since the Butler–Volmer equation is derived by arguments based on Eq. (5.4), and the present discussion is based on application of the Butler–Volmer equation, this would appear to indicate that either these reactions are not electrochemical, as defined by workers in this field, or that other factors need to be considered.

The apparent formation of a passivating layer of a product during the initial stage of electrolytic leaching of natural chalcopyrite had been reported previously. Biegler and Swift[309] studied the anodic behaviour of chalcopyrite in 1 M H_2SO_4 and 1 M HCl at 25°C by linear sweep voltammetry and electrolytic oxidation at constant potential. Using a rotating disc and a sweep rate of 20 mVs^{-1}, the first sweep to positive potentials showed three distinct regions. From the rest potential of about 400 mV versus SHE to 800 or 950 mV, depending on whether the electrode surface was polished or ground, there was a broad "prewave" in which the current, $\sim$ 0.1–1 mA cm^{-2}, passed through a peak or plateau, depending on the conditions used. Between 950 mV and 1.2 V the current rose steadily from about 1 to 100 mA cm^{-2}. Above this potential the current–voltage relationship was linear before reaching an apparent potential of 1.8–3.5 V, depending on the chalcopyrite electrode and the sweep rate. When the potential was cycled again the prewave was absent from the second and subsequent scans.

The prewave was attributed to surface oxidation, and the current passed in completing the process was of the order of 1 mC per real cm^2 of surface. This corresponds to oxidation of a layer of mineral $4.6/n$ nm thick, where n is the number of Faradays per mol, of chalcopyrite reacting. The range of potential where the prewave occurs covers that over which chemical oxidation of the mineral by Fe^{III} takes place in leaching processes, and it was considered that formation of the oxidised layer might be associated with the progressively decreasing rate of leaching commonly observed, now attributed to the porous layer of sulphur produced and to the decreasing particle size of particulate mineral.

In an attempt to grow a thicker layer of the product a polished chalcopyrite electrode was held at constant potentials up to 950 mV for periods of 1 hour, by which time the current density had fallen to about 1 μA cm^{-2}. The electrode surface still had the appearance of polished chalcopyrite. After maintaining the electrode at potentials above 1 V, in the main wave region, attempts were made to identify solid products of the reaction but none except sulphur was found. It was concluded that in this steep region of the current–potential curve the rate of oxidation of the chalcopyrite is determined by the rate of nucleation and growth of 3-dimensional anodic dissolution sites. The applied potential in the main wave region was sufficiently high to overcome any passivating effect.

These observations were confirmed.[310,311] At constant potentials in the range 400–850 mV versus SHE the current flowing steadily decreased with time and it was concluded that the rate of oxidation of the mineral was controlled by a surface layer of product which built up as the reaction continued. When elemental sulphur formed on the electrode surface it played no part in controlling the rate of the reaction. The film of product was said to break down when treated with carbon disulphide, acetone or acidified water, or when the electrode was left under dry nitrogen. If the applied potential was removed while the electrode was being oxidised in hydrochloric acid at a time when the current flowing had decreased significantly, and was reapplied shortly afterwards the large current flowed again and decreased with time as before. The reactivation process was found to be independent of stirring of the solution, more rapid at higher temperatures, and its extent was dependent on the time elapsed after removal of the applied potential. This behaviour is consistent with that reported by Majma *et al.* A stable current is reached by the anodic oxidation process so that the amount of surface product formed increases during the period when the current falls and then must remain effectively constant. Such behaviour is consistent with the formation of a surface layer of the mineral which is metal-deficient due to diffusion of cations under the influence of the applied electric field. The distinct difference in the morphology of the surfaces of chalcopyrite leached electrochemically and chemically, using $FeCl_3$, has been demonstrated.[294]

Linge[312] leached particulate chalcopyrite in iron (III) nitrate solutions and measured the reaction rate by determining the amount of Fe^{III} reduced. The pH and oxidation potential were held constant. He presents strong evidence for an initial reaction, persisting for a period of a few minutes, during which more iron than copper was dissolved from the lattice leaving a metal-deficient surface layer. He found a dissolution rate after 1 hour equivalent to an anodic current of only 0.02 μA cm^{-2}, consistent with the electrochemical measurements of Biegler and Swift.[309] the limiting thickness of the metal-deficient layer, which built up rapidly, was about 5 nm.

Further evidence that the first stages of oxidative leaching of chalcopyrite in acidic solutions lead to the formation of a surface layer of the mineral which is rich in copper had been presented.[309–311,313] It had also been suggested that the passive surface layer is a copper polysulphide of the type CuS_2,[310] or a solid electrolyte interphase.[311] Its effect on the kinetics of the oxidation of the chalcopyrite has been described in terms of diffusion of metal ions in the solid,[312] passive film growth,[313] electron transport properties,[283] or electrolytic conduction of metal ions.[311] Further work using linear sweep voltammetry[314] confirmed that the anodic prewave reported previously[309] for freshly prepared chalcopyrite electrodes in acid solutions represents a surface oxidation reaction, which is distinct in mechanism from the bulk oxidation of the mineral which occurs at higher potentials, and is not due to oxidation of the surface in air before the measurements are made. The product was considered to be the layer of material which inhibits the chemical leaching of chalcopyrite with oxidants such as Fe^{III}. The area of the prewave is roughly independent of the sweep rate and acid electrolyte used but is increased by grinding to an extent greater than can be attributed to the increased surface roughness alone. The layer can accumulate progressively and at 25°C remains stable for many hours at open circuit, even when the electrode is removed from the cell. The charge passed in forming the prewave layer was confirmed as about 2 mC cm^{-2} for a polished chalcopyrite surface in 1 M HCl. This gives the thickness of the chalcopyrite layer which has reacted as 3.7 nm and a product layer thickness as 2.9 nm if certain assumptions are made. The relative amounts of Fe^{II} and Cu^{II} which pass into solution are in the ratio 4:1 and the assumptions led the authors to the conclusion that the solid product consisted of CuS and S.

The formation of CuS during oxidation of chalcopyrite electrodes in acid chloride solutions had been reported previously by Ammou-Chokroum *et al.*[315] They proposed a mechanism to explain the decrease in the rate of electrolyte dissolution of the mineral as being due to enrichment of its surface in copper due to accumulation of CuS in the porous layer of sulphur. As it blocks the pores the surface area available for reaction is decreased as also is the rate of oxidation. The next step in the proposed sequence of dissolution reactions is oxidation of the CuS at the interface between solid and the solution. It is not accepted here that the formation of CuS can be a primary reason for the progressive lowering of the rate of leaching of chalcopyrite in acidic solutions as long as the oxidation potential in the aqueous phase close to the solid surface is sufficiently high to maintain $\{S^{2-}\}$ at a very low value. The authors report that H$_2$S could be easily detected by smell in some of their suspensions of mineral particles and that the rest potential of an electrode changed markedly during the 30 minutes after its immersion in the acid electrolyte, containing no oxidant before the anodic potential was applied.

The nature of the passivating layer of product formed during electrolytic leaching of chalcoprite is not clearly established. It is considered briefly again in Sec. 5.3.6 in terms of the diffusion of iron and copper in the mineral. The layer does not appear to form during chemical leaching with Fe^{III} sulphate or chloride, or with CuCl$_2$ as oxidants. Decreasing reaction rates in these solutions can be analysed quantitatively in terms of progressive decrease of mineral surface area, or formation of a layer of porous sulphur surrounding a core of remaining mineral. Adhesion of this layer to the underlying chalcopyrite is greatest when it forms a continuous envelope surrounding a particle, and is much less in the case of a flat surface of an electrode. The relevance of the passivating layer to the hydrometallurgical processing of chalcopyrite does not seem to be significant. This is not the case when the surface chemistry of a mineral is important, as it is when froth flotation is used as a mineral benefication step for example. In such systems the pH is usually in the range weakly acid to moderately alkaline.

5.3.6.   *Diffusion of Cations in Copper-Containing Sulphides*

Copper–Iron Sulphides

The rate at which copper and iron can be leached from α-chalcopyrite in acidic solutions using Fe^{III} or, in solutions containing sufficient chloride, $CuCl_2$ as oxidants is very slow even at elevated temperatures. The question arises as to whether this is due to limiting effects of solid state diffusion of the metals within the mineral. Use of electrochemical experimental methods and application of the Sand Eq. (4.20) or the Cottrell Eq. (4.21) permits measurement of the diffusion coefficient of the electroactive species under the driving force due to the applied electric field. Chronopotentiometry, or chronoamperometry, or linear or cyclic voltammetry may be used, and application of the method to the leaching of bornite is outlined in Sec. 5.3.4. The rate at which the electroactive species can reach the reacting interface controls the current which can flow under the conditions applied to the system.

When chemical as distinct from electrochemical leaching is taking place the only driving force causing diffusion is random thermal movement of ions and the concentration gradient within the solid boundary layer caused by removal of atoms from the reacting layer at the surface of the solid in contact with the solution. If a nonporous layer of an unreactive product forms, isolating the mineral from the solution, there is no driving force to make solute species penetrate the layer to reach the mineral. It is well known that at temperatures above its melting point sulphur produced by oxidative leaching of some sulphide minerals wets the mineral surface and the leaching reaction stops. If a porous layer of solid sulphur forms on the surface soluble reactant and product species diffuse to and from the surface of the remaining mineral and reaction continues at a reduced rate.

Under chemical leaching conditions self-diffusion drives the solid system towards homogeneity. Measurement of self-diffusion coefficients in solids involves the formation of an interface between two pieces of solid, one containing a higher concentration of the diffusing species being studied than the other, which is the solid in which the diffusion coefficient is to be measured. The most common procedure is to section the solid and measure the amount of the diffusing species present in each slice after a given time. Thus each sample of the solid provides a single value of D. If the diffusing species has a suitable radioactive isotope, movement of this into the solid can be used to measure D. Harvey and Chen[316] used ^{64}Cu and ^{59}Fe to measure the diffusion coefficients of both metals in two specimens of chalcopyrite, one from Messina, Transvaal, S. Africa; the other from Tintic, Utah, USA. The massive samples were first cut into large slabs, 5 mm thick, and one large surface of each was ground flat and polished, finishing with 1 μm diamond paste. Each slab was then cut to form 5 mm cubes or $10 \times 5 \times 5$ mm parallelepipeds. Any inclusions of silicates exposed at the newly formed surfaces were removed by hydrofluoric acid.

The three elements forming chalcopyrite produce radioactive isotopes on neutron activation. ^{35}S is a β-emitter, half-life 88 days, but because of the low energy of the electrons they are absorbed within the mineral so that diffusion of ^{35}S could not be measured using the nondestructive technique. ^{64}Cu emits γ-rays of energy 0.511 MeV during decay by positron annihalation, half-life 12.9 h. ^{59}Fe emits γ-rays, the two highest intensities corresponding to 1.099 and 1.292 MeV, half-life 45.1 days. Two thermal neutron sources were used to produce the radioactive isotopes within the parallelepipeds of the chalcopyrite specimens. The conditions of irradiation, and when necessary the times of storage to decay to the desired levels of activity, are given in the paper. The diffusion couple consisted of one radioactive and one initially nonradioactive parallelepiped with their polished faces in contact, clamped between two stainless steel discs of diameter 2.5 cm. Tantalum foils were inserted between the samples and the discs to prevent reaction with them. The diffusion took place in a tube furnace with the temperature controlled to within 1°C. A slight positive pressure of helium was maintained within the tube to prevent oxidation of the sulphide. Decrepitation, probably caused by fluid inclusions, generally occurred in the chalcopyrite in the temperature range 310–350°C, imposing an upper temperature limit for the measurements. Usually reactivation was required for each diffusion measurement. Separation of the two halves of the diffusion couple was easy since there was no tendency for the faces to fuse together. The γ-rays from the ^{64}Cu and ^{59}Fe were measured using a well shaped NaI scintillation detector with a hundred-channel pulse height analyser. The activity of the sample was determined from the area of the appropriate γ-ray peak after subtraction of background.

For unidirectional diffusive flow of labelled species in a diffusion couple the diffusion coeffiecient at a given temperature is given by

$$D = \left(\frac{A}{A_o}\right)^2 \frac{\pi h^2}{t}$$

where A is the activity of the initially nonradioactive member of the couple after diffusion

A_o is the total activity of the diffusion couple

h is the length of the neutron-activiated half of the diffusion couple

t is the diffusion time.

The derivation of this equation is given in the paper together with a discussion of its applicability to the geometry used in the measurements.

The Tintic chalcopyrite contained a trace level of silver, which formed a number of isotopes during the irradiation with neutrons. Those with short half-lives could not be detected in the presence of the much higher activities of ^{64}Cu and ^{59}Fe, but the metastable isomer 110^m Ag has a half-life of 253 days and emits γ-rays, the two highest intensities corresponding to 0.558 and 0.885 MeV. These could be detected after the ^{64}Cu and ^{59}Fe had decayed sufficiently and were used to obtain two values of D for silver as a byproduct of the measurements of iron diffusion. Self-diffusion of iron in pyrite was also measured using the same experimental procedure with four sets of diffusion couples. The massive sample of the mineral was from Ambasaguas, Spain.

Values of the diffusion constants obtained by Harvey and Chen are given in Table 5.22. Derived preexponential factors, D_o, are diffusion coefficients extrapolated to $1/T = 0$ according to the equation $D = D_o \exp(-E_A/RT)$. This is equivalent to the Arrhenius equation. The activation energy E_A is commonly writen as Q in North America, and in chemical kinetics D_o is known as the frequency factor. The activation energy values for copper self-diffusion measured using the specimens of chalcopyrite from the two areas are identical, within the estimated range of reliability, ± 6 kJ mol^{-1}. The same is true for iron self-diffusion. Although the uncertainty in the activation energy for silver diffusion in chalcopyrite is greater than for copper and iron self-diffusion, the values can reliably be placed in the order Ag > Cu > Fe. Since the values of the temperature dependency of self-diffusion of iron and copper are very different it can be inferred that copper and iron atoms need not move cooperatively through the chalcopyrite lattice. Differences in the values of D_o for the three diffusing species are orders of magnitude greater than the differences between samples so that the order of variation of the preexponential factors is

Table 5.22. Diffusion constants (from Ref. 316).

Solid phase	Diffusing isotrope	D_o cm^2 s^{-1}	E_A kJ mol^{-1}
CuFeS$_2$ (Messina)	^{64}Cu	9.4×10^{-7}	
			50.6 ± 6
CuFeS$_2$ (Tintic)	^{64}Cu	2.8×10^{-7}	
CuFeS$_2$ (Messina)	^{59}Fe	5.4×10^{-12}	
			26.8 ± 6
CuFeS$_2$ (Tintic)	^{59}Fe	1.3×10^{-12}	
CuFeS$_2$ (Tintic)	^{110}Ag	8.6×10^{-4}	73
FeS$_2$ (Ambasaguas)	^{59}Fe	2.5×10^{-12}	42

Ag > Cu > Fe. This is the same as the order of the values of E_A and the authors conclude that this consistency suggests that the diffusing silver is present as a substitutional impurity in the specimens of chalcopyrite in which it was studied.

The larger activation energy of iron in pyrite relative to that in chalcopyrite is explained in terms of the density of packing of atoms in the unit cells of the two solids. That of pyrite has a volume of $(5.4 \text{ Å})^3$ and contains 4 formula units ("molecules") while one half of the chalcopyrite unit cell has a comparable volume $(5.24 \text{ Å})^2 \times (5.15 \text{ Å})$, containing only 2 "molecules". Thus for the two cells of approximately equal volume, there are twice as many sulphur atoms in pyrite as there are in chalcopyrite, with an equal number of metal atoms.

Harvey and Chen[316] interpreted their diffusion measurements in terms of the lattice structure of chalcopyrite, using data from the X-ray crystal structure refinement study of Hall and Stewart.[317] They state that there is sterochemical evidence that the structure exists in a strong covalently bonded configuration which has an effective ionic state between Cu$^+$ Fe^{3+} S$_2^{2-}$ and Cu^{2+} Fe^{2+} S$_2^{2-}$. The sulphur atoms of CuFeS$_2$ are arranged in an essentially face-centred cubic array. Each S atom is bonded tetrahedrally to two Cu and two Fe atoms which occupy holes in the close-packed sulphur lattice. The average S–S distance between nearest neighbours is 3.74 Å, which is close to the internuclear distance between two S^{2-} ions in contact, 3.68 Å. This does not imply that bonding in CuFeS$_2$ is largely ionic. Pauling[318] gave the radius of sulphur atoms whose covalent bonding orbitals are fully occupied and are making van der Waals contacts with one another, as 1.85 Å. Thus the sulphur atoms in the chalcopyrite lattice can be described as being in close-packed array with metal atoms distributed in an ordered fashion in interstitial holes. Alternate planes of tetrahedral holes are occupied by metal atoms, and octahedral holes are unoccupied. Harvey and Chen consider therefore that it is very probable that diffusion of metal in chalcopyrite occurs interstitially. In addition they show that there is good linear correlation between activation energy for diffusion and ionic radius for the species Fe^{3+}, Cu^+ and Ag^+. Correlation is poor for the species Fe^{2+}, Cu^2 and Ag^+, and also for the uncharged atoms since Cu and Fe have nearly the same radius. They conclude that the metal atoms diffuse in chalcopyrite as the species Fe^{3+}, Cu^+ and Ag^+. This is consistent with the results of magnetic susceptibility,[319] neutron diffraction[320] and Mössbauer spectroscopy[321,322] measurements which indicated that the $3d$ electron configurations of copper and iron in chalcopyrite are closer to those of Cu^I and Fe^{III} than to those of Cu^{II} and Fe^{II}. The chalcopyrite structure can accommodate excess metal which was shown by lattice parameter and density measurements to occupy interstitial positions. Harvey and Chen suggest that in the diffusion process metal atoms move from their normal tetrahedral lattice positions into adjacent octahedral holes as positive ions. Migration can then occur along series of unoccupied, interconnected tetrahedral and octahedral holes. Because the total concentration of unoccupied holes is always approximately equal to the total atom concentration in chalcopyrite, they point out that self-diffusion rates should not be sensitive to small deviations from the stoichiometric composition. This prediction could not be confirmed because of the unavailability of suitable specimens for measurement. Disorder in the sulphur atom lattice would modify the situation considerably.

The fastest rates of chemical leaching reactions of chalcopyrite which might be maintained by solid state diffusion of metal atoms to the mineral surface at the solid–liquid interface were estimated by Harvey and Chen, using the diffusion coefficients they obtained. Since the copper ion moves faster than iron it was assumed that it is removed by reaction as rapidly as it can be supplied from the unreacted bulk of the mineral specimen or particle to the reaction zone by diffusion. This must lead to the formation of a layer of solid which is depleted in copper, the boundary or diffusion layer within the solid. They take the extreme lower limit of this as one unit cell dimension, $a \sim 10$ Å, so that maximum concentration gradient of diffusing copper ions would be n/a where n is the concentration of Cu or Fe atoms in solid $CuFeS_2$ and $n = 1.4 \times 10^{22}$ cm^{-3}. By extrapolation of measured copper diffusion coefficients, the corresponding diffusion velocities $\nu = D/a$ would be 7×10^{-8}, 9×10^{-7}, and 5×10^{-6} cm s^{-1} at 50°C, 100°C and 150°C respectively. A leach rate of 10^{-6} cm s^{-1} would lead to complete consumption of a 100 μm spherical grain in about 3 hours. Thus it may be considered possible that the rates of some slow leaching reactions of α-chalcopyrite could be limited by diffusion through the solid. However the depleted diffusion layer in the solid would have to be very thin indeed. Jones and Peters[282] leached a flat surface of chalcopyrite in 1 M Fe$_2$(SO$_4$)$_3$ + 0.2 M H$_2$SO$_4$ at 90°C and their published graph indicated that after 2 days the leaching rate became constant. During a subsequent 40 day period of constant leaching rate 3×10^{-3} g atom cm^{-2} dissolved. This equals 0.191 g of Cu dissolved cm^{-2} of solid of density 4.3 g ml^{-1}. Thus the thickness of the layer of copper dissolved was 0.0443 cm during a period of 3.46×10^6 s, a rate of 1.28×10^{-8} cm s^{-1}. This is a little slower than the diffusion velocity at 100°C, consistent with the possibility that the boundary layer within the solid is somewhat thicker than 10 Å. However the copper ions in the solid surface at its interface with the leaching solution are merely available for reaction and so for transfer into it. The rate at which they actually do so depends on concentrations of reactants in the solution and the value of E in the case of chemical leaching, and on the value of E for electrochemical leaching in the absence of added oxidants, if no reagents which significantly affect the electrochemical reaction mechanism have been added.

The semiquantitative estimate of the thickness of the chalcopyrite layer which had reacted to form the protective layer on the mineral surface during electrochemical leaching of an electrode was 3.7 nm.[314] Using Fe(NO$_3$)$_3$ as the oxidant to leach particulate chalcopyrite, Linge found that during the first few minutes *iron* dissolved preferentially from the solid and that the metal-deficient layer which built up rapidly was about 5 nm thick.[312] The rates of the chemical and electrochemical leaching reactions were then similar. If they were controlled by diffusion of metals across a boundary layer of that thickness of solid in which their diffusion coefficients were similar to the values determined by Harvey and Chen, then the rates would be very slow.

The suggestion proposed by those authors that cations can diffuse through interstitial positions in the lattice of chalcopyrite leads to questions concerning the behaviour of the other minerals in the chalcopyrite group. The relative volumes of their unit cells increase with an increasing value of metal to sulphur ratio, indicating that the solids are metal-rich rather than sulphur-poor. The basic structures of talnakhite, mooihoekite and haycockite are due to the presence of extra metal atoms at interstitial sites of the cubic close-packed sulphur lattice of chalcopyrite. Hall describes the interstitial metals in terms of their forming metal coordination octahedra which vary in size according to the packing arrangements.

If diffusion of cations in α-chalcopyrite occurs through intersitial holes in the lattice it might reasonably be expected that the extra metals present in $Cu_9Fe_8S_{16}$, $Cu_9Fe_9S_{16}$ and $Cu_8Fe_{10}S_{16}$ could diffuse out of the solids under acidic oxidative conditions if the solid structure could maintain charge balance. There appears to be no information concerning the behaviour of talnakhite and haycockite, but this is the behaviour of mooihoekite which is discussed above. It is interesting to note that the changes in the X-ray diffraction pattern of mooihoekite as leaching proceeded do not lead to a structure resembling that of talnakhite or haycockite at any stage.

Adams *et al.*[323] carried out several series of experiments to clarify aspects of the work of Hiller and Probsthain[273] and establish composition limits for some phases in the chalcopyrite series. The required quantities of the elements were heated in sealed silica tubes and held at 800°C in the usual way. The cooled sintered products were homogenised by grinding in agate under N_2, rigorously excluding air, and reheated in evacuated silica tubes at 800°C for 4 days. This procedure was carried out three times. 5 mg samples of products were heated in a dynamic vacuum and weight loss was measured as a function of temperature. The tetragonal α-chalcopyrite was found to lose sulphur at 390°C, rather than 550°C as reported previously.[273] Cell parameters of the products were measured using a diffractometer and, for the density of the solid, a density bottle and 1, 1, 2, 2,-tetrachloroethane after thorough degassing. For samples $CuFeS_{2-x}$:

With $x = 0$, $a_o = 5.292 \pm (5)$; $c_o = 10.407 \pm (5)$ Å. Theoretical density $4.18 \pm (1)$ g cc^{-1}, $[4(CuFeS_2)]$; α-tetragonal.

With $x = 0$ to $x = 0.16$, mixed phase, $\alpha + \gamma$-tetragonal. The cubic β phase was possibly also present.

With $x = 0.17$ to $x = 0.24$, $a_o = 10.598 \pm (5)$; $c_o = 5.380 \pm (5)$ Å (values for $x = 0.2$); theoretical density $4.32 \pm (1)$ [8 $Cu_{1.11}$ $Fe_{1.11}$ $S_{2.00}$] experimental density $4.30 \pm (1)$; γ-tetragonal. All theoretical densities are based on close-packed sulphur with excess metal occupying additional lattice sites.

Attempts to reproduce the cubic β-phase of Hiller and Probsthain by direct combination of the elements gave only γ-phase. The β-phase is equivalent to talnakhite, $Cu_9Fe_8S_{16}$, and it was shown that this can be prepared by heating

$CuFeS_2$ in a stream of hydrogen at 450°C. The synthetic material with $x = 0.2$ had values of a_o and c_o in agreement with those of Hiller and Probsthain for the same composition. However whereas those authors reported this phase to be unstable at low temperatures, Adams *et al.* found the γ-phase to be stable when quenched from any temperature between ambient and 800°C, and to be stable at room temperature for 450 days.

Attempts to produce polycrystalline samples of compositions $Cu_{1-x}Fe_{1+x}$· S_2 and $Cu_{1+x}Fe_{1-x}S_2$ were unsuccessful. In each case α-$CuFeS_2$ and other unidentified phases appeared in the diffraction pattern. However a single phase could be obtained for compounds having the general formula $Cu_{1-x}Fe_{1+x}S_{2-y}$ for $x = 0.25 > x > 0.08$ with y chosen as 0.2 (i.e in the γ-chalcopyrite region). The stability range of this phase with respect to sulphur was not determined. The compounds in this composition range have a cubic structure and changes in lattice dimensions as the Cu:Fe ratio is varied from $x = 0$ to $x = 0.25$ are given. The tetragonal γ-phase exists only when the value of the Cu:Fe ratio approaches unity. Efforts to prepare $Cu_{1+x}Fe_{1-x}S_{1.8}$ were unsuccessful; the phases identified by X-ray diffraction were $CuFeS_2$ or $CuFeS_{2-x}$ in addition to bornite solid solution, Cu_5FeS_4.

Some of the α-chalcopyrite samples prepared by Adams *et al.* contained ferromagnetic impurites which were detected by magnetic and density measurements, but not by X-ray diffraction. The densities of these samples were greater than the value for the pure α-phase and when annealed with sulphur to correct the deficiency the magnetic susceptibility showed no field dependency, indicating removal of the impurities. Attempts to synthesise α-chalcopyrite by melting the sintered product formed by reacting together the elements as described in the Introduction to this section also resulted in solids containing excess metal due to incomplete absorption of sulphur.[231] Material shown by chemical analysis to have the composition $Cu_{1.12}$ $Fe_{1.09}$ S_2 gave no X-ray diffraction lines other than these of α-chalcopyrite. Samples of 100–150 μm solid were leached in 0.1 M $Fe_2(SO_4)_3$ solution having pH 1.0 at 80°C for about 70 hours, the experiments being started 6, 40 and 78 days after completion of the synthesis of the material. The 6 days old solid gave a parabolic leaching curve during the first 10 hours, which then became linear. The 40 and 78 days old solids gave identical, superimposed parabolic leaching curves during the first 10 hours, with a faster rate than the 6 days old materials, becoming linear after about 20 hours and considerably slower than the latter, and the 40 day slightly faster than the 78 day old solid. After being stored for

9 months the unleached solid had fractured and segregation had occurred in many places at outside edges. Some holes had formed in the solid and electron microprobe analyses showed a higher copper content in solid close to these edges and around the holes than in the bulk of the solid. The same procedure had shown the material to be homogeneouns shortly after its preparation. After storing it for 14 months a sample of the solid was leached under the same conditions as before. 11–12% of its copper content was leached out relatively rapidly, corresponding to the 10.7% stoichiometric excess over that required for the composition $CuFeS_2$ which had been present in the freshly prepared homogeneous solid. Analysis of the leach residue showed that none of the 8.3% stoichiometric excess of iron present in that solid had been lost during the same period of leaching.

Copper Sulphides

Several investigations have been carried out on the diffusion of copper in chalcocite and digenite, using both the high and low temperature forms. Self-diffusion at 140–150°C in synthetic Cu_2S, purified by melting in vacuo, was measured using a cylindrical specimen, 1.5 cm in diameter, incorporating ^{64}Cu in it from a solution of the nitrate, and heating in vacuo for 90–375 minutes.[325] The diffusion coefficient at 134°C was approximately 6×10^{-7} cm^2 s^{-1} and the value did not increase significantly at higher temperatures within the range used, E_A being estimated to be about 20 kJ mol^{-1}. The results were valid for the intermediate β-phase, stable between about 103°C and 435°C. The same result was obtained for the diffusion of Cu in $Cu_{1.77}S$ and D does not depend on composition over the range $Cu_{1.77}S$ and $Cu_{1.99}S$.

Buerger and Bernhardt[326] quoted results of Jensen[327] who observed that the diffusion rate of copper in low chalcocite increased rapidly as the specimen approached the transformation temperature, $\sim$ 103°C and was much higher throughout the high chalcocite range, the activation energy being extremely low. They explained this by assuming that the copper atoms are mobile through the interstices of the hexagonal close-packed sulphur atoms. This mobility was attributed to the fact that these atoms can assume tetrahedral, trigonal, and linear coordination. They suggested that in the low temperature form the thermal energy is not sufficient to maintain such high mobility and the copper atoms are in at least two of the three coordination positions in the hcp array of S atoms which occur in the high temperature form.

Etienne[118] employed a method based on chronopotentiometry to measure the diffusion coefficients of copper in chalcocite and digenite, see Sec. 4.2.5. She mounted a copper metal rod, about 0.9 cm in diameter, to expose one polished circular planar face to a 0.2 M Na_2SO_4 + 0.2 M H_2SO_4 solution which was saturated with H_2S, a continuous stream of the gas being bubbled through it. The liquid was stirred vigorously and maintained at constant temperature.

The electrode potential (versus saturated calomel electrode) required to maintain a constant current was measured. The reaction is expressed as

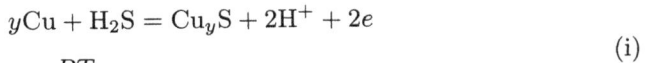

$$yCu + H_2S = Cu_yS + 2H^+ + 2e$$

$$E = E^\circ - \frac{RT}{2F} \ln\{Cu\}^y\{H_2S\}/\{Cu_yS\}\{H^+\}^2$$

(i)

$\{Cu\}$ at the solid–liquid interface reflects the conditions imposed on the system and the transport properties of the copper sulphide "scale" being produced. $\{Cu_yS\}$ was taken as unity.

In defined geometric and stirring conditions, with constant current flowing, $\{H_2S\}$ and $\{H^+\}$ are constant close to the electrode. In steady state conditions the activation overvoltage for a given sulphide is a constant. Therefore the variation of the electrode potential with time follows the change of the copper activity according to (ii), derived form (i):

$$dE/dt = -\frac{RT}{2F}\frac{d}{dt}\ln A\{Cu\}^y$$

and if y is constant

$$dE/dt = -y\frac{RT}{2F}\frac{d}{dt}\ln\{Cu\}.$$

(ii)

If the sulphide scale grows uniformly on the electrode surface the copper activity at the liquid–solid interface can be directly linked to the transport properties of the sulphide layer. The theoretical model was then developed.

Appendix

Appendix. Some thermodynamic data for elevated temperatures Sources: Data for 298 K Ref. 50 with some amendments. Cp values Ref. 53.

Formula	Description	Sate	ΔH° kJ mol^{-1}	ΔG° kJ mol^{-1}	S° J mol^{-1} K^{-1}	Cp J mol^{-1} K^{-1}	ΔCp_{298}^{373} J mol^{-1} K^{-1}	ΔCp_{298}^{423} J mol^{-1} K^{-1}	ΔG° kJ mol^{-1} 423 K
			298 K (25°C) and 0.1 MPa (1 bar)						
H^+		aq	0	0	0	0	130	138	-3.198
H_2O		l	-285.830	-237.129	69.91	75.291	75.44	75.90	-247.63
H_2		g	0	0	130.684	28.824	28.9	28.9	-17.01
OH^-		aq	-229.994	-157.244	-10.75	-148.5	-243	-255	-26.45
O_2		g	0	0	205.138	29.355	34.7	34.7	
Fe		cr	0	0	27.28	25.10			
Fe^{2+}		aq	-89.1	-78.90	-137.7		276	285	-68.29
Fe^{3+}		aq	-48.5	-4.7	-315.9		389	402	25.47
$Fe_{0.947}O$	wustite	cr	-266.27	-245.12	57.49	48.12			
Fe_2O_3	hematite	cr	-824.2	-742.2	87.40	103.85	110.9	113.8	-755.76
Fe_3O_4	magnetite	cr	-1118.4	-1015.4	146.4	143.43	159.4	164.4	-1037.5
$Fe(OH)_3$	precipitated	cr	-823.0	-696.5	106.7				
$Fe_{1.000}S$	pyrrhotite-α	cr	-100.0	-100.4	60.29	50.54	63.6	64.9	-109.44
FeS_2	pyrite	cr	-178.2	-166.9	52.93	62.17	63.6	65.3	-175.0
S	rhombic	cr	0	0	31.80	22.64	24.7	26.8	-4.60
S^{2-}		aq		111.4	-14.6				
HS^-		aq	-39.7	12.08	62.8		-243	-259	10.23
H_2S		aq		-27.83	121	-84	268	255	-48.9
HSO_4^-		aq	-887.34	-755.91	131.8		-42	-75	-770.6
SO_4^{2-}		aq	-909.27	-744.53	20.1	-293	-452	-439	-736.9
Cu		cr	0	0	33.150	24.435	24.73	24.85	-4.72
Cu^+		aq	71.67	49.98	40.6		218	222	39.76
Cu^{2+}		aq	64.77	65.49	-99.6		268	276	71.54
CuO		cr	-157.3	-129.7	42.63	42.30	45.48	46.02	-136.4
Cu_2O		cr	-168.6	-146.0	93.14	63.64	70.33	70.92	-159.3
$HCuO_2^-$		aq		-258.5	41.8		-530.5	-509.0	-252
CuO_2^{2-}		aq	-183.6	-183.6	-96.2		-699	-680	-156
CuS		cr	-53.1	-53.6	66.5	47.82	48.07	48.32	-63.03
Cu_2S		cr	-79.5	-86.2	120.9	76.32	81.59	89.96	-103.4

References

1. A.F. Wells, *Structural Inorganic Chemistry*, 5th edition, Oxford University Press, Oxford (1984).
2. E.J. Fasika, Structural aspects of the oxides and oxyhydroxides of iron, *Corrosion Science* **7**, 833–839 (1967).
3. J.D. Bernal, D.R. Dasgupta and A.L. MacKay, The oxides and hydroxides of iron and their structural inter-relationships, *Clay Min. Bull.* **4**, 15–30 (1959).
4. P. Ramdohr, *The Ore Minerals and their Intergrowths* (English translation of 3rd edition), Pergamon Press, Oxford (1969).
5. D.J. Vaughan and J.R. Craig, *Mineral Chemistry of Metal Sulphides*, Cambridge University Press, Cambridge (1978).
6. P. Debye and E. Hückel, The theory of electrolytes. (1) Lowering of freezing point and related phenomena, *Physik. Z.* **24**, 185–206 (1923). (2) The limiting law of electrical conductivity, *Ibid.* **24**, 305–325 (1923). (3) Osmotic equation of state and the activity of strong electrolytes in dilute solutions, *Ibid.* **25**, 97–107 (1924).
7. H.S. Harned and B.B. Owen, *The Physical Chemistry of Electrolytic Solutions*, 3rd edition, Reinhold Publishing Corporation, New York (1958).
8. R.A. Robinson and R.H. Stokes, *Electrolyte Solutions*, 2nd edition (revised), Butterworths, London (1965).
9. M.D. Cohen, R.C. Flagan and J.H. Seinfeld, Studies of concentrated electrolyte solutions using the electrodynamic balance. (1) Water activities for single-electrolyte solutions, *J. Phys. Chem.* **91**, 4563–4574 (1987).
10. *Idem.* Studies of concentrated electrolyte solutions using the electrodynamic balance. (2) Water activities for mixed-electrolyte solutions, *Ibid.* **91**, 4575–4583 (1987).
11. *Idem.* Studies of concentrated electrolyte solutions using the electrodynamic balance. (3) Solute nucleation, *Ibid.* **91**, 4583–4590 (1987).
12. E.A. Guggenheim, *Thermodynamics*, North Holland Publishing, Amsterdam (1949).

13. C.W. Davies, The extent of dissociation of salts in water. (8) An equation for the mean ionic activity coefficient of an electrolyte in water, and a revision of the dissociation constants of some sulphates, *J. Chem. Soc.* (London), 1938, 2093–2098.

14. G. Scatchard, Concentrated solutions of strong electrolytes, *Chem. Rev.* **19**, 309–327 (1936).

15. J.F. Zemaitis Jr., D.M. Clark, M. Rafal and N.C. Scrivner, *Handbook of Aqueous Electrolyte Thermodynamics*, Design Institute for Physical Property Data, Sponsored by AIChE, New York (1986).

16. A.L. Horvath, *Handbook of Aqueous Electrolyte Solutions, Physical Properties, Estimation and Correlation Methods*, Ellis Horwood Ltd., Chichester (1985).

17. R.H. Davies, Aqueous solutions of dilute and concentrated electrolytes, in *Chemical Thermodynamics in Industry: Models and Computations*, Ed. T.I. Barry, SCI: Blackwell Scientific Publications, Oxford (1985).

18. K.S. Pitzer, Thermodynamics of electrolytes. (1) Theoretical basis and general equations, *J. Phys. Chem.* **77**, 268–277 (1973).

19. K.S. Pitzer and G. Mayorga, Thermodynamics of electrolytes. (2) Activity and osmotic coefficients for strong electrolytes with one or both ions univalent, *J. Phys. Chem.* **77**, 2300–2308 (1973).

20. K.S. Pitzer and G. Mayorga, Thermodynamics of electrolytes. (3) Activity and osmotic coefficients for 2-2 electrolytes, *J. Solution Chem.* **3**, 539–546 (1974).

21. K.S. Pitzer and J.J. Kim, Thermodynamics of electrolytes. (4) Activity and osmotic coefficients for mixed electrolytes, *J. Am. Chem. Soc.* **96**, 5701–5707 (1974).

22. K.S. Pitzer, Thermodynamics of electrolytes. (5) Effects of higher-order electrostatic terms, *J. Solution Chem.* **4**, 249–265 (1975).

23. K.S. Pitzer and L.F. Silvester, Thermodynamics of electrolytes. (6) Weak electrolytes including H_3PO_4, *J. Solution Chem.* **5**, 269–278 (1976).

24. K.S. Pitzer, N.R. Rabindra and L.F. Silvester, Thermodynamics of electrolytes. (7) Sulphuric acid, *J. Am. Chem. Soc.* **99**, 4930–4936 (1977).

25. L.F. Silvester and K.S. Pitzer, Thermodynamics of electrolytes. (8) High temperature properties including enthalpy and heat capacity, with application to sodium chloride, *J. Phys. Chem.* **81**, 1822–1828 (1977).

26. K.S. Pitzer, J.R. Peterson and L.F. Silvester, Thermodynamics of electrolytes. (9) Rare earth chlorides, nitrates and perchlorates, *J. Solution Chem.* **7**, 45–56 (1978).

27. L.F. Silvester and K.S. Pitzer, Thermodynamics of electrolytes. (10) Enthalphy and the effect of temperature on the activity coefficients, *J. Solution Chem.* **7**, 327–337 (1978).

28. D.J. Bradley and K.S. Pitzer, Thermodynamics of electrolytes. (12) Dielectric properties of water and Debye–Hückel parameters to 350°C and 1 Kbar, *J. Phys. Chem.* **83**, 1599–1603 (1979).

29. Y.C. Wu, R.M. Rush and G. Scatchard, Osmotic and activity coefficients for binary mixtures of sodium chloride, sodium sulphate, magnesium sulphate and

magnesium chloride in water at 25°C. (1) Isopiestic measurements on the four systems with common ions, *J. Phys. Chem.* **72**, 4048–4057 (1968).

30. *Idem.* (2) Isopiestic and electromotive force measurements on the two systems without common ions, *Ibid.* **73**, 2047–2053 (1969).

31. C.J. Downes and K.S. Pitzer, Thermodynamics of electrolytes. Binary mixtures formed from aqueous sodium chloride, sodium sulphate, copper (II) chloride and copper sulphate at 25°C, *J. Solution Chem.* **5**, 389–398 (1976).

32. P.S.Z. Rogers and K.S. Pitzer, Volumetric properties of aqueous sodium chloride solutions, *J. Phys. Chem. Ref. Data* **11**, 15–81 (1982).

33. K.S. Pitzer and J.S. Murdzek, Thermodynamics of aqueous sodium sulphate, *J. Solution Chem.* **11**, 409–413 (1982).

34. P.S.Z. Rogers and K.S. Pitzer, High-temperature thermodynamic properties of aqueous sodium sulphate solutions, *J. Phys. Chem.* **85**, 2886–2895 (1981).

35. T.J. Edwards, G. Maurer, J. Newman and J.M. Prausnitz, Vapour–liquid equilibria in multicomponent aqueous solutions of volatile weak electrolytes, *AIChE J.* **24**, 966–976 (1978).

36. D. Beutier and H. Renon, Representation of NH_3–H_2S–H_2O, NH_3–CO_2–H_2O and NH_3–SO_2–H_2O vapor–liquid equilibriums, *Ind. Eng. Chem. Process Des. Dev.* **17**, 220–230 (1978).

37. C.-C. Chen, H.I. Britt, J.F. Boston and L.B. Evans, Extension and modification of the Pitzer equation for vapor–liquid equilibriums of aqueous electrolyte systems with molecular solutes, *AIChE J.* **25**, 820–831 (1979).

38. *Idem.* Two new activity coefficient models for the vapor–liquid equilibrium of electrolyte systems, in *Thermodynamics of Aqueous Systems with Industrial Applications*, ACS Symposium Series 133, Ed. S.A. Newman, American Chemical Society, Washington D.C. (1980).

39. *Idem.* Local composition model for excess Gibbs energy of electrolyte solutions, Part 1 — Single solvent, single completely dissociated electrolyte systems, *AIChE J.* **28**, 588–596 (1982).

40. L.A. Bromley, Thermodynamic properties of strong electrolytes in aqueous solutions. *AIChE J.* **19**, 313–320 (1973).

41. E.A. Washburn (ed)., *International Critical Tables*, McGraw Hill, New York (1926).

42. J.A. Rard and D.G. Miller, Isopiestic determination of the osmotic coefficients of aqueous sodium sulfate, magnesium sulfate and sodium sulfate–magnesium sulfate at 25°C, *J. Chem. Eng. Data* **26**, 33–38 (1981).

43. J.A. Rard, A. Habenschuss and F.H. Spedding, A review of the osmotic coefficients of aqueous calcium chloride at 25°C, *J. Chem. Eng. Data* **22** 180–186 (1977).

44. J.A. Rard, Isopiestic determination of the osmotic and activity coefficients of aqueous manganese II chloride, manganese II sulphate and rubidium chloride at 25°C, *J. Chem. Eng. Data* **29**, 443–450 (1984).

45. W. Kangro and A. Groeneveld, Concentrated aqueous solutions, *I Z. Physik. Chem.* **32**, 110–126 (1962).

46. L.G. Sillén and A.E. Martell, *Stability Constants of Metal–Ion Complexes*, 2nd edition Special Publication No. 17, The Chemical Society, London (1964).

47. L.G. Sillén, E. Högfeldt and A.E. Martell, *Stability Constants of Metal–Ion Complexes*, Supplement No. 1, Special Publication No. 25, The Chemical Society, London (1971).

48. R.M. Smith and A.E. Martell, *Critical Stability Constants*, Plenum Press, New York, Vol. 1 Amino Acids (1974); Vol. 2 Amines (1975); Vol. 3 Other Organic Ligands (1977); Vol. 4 Inorganic Complexes (1976); Vol. 5 First Supplement (1982).

49. M. Pourbaix, *Atlas of Electrochemical Equilibria in Aqueous Solutions*, Pergamon Press, Oxford (1966).

50. D.D. Wagman *et al.*, *The NBS Tables of Chemical Thermodynamic Properties*, Journal of Physical and Chemical Reference Data, 11, Supplement No. 2, American Chemical Society and American Institute of Physics, New York (1982).

51. F.M.M. Morel, *Principles of Aquatic Chemistry*, John Wiley and Sons, New York (1983), p. 317.

52. G. Valensi, Contribution to the potential-pH diagram of sulphur, *Compt. Rend. 2me Réunion Comite Int. Thermo. Kinetics Electrochem*, Milan (1950), 51–67.

53. R.C.H. Ferreira, High temperature E-pH diagrams for the systems S–H$_2$0, Cu–S–H$_2$O and Fe–S–H$_2$O, in *Leaching and Reduction in Hydrometallurgy*, Ed. A.R. Burkin, The Institution of Mining and Metallurgy, London (1975), 67–83.

54. G.N. Lewis and M. Randall, *Thermodynamics*, revised by Eds. K.S. Pitzer and L. Brewer, 2nd edition, McGraw Hill Book Co. New York (1961), p. 523.

55. C.M. Criss and J.W. Cobble, The thermodynamic properties of high temperature aqueous solutions. (IV) Entropies of the ions up to 200° and the correspondence principle, *J. Am. Chem. Soc.* **86**, 5385–5390 (1964).

56. C.M. Criss and J.W. Cobble, The thermodynamic properties of high temperature aqueous solutions. (V) The calculation of ionic heat capacities up to 200°C. Entropies and heat capacities above 200°, *J. Am. Chem. Soc.* **86**, 5390–5393 (1964).

57. F.A. Schaufelberger, Precipitation of metal from salt solution by reduction with hydrogen, *J. Metals* **8**, 695–704 (1956); *Trans. Am. Inst. Mining. Eng.* **205**, 539–548 (1956); Also published in *Mining Eng.* **8**, 539–548 (1956).

58. B. Meddings and V.N. Mackiw, The gaseous reduction of metals from aqueous solutions, in *Symposium on Unit Processes in Hydrometallurgy*, Ed. M.E. Wadsworth and F.T. Davis, Gordon and Breech, New York (1964), 345–384.

59. J. Bjerrum, *Metal Ammine Formation in Aqueous Solution*, P. Haase and Son, Copenhagen (1957).

60. A.J. Monhemius, Precipitation diagrams for metal hydroxides, sulphides, arsenates and phosphates, *Trans. Inst. Mining Metall.* (*Section C. Mineral Processing and Extractive Metallurgy.*) **86**, C202–206 (1977).

61. A.J. Ellis and N.B. Milestone, The ionization constants of hydrogen sulphide from 20° to 90°, *Geochim. Cosmochim. Acta* **31**, 615–620 (1967).

62. W. Giggenbach, Optical spectra of highly alkaline sulfide solutions and the second dissociation constant of hydrogen sulfide, *Inorg. Chem.* **10**, 1333–1338 (1971).

63. A.J. Ellis and W. Giggenbach, Hydrogen sulphide ionization and sulphur hydrolysis in high temperature solution, *Geochim. Cosmochim. Acta* **35**, 247–260 (1971).

64. B. Meyer, K. Ward, K. Koshlap and L. Peter, Second dissociation constant of hydrogen sulfide, *Inorg. Chem.* **22**, 2345–2346 (1983).

65. S. Licht, pH measurement in concentrated alkaline solution, *Anal. Chem.* **57**, 514–519 (1985).

66. S. Licht and J. Manassen, The second dissociation constant of H_2S, *J. Electrochem. Soc.* **134**, 918–921 (1987).

67. L. Burkhart and J. Voigt, Aqueous precipitation in hydrometallurgy, in *Hydrometallurgical Reactor Design and Kinetics*, Eds. R.G. Bautista, R.J. Wesley and G.W. Warren, AIME, New York (1986), 441–456.

68. I.N. Tang and H.R. Munkelwitz, An investigation of solute nucleation and levitated solution droplets, *J. Colloid Interface Sci.* **98**, 430–438 (1984).

69. J.E. Dutrizac, The physical chemistry of iron precipitation in the zinc industry, in *Lead–Zinc–Tin '80, Conf. Proc.* Eds. J.M. Cigan, T.S. Mackey and T.J. O'Keefe, AIME, New York (1979), 532–564.

70. R.N. Sylva, The hydrolysis of iron (III), *Pure Appl. Chem.* **22**, 115–132 (1972).

71. Y.B. Yakovlev, F.Y. Kul'ba, A.G. Pus'ko and M.N. Gerchikova, Hydrolysis of iron (III) sulphate in zinc sulphate solutions at 25°C, 50°C and 80°C, *Russ. J. Inorg. Chem.* **22**, 27–29 (1977).

72. M. Kiyama and T. Takada, The hydrolysis of ferric complexes, Magnetic and spectrophotometric studies of aqueous solutions of ferric salts, *Bull. Chem. Soc. Jap.* **46**, 1680–1686 (1973) (in English).

73. S. Agatzini and A.R. Burkin, Statistical approach to the precipitation of iron as "goethite", *Trans. Inst. Mining Metall (Section C. Mineral Processing and Extractive Metallurgy)* **94**, C105–114 (1985).

74. C.J. Haigh, The hydrolysis of iron in acid solutions, *Proc. Aust. Inst. Min. Metall* 49–56 (September 1967).

75. E. Posnjak and H.E. Merwin, The system Fe_2O_3–SO_3–H_2O, *J. Am. Chem. Soc.* **44**, 1965–1994 (1922).

76. R.G. Bates, *Determination of pH; Theory and Practice*, 2nd edition, John Wiley and Sons, New York (1964).

77. R.G. Bates, Revised standard values for pH measurements from 0 to 95°, *J. Res. Natl. Bur. Std.* **A66**, 179–184 (1962).

78. G.D. Manning, pH calibration of tetroxalate, tartrate and phthalate buffer solutions at above 100°C, *J. Chem. Soc. Faraday Trans. 1*, **74**, 2434–2451 (1978).

79. M.L. Gavrish and I.S. Galinker, Solubility at elevated temperatures in the AgF–H_2O, Ag_2CrO_4–H_2O and Ag_2O–H_2O systems, *Zh. Neorg. Khim.* **15**, 1979–1981 (1970).

80. R.G. Robins, Hydrolysis of uranyl nitrate solutions at elevated temperatures, *J. Inorg. Nucl. Chem.* **28**, 119–123 (1966).

81. T.R. Scott, Alumina by acid extraction, *J. Metals* **14**, 121–125 (1962).

82. P.T. Davey and T.R. Scott, The hydrolysis of aluminum sulphate solutions at elevated temperatures, *Aust. J. Appl. Sci.* **13**, 229–241 (1962).

83. J. J. Fritz, Chloride complexes of CuCl aqueous solution, *J. Phys. Chem.* **84**, 2241–2246 (1980).

84. D.M. Muir, The application of thermodynamics to extractive metallurgy with chloride solutions — A review, Warren Spring Laboratory, Stevenage, SG1 2BX, UK (1984).

85. U. Strahm, R.C. Patel and E. Matijevic, Thermodynamics and kinetics of aqueous iron (III) chloride complexes formation, *J. Phys. Chem.* **83**, 1689–1695 (1979).

86. T. Sato and T. Kato, The stability constants of the chloro complexes of Cu(II) and Zn(II) determined by tri-n-octylamine extraction, *J. Inorg. Nucl. Chem.* **39**, 1205–1208 (1977).

87. T. Sato and T. Nakamura, The stability constants of the aqueous chloro complexes of divalent Zn, Cd and Hg determined by solvent extraction with tri-n-octylphosphine oxide, *Hydrometallurgy* **6**, 3–12 (1980).

88. D.K. Nordstrom *et al.* A comparison of computerized models for equilibrium calculations in aqueous systems, in *Chemical Modeling in Aqueous Systems*, ACS Symposium Series No. 93, Ed. E.A. Jenne, American Chemical Society, Washington, D.C. (1979), 857–892.

89. G. Eriksson, An algorithm for the computation of aqueous multicomponent multiphase equilibria, *Anal. Chim. Acta* **112**, 375–383 (1979).

90. F.J. Zeleznik and S. Gordon, Calculation of complex chemical equilibria, *Ind. Eng. Chem.* **90**, 27–57 (1968).

91. F. Van Zeggeren and S.H. Storey, *The Computation of Chemical Equilibria*, Cambridge University Press, London (1970).

92. F.J. Millero and D.R. Schreiber, The use of the ion pairing model to estimate activity coefficient of the ionic components of natural waters, *Am. J. Sci.* **282**, 1508–1540 (1982).

93. C.W. Bale and G. Eriksson, Metallurgical thermochemical databases — A review, *Can. Metall. Quart.* **29**, 105–132 (1990).

94. R.H. Davies, A.T. Dinsdale, T.G. Chart, T.I. Barry and M.H. Rand, Application MTDATA to the modelling of multicomponent equilibria. High Temperature Science, 1988–1989 (Published 1990) **26**, 251–262.

95. T.I. Barry and T.G. Chart, New approach to materials design: Calculated phase equilibria for composition and structural control, in *Research and Development of High Temperature Materials for Industry*, Conf. Proc. Ed. E. Bullock, Elsevier Applied Science (1989), 565–592.

96. C.A. Johnson, The regulation of trace element concentrations in river and estuarine waters contaminated with acid mine drainage, The adsorption of Cu and Zn on amorphous Fe oxyhydroxides, *Geochim, Cosmochim. Acta.* **50**, 2433–2438 (1986).

97. N.K. Adam, *The Physics and Chemistry of Surfaces*, 2nd edition, Oxford University Press, Oxford (1938) p. 260.

98. H. Freundlich, *Colloid and Capillary Chemistry* (English translation from 2nd and 3rd German editions by H.S. Hatfield), Methuen and Co., London (1926) 169–239.

99. J.E. Dutrizac, The behaviour of impurities during jarosite precipitation, in *Hydrometallurgical Process Fundamentals, Conf. Proc.* Ed. R.G. Bautista, Plenum Press, New York (1984), 125–169.

100. A.C. Riddiford and L.L. Bircumshaw, The kinetics of the solution of zinc in aqueous iodine solutions. (III and IV), *J. Chem. Soc.* 1952, 698–701; 701–704.

101. P.S. Roller, The physical and chemical relations in fluid-phase heterogeneous reaction, *J. Phys. Chem.* **39**, 221–237 (1935).

102. A. Fage and H.C.H. Townend, An examination of turbulent flow with an ultramicroscope, *Proc. Roy. Soc.* (London) **A135**, 656–667 (1932).

103. L.L. Bircumshaw and A.C. Riddiford, Transport control in heterogeneous reactions, *Quarterly Rev.* (London) **6**, 157–185 (1952).

104. V.G. Levich, *Physicochemical Hydrodynamics*, Prentice Hall, Englewood Cliffs, New Jersey (1962).

105. V.G. Levich, Theory of concentration polarization, *J. Phys. Chem.* (USSR) **18**, 335–355 (1944).

106. C.N. Hinshelwood, *Kinetics of Chemical Change*, Oxford University Press, Oxford (1940).

107. M. Kameda, Fundamental studies on solution of gold in cyanide solutions, (II) Equations of reactions and effects of cyanide strength and other variables on solution rate, Science Reports Research Institutes Tohok. University Ser. **A1**, 223–230 (1949); *Chem. Abs.* **45**, 5583c (1951).

108. V. Lund, Corrosion of silver by potassium cyanide solutions and oxygen, *Acta. Chem. Scand.* **5**, 555–567 (1951).

109. G.A. Deitz and J. Halpern, Reaction of silver with aqueous solutions of cyanide and oxygen. *J. Metals* **5**, *AIME Trans.* **197**, 1109–1116 (1953).

110. I.A. Kakovsky, A study of the kinetics and mechanism of some hydrometallurgical processes, *6me Congrés International de la Preparation des Minerais, 1963. Compte Rendu Scientifique*, Société de l'Industrie Minérale, Saint–Etienne (1964), 157–180 (in French).

111. J.A. King, A.R. Burkin and R.C.H. Ferreira, Leaching of chalcocite by acidic ferric chloride solutions, in *Leaching and Reduction in Hydrometallurgy*, Ed. A.R. Burkin, Institution of Mining and Metallurgy London (1975), 36–45.

112. J.D. Sullivan, Chemistry of leaching chalcocite, *Technical Papers, US Bur. Mines* 473 (1930), 24 pages. Chemistry of leaching covellite, *Technical Papers, US Bur. Mines*, 487 (1930), 18 pages.

113. P.F. Thompson, The dissolution of gold in cyanide solutions, *Trans. Electrochem. Soc.* **91**, 41–71 (1947).

114. V. Kudryk and H.H. Kellogg, Mechanism and rate-controlling factors in the dissolution of gold in cyanide solutions, *J. Metals* **6**, *AIME Trans.* **200**, 541–548 (1954).

115. P. Delahay, *New Instrumental Methods in Electrochemistry*, Interscience, New York (1954).

116. P. Delahay, *Double Layer and Electrode Kinetics*, Interscience, New York (1965).

117. A.S. Bard and L.R. Faulkner, *Electrochemical Methods and Applications*, John Wiley and Sons, New York (1980).

118. A. Etienne, Electrochemical method to measure the copper ionic diffusivity in a copper sulfide scale, *J. Electrochem. Soc.* **117**, 870–874 (1970).

119. D.D. Macdonald, *Transient Techniques in Electrochemistry*, Plenum Press, New York (1977).

120. M.E. Wadsworth, Hydrometallurgical processes. Principles of leaching, in *Rate Processes of Extractive Metallurgy*, Eds. H.Y. Sohn and M.E. Wadsworth, Plenum Press, New York (1979), 133–186.

121. F.M. Doyle, N. Ranjan and E. Peters, Mathematical modelling of zinc oxide leaching in dilute acid solutions, *Trans. Inst. Min. Metall. (Section C: Mineral Processing and Extractive Metallurgy.)* **96**, C69–78 (1987); Also published in *Hydrometallurgical Reactor Design and Kinetics*, Eds. R.G. Bautista, R.J. Wesely and G.W. Warren, Metallurgy Society, Warrendale, PA (1986), 49–66.

122. V.N. Mackiw, W.C. Lin and W. Kunda, Reduction of nickel by hydrogen from ammoniacal nickel sulfate solutions, *J. Metals* **9**, *AIME Trans.* **209**, 786–793 (1957).

123. F.A. Schaufelberger, Precipitation of metal from salt solution by reduction with hydrogen, *J. Metals* **8**, 539–548 (1956); *Trans. Am. Inst. Mining Metall. Eng.* **205**, 539–548 (1956).

124. W. Kunda, D.J.I. Evans and V.N. Mackiw, Low density nickel powder by hydrogen reduction from the aqueous ammonium carbonate system, *Planseeber. Pulvmetall.* **12**, 153–171 (1964).

125. O. Knacke, F. Pawlek and E. Sussmuth, Beiträge zur Metallgewinnung durch Druck reduktion, *Erzmetall.* **9**, 566–574 (1956).

126. S.C. Sircar and D.R. Wiles, Hydrogen precipitation of nickel from buffered acid solutions, *Trans. Am. Inst. Min. Eng.* **218**, 891–893 (1960).

127. G.N. Dobrokhotov and N.I. Onuchkina, Kinetics of nickel reduction with hydrogen from ammonium sulfate solutions, *Izv. Vyssh. Ucheb. Zaved. Esvetn, Metall.* **5**(5), 72–78 (1962); *Chem. Abs.* **58**, 9925 (1963).

128. C.R.S. Needes and A.R. Burkin, Kinetics of reduction of nickel in aqueous ammoniacal ammonium sulphate solutions by hydrogen, in *Leaching and Reduction in Hydrometallurgy*, Ed. A.R. Burkin, The Institution of Mining and Metallurgy, London (1975), 91–96.

129. L.F. Epstein, Constants in the equation of state for a gas, *J. Chem. Phys.* **20**, 1981–1982 (1952).

130. W.E. Putnam and J.E. Kilpatrick, The general relation between the density virial coefficients and the pressure virial coefficients, *J. Chem. Phys.* **21**, 951 (1953).

131. K.E. Gubbins, K.K. Bhatia and R.D. Walker Jr., Diffusion of gases in electrolyte solutions, *AIChE J.* **12**, 548–552 (1966).

132. W. Kunda, J.P. Warner and V.N. Mackiw, The hydrometallurgical production of cobalt, *Trans. Can. Inst. Min. Metall.* **65**, 21–25 (1962).

133. R.T. Wimber and M.E. Wadsworth, Kinetics of the platinum-catalyzed hydrogen reduction of aqueous cobalt sulfate–ammonium acetate solutions, *Trans. AIME* **221**, 1141–1148 (1961).

134. T.M. Kaneko and M.E. Wadsworth, The catalytic reduction of cobalt from ammoniacal cobalt sulfate solution, *J. Phys. Chem.* **60**, 457–462 (1956).

135. W.G. Courtney, The catalytic reduction of cobalt with hydrogen from ammoniacal cobalt sulfate solution, *J. Phys. Chem.* **61**, 693–694 (1957).

136. C.R.S. Needes and A.R. Burkin, Kinetics of reduction of cobalt in aqueous ammoniacal ammonium sulphate solutions by hydrogen, in *Leaching and Reduction in Hydrometallurgy*, Ed. A.R. Burkin, The Institution of Mining and Metallurgy, London (1975), 97–101.

137. W. Kunda and R. Hitesman, The reduction of cobalt from its aqueous ammine ammonium sulphate system using hydrogen under pressure, *Hydrometallurgy* **4**, 347–375 (1979).

138. H.A. Pray, C.E. Schweickert and B.H. Minnich, Solubility of hydrogen, oxygen nitrogen and helium in water at elevated temperatures, *Ind. Eng. Chem.* **44**, 1146–1151 (1952).

139. R.C. Mehrotra and R. Bohra, *Metal Carboxylates* Academic Press, London (1983).

140. R.C. Dakers and J. Halpern, Kinetics of the homogenous reaction between cupric acetate and molecular hydrogen in aqueous solution, *Can. J. Chem.* **32**, 969–978 (1954).

141. T.F. Mason and A.R. Burkin, Kinetics of reduction of aqueous cupric acetate solution by hydrogen — A spectrophometric study, in *Leaching and Reduction in Hydrometallurgy*, Ed. A.R. Burkin, Institution of Mining and Metallurgy, London (1975), 102–109.

142. J. Halpern, E.R. MacGregor and E. Peters, The nature of the activated intermediate in the homogeneous catalytic activation of hydrogen by cupric salts, *J. Phys. Chem.* **60**, 1455–1456 (1956).

143. E.R. MacGregor and J. Halpern, The reduction of cupric salts in aqueous perchlorate and sulfate solutions by molecular hydrogen, *Trans. Metall. Soc. AIME.* **212**, 244—247 (1958).

144. H.F. McDuffie, E.L. Compere, H.H. Stone, L.F. Woo and C.H. Secoy, Homogeneous catalysis for homogeneous reactors. Catalysis of the reaction between hydrogen and oxygen, *J. Phys. Chem.* **62**, 1030–1036 (1958).

145. E.A. von Hahn and E. Peters, The kinetics and mechanism of the Cu^{++} — Catalyzed reduction of Cr^{VI} by hydrogen in aqueous solutions, *Can. J. Chem.* **39**, 162–170 (1961).

146. W.J. Dunning and P.E. Potter, Homogeneous catalytic activation of molecular hydrogen by cuprous ions in aqueous solution, *Proc. Chem. Soc.* 1960, 244–245.

147. E. Peters and E.A. von Hahn, The precipitation of copper from aqueous solutions by hydrogen reduction, in *Symposium on Unit Processes in Hydrometallurgy*, Eds. M.E. Wadsworth and F.T. Davis, Gordon and Breach, New York (1964), 204–226.

148. E.A. von Hahn and E. Peters, Kinetics of copper (II) and copper (I) catalyzed deuterium exchange in sulfuric and perchloric acid solutions, *J. Phys. Chem.* **75**, 571–579 (1971).

149. E.A. von Hahn and E. Peters, The role of copper (I) in the kinetics of hydrogen reduction of aqueous cupric sulfate solutions, *J. Phys. Chem.* **69**, 547–552 (1965).

150. R.T. McAndrew and E. Peters, Kinetics of replacement of silver from acetate and perchlorate solution by carbon monoxide, *Can. Metall. Quart.* **3**, 153–173 (1964).

151. J.J. Byerley and E. Peters, The reduction of cupric salts in aqueous solutions by carbon monoxide, in *Symposium on Unit Processes in Hydrometallurgy*, Eds. M.E. Wadsworth and F.T. Davis, Gordon and Breach, New York (1964), 183–203.

152. J.J. Byerley and E. Peters, Kinetics and mechanisms of the reaction between carbon monoxide and copper (II) in aqueous solution, *Can. J. Chem.* **47**, 313–321 (1965).

153. J.J. Byerley and Y.H. Lee Jolland, Copper (II) catalyzed oxidation of carbon monoxide by molecular oxygen, *Can. J. Chem.* **45**, 3025–3030 (1967).

154. E. Hirsch and E. Peters, Kinetics of nickel reduction from ammine solution by carbon monoxide, *Can. Metall. Quart.* **3**, 137–151 (1964).

155. H.N. Halvorson and E. Peters, Kinetic study of the reduction of cobalt (II) by carbon monoxide in an aqueous system, *Can. J. Chem.* **47**, 2535–2543 (1969).

156. M.L. Hair, *Infrared Spectroscopy in Surface Chemistry*, Marcel Dekker, New York (1967).

157. I.H. Warren and E. Devuyst, Leaching of metal oxides, *Int. Symp. Hydrometallurgy*, Chicago 1973, Eds. D.J.I. Evans and R.S. Shoemaker, AIME, New York (1973) 229–264.

158. I.H. Warren, M.D. Bath, A.P. Prosser and J.T. Armstrong, Anisotropic dissolution of hematite, *Trans. Inst. Min. Metall. (Section C: Mineral Processing and Extractive Metallurgy.)* **78**, C21–27 (1969).

159. K. Azuma and H. Kametani, Kinetics of dissolution of ferric oxide, *Trans. Metall. Soc. AIME* **230**, 853–862 (1964).

160. I.H. Warren and E.A. Devuyst, Fundamentals of the leaching of simple oxides, in *Hydrometallurgy*. I. Chemical, E. Symposium Series No. 42, Eds. G.A. Davies and J.B. Scuffham, Institution of Chemical Engineers, London (1975), Paper 7.

161. J.K. Wood and V. K. Black, Amphoteric metallic hydroxides. 3 Chromium hydroxide, *J. Chem. Soc.* (London) (1916), 164–171.

162. V.V. Ipatiev Jr. and V.G. Tronev, The theory of the oxidation of chromium oxide by atmospheric oxygen under pressure and in aqueous solution, *Zh. Prikl. Khim.* **6**, 832–838 (1933) (in Russian text).

163. V.V. Ipatiev Jr. and M.N. Platonova, Oxidation of chromium hydroxide and chromite by atmospheric oxygen in an alkaline medium, *Zh. Prikl. Khim.* **4**, 633–636 (1931) (in Russian text); Also published in *Ber. Dt. Chem. Ges.* **B65**, 572–579 (1932) (in German text).

164. Société Bozel-Maletra, French Patents 683 602 (1930); 683 604 (1930); 768 366 (1934).

165. C.J. Farrow and A.R. Burkin, Alkali pressure leaching of chromium (III) oxide and of chromite mineral, in *Leaching and Reduction in Hydrometallurgy*, Ed. A.R. Burkin, The Institution of Mining and Metallurgy, London (1975), 20–27.

166. A.R. Burkin, The effects and mechanisms of oxidation reactions at solid surfaces during leaching, in *Unit Processes in Hydrometallurgy, Conf. Proc.* Eds. M.E. Wadsworth and F.T. Davis, Gordon and Breach, New York (1964) pp. 80–94.

167. A.R. Burkin, Solid-state transformations during leaching, *Min. Sci. Eng.* **1**, 4–14 (1969).

168. F.A. Forward and V.N. Mackiw, Chemistry of the ammonia pressure process for leaching Ni, Cu and Co from Sherritt–Gordon sulphide concentrates, *J. Metals* **7**, 457–463 (1955).

169. S. Nashner, Refining at Fort Saskatchewan, *Trans. Can. Inst. Min. Metall.* **58**, 212–226 (1955).

170. P.A. Laxen, The dissolution of UO_2 as an electron transfer reaction, in *The Recovery of Uranium*, International Atomic Energy Agency, Vienna (1971), pp. 321–330.

171. T.L. Mackay and M.E. Wadsworth, A kinetic study of the dissolution of UO_2 in sulfuric acid, *Trans. Am. Inst. Min. Eng.* **212**, 597–603 (1958).

172. E.A. Kanevskii, A.P. Filippov and G.M. Nesmeyanova, Heterogeneous oxidation of UO_2 and uranium leaching processes in acid solutions, *Proc. 3rd Int. Conf. Peaceful Uses Atomic Energy*, Geneva, 1964, United Nations, New York (1965), Vol. 12, 242–249 (Russian text; English summary).

173. P.A. Laxen, A kinetic study of the dissolution of uraninites in sulphuric acid, in *Research in Chemical and Extraction Metallurgy*, Australian Institute of Mining and Metallurgy, Melbourne (1967), 181–192.

174. V.I. Spitsyn, G.M. Nesmeyanova and E.A. Kanevskii, Thermodynamics and kinetics of the dissolution of uranium oxides in acid, *Russ. J. Inorg. Chem.* **5**, 945–947 (1960).

175. M.J. Nicol, C.R.S. Needs and N.P. Finkelstein, Electrochemical model for the leaching of uranium dioxide: 1-acid media, in *Leaching and Reduction in Hydrometallurgy*, Ed. A.R. Burkin, The Institution of Mining and Metallurgy, London (1975), 1–11.

176. C.R.S. Needes, M.J. Nicol and N.P. Finkelstein, Electrochemical model for the leaching of uranium dioxide: 2 — Alkaline carbonate media, in *Leaching and Reduction in Hydrometallurgy*, Ed. A.R. Burkin, The Institution of Mining and Metallurgy, London (1975), 12–19.

177. National Institute for Metallurgy, *Int. NIM Report No. 7079* (1973), 60 pages.

178. R.K. Iler, *The Chemistry of Silica Solubility, Polymerisation, Colloid and Surface Properties and Biochemistry*, Wiley, New York (1979).

179. B. Terry, The acid decomposition of silicate minerals. Part 1 — Reactivities and modes of dissolution of silicates, *Hydrometallurgy*, **10**, 135–150 (1983).

180. B. Terry, The acid decomposition of silicate minerals. Part 2 — Hydrometallurgical applications, *Hydrometallurgy* **10**, 151–171 (1983).

181. K.J. Murata, Internal structure of silicate minerals that gelatinize with acid, *The American Mineralogist* **28**, 545–562 (1943).

182. B. Terry, Specific chemical rate constants for the acid dissolution of oxides and silicates, *Hydrometallurgy* **11**, 315–344 (1983).

183. A.R. Burkin, Pressure leaching of some silicate minerals in alkali solutions, *Proc. Int. Min. Proces. Congress, 1960*, Institution of Mining and Metallurgy, London (1960), 857–862.

184. G.H. Kelsall and I. Thompson, The redox chemistry of H_2S oxidation in the British Gas Stretford process. (I) Thermodynamics of sulphur–water systems at 298 K, *J. Appl. Electrochem.* **23**, 279–286 (1993).

185. T.I. Barry, R.H. Davies, J.A. Gisby, S.M. Hodson and N.J. Pugh, *MT DATA Handbook: Documentation for the NPL Metallurgical and Thermochemical Databank*, National Physical Laboratory. Teddington, UK (1989).

186. Idem. *Proc. 6th Conf. High Temperatures Chem. Inorg. Mat.*, 3–7 April 1989, NIST, Gaithersburg, Maryland, USA.

187. J.F. Stenhouse and W.M. Armstrong, The aqueous oxidation of pyrite, *Trans. Can. Inst. Min. Metall.* **55**, 38–42 (1952).

188. A.R. Burkin and A.M. Edwards, The formation of insoluble iron oxides during the alkali pressure leaching of pyrite, in *Mineral Processing, Proc. 6th Int. Congress*, Cannes, 1963, Ed. A. Roberts, Pergamon, Oxford (1965), 159–169; Also published as Insoluble oxide coating during alkaline pressure leaching of pyrite, *6th Congrés International de la Préparation des Minérais, Compte Rendu Scientifique*, Société de l'Industrie Minerale, St. Etienne (1963), 199–209 (in French).

189. A.N. Buckley, I.C. Hamilton and R. Woods, Studies of the surface oxidation of pyrite and pyrrhotite using X-ray photoelectron spectroscopy and linear potential voltammetry, in *Electrochemistry in Mineral and Metal Processing. (II) Conf. Proc.* Eds. P.E. Richardson and R. Woods, The Electrochemical Society Inc., Pennington (1988) 234–246.

190. S. Djurle, An X-ray study on the system Cu–S, *Acta. Chem. Scand.* **12**, 1415–1426 (1958).

191. E.H. Roseboom, Jr., An investigation of the system Cu–S and some natural copper sulfides between 25°C and 700°C, *Econ. Geol.* **61**, 641–672 (1966).

192. N. Morimoto and K. Koto, Phase relations of the Cu–S system at low temperatures — Stability of anilite, *Am. Mineral.* **55**, 106–117 (1970).

193. W.R. Cook, Jr., The Cu–S phase diagram, Thesis, Gould Laboratories, Cleveland, Ohio 1971.

194. W.R. Cook, Jr., Phase changes in copper (I) sulfide as a function of temperature, in *Solid State Chemistry*, Eds. R.S. Roth and S.J. Schneider, Jr., *Special Publication, US Natn. Bur. Stand.*, No. 364, 703–712 (1972).
195. R.W. Potter II and H.T. Evans, Jr., Definitive X-ray powder data for covellite, anilite, djurleite and chalcocite, *J. Res. US Geol. Survey* **4**, 205–212 (1976).
196. R.W. Potter II, An electrochemical investigation of the system copper–sulphur, *Econ. Geol.* **72**, 1524–1542 (1977).
197. N.T. Evans, Jr., Crystal structure of low chalcocite, *Nature: Phys. Sci. V.* **232**(29), 69–70 (1971).
198. H. Takeda, J.D.H. Donnay, E.H. Roseboom, Jr., and D.E. Appleman, The crystallography of djurleite, $Cu_{1.97}S$, *Zeitschr. Kristallogr.* **125**, 404–413 (1967).
199. L.G. Berry, The crystal structure of covellite, CuS and klockmannite, CuSe, *Am. Mineral.* **39**, 504–509 (1954).
200. N. Morimoto and G. Kullerud, Polymorphism in digenite, *Am. Mineral.* **48**, 110–123 (1963).
201. G.H. Moh, Low-temperature sulfide synthesis, *Carnegie Institute in Washington Year Book 1962*, 214–215 (1963).
202. G.H. Moh, Blaubleibend covellite, *Carnegie Institute in Washington Year Book 1963*, 208–209 (1964).
203. G. Frenzel, The blue-remaining covellite: A contribution to the system CuS–Cu_2S, *Neues Jahrb. Mineral. Abhandl.* **93**, 115–132 (1959).
204. G. Frenzel, The copper excess of blue remaining covellite, *Neues Jahrb. Mineral. Monatsh.*, 1961, 199–204.
205. N. Morimoto, K. Koto and Y. Shimazaki, Anilite, Cu_7S_4, a new mineral, *Am. Mineral.* **54**, 1256–1268 (1969).
206. K. Koto and N. Morimoto, The crystal structure of anilite, *Acta Cryst.* **B26**, 915–924 (1970).
207. R.J. Goble and G. Robinson, $Cu_{1.60}S$, a new copper sulfide from Dekalb Township, New York, *Can. Mineral.* **18**, 519–523 (1980).
208. R.J. Goble, L.S. Whiteside and A.M. Ghaz, Geerite-type structures and the flotation of sphalerite, *CIM Bull.* **81**(911), 110–114 (1988).
209. R.J. Goble, Copper sulfides from Alberta: Yarrowite Cu_9S_8 and spionkopite $Cu_{39}S_{28}$, *Can. Mineral.* **18**, 511–518 (1980).
210. R.J. Goble, The relationship between crystal structure, bonding and cell dimensions in the copper sulfides, *Can. Mineral.* **23**, 61–76 (1985).
211. I.H. Warren, A study of the acid pressure leaching of chalcopyrite, chalcocite and covellite, *Aust. J. Appl. Sci.* **9**, 36–51 (1958).
212. G. Thomas and T.R. Ingraham, Kinetics of dissolution of synthetic covellite in aqueous acidic ferric sulphate solutions, *Can. Metall. Quart.* **6**, 153–165 (1967).
213. G. Thomas and T.R. Ingraham and R.J.C. MacDonald, Kinetics of dissolution of synthetic digenite and chalcocite in aqueous acidic ferric sulphate solutions, *Can. Metall. Quart.* **6**, 281–292 (1967).
214. W.W. Fisher and R.J. Roman, The dissolution of chalcocite in oxygenated sulfuric acid solution, *Circ. New. Mexico St. Bur. Mines Min. Resour.* 112, 1971, 28 pages.

215. B.Z. Ustinskii and D.M. Chizhikov, Potentials of some metal sulphides and their alloys, *Zh. Prikl. Khim.* **22**, 1249–1252 (1949) (in Russian text).

216. A.G. Loshkarev and A.F. Vozisov, Anodic solution of copper sulphide, *Zh. Prikl. Khim.* **26**, 55–62 (1953) (in Russian text).

217. F. Habashi and N. Torres Acuña, The anodic dissolution of copper (I) sulfide and the direct recovery of copper and elemental sulfur from white metal, *Trans. Am. Inst. Min. Eng.* **242**, 780–787 (1968).

218. P. Cavallotti and G. Salvago, Electrodic behaviour of copper sulphides in aqueous solutions, *Electrochim. Metallorum.* **4**, 181–210 (1969).

219. J.A. King, PhD Thesis, University of London, 1966.

220. W.R. Cook Jr., *JCPDS Powder Diffraction File* 23–959. From (193).

221. W.R. Cook Jr., *JCPDS Powder Diffraction File* 23–962. From (193).

222. N. Morimoto, K. Koto and Y. Shimazaki, Data from (205); Other data in *JCPDS Powder Diffraction File* 33–489. From (195).

223. R.J. Goble and G. Robinson, *JCPDS Powder Diffraction File* 33–492. From (207).

224. R.J. Goble, *JCPDS Powder Diffraction File* 36–380. From (209).

225. R.J. Goble, *JCPDS Powder Diffraction File* 36–379. From (209).

226. H.E. Swanson, R.K. Fuyat and G.M. Ugrinic. *JCPDS Powder Diffraction File* 6–464. From: Standard X-ray diffraction powder patterns. Copper (II) sulfide (covellite) CuS (hexagonal), *Circ. US Natn. Bur. Stand.*, No. 539 (1955).

227. *Powder Diffraction File Search Manual, Hanawalt Method, Inorganic 1991.* International Centre for Diffraction Data, Swarthmore, PA (1991).

228. R.J. Goble, The relationship between crystal structure, bonding and cell dimensions in the copper sulfides, *Can. Mineral.* **23**, 61–76 (1985).

229. L.S. Whiteside and R.J. Goble, Structural and compositional changes in copper sulfides during leaching and dissolution, *Can. Mineral.* **24**, 247–258 (1986).

230. J.E. Dutrizac, R.J.C. MacDonald and T.R. Ingraham, The kinetics of dissolution of synthetic chalcopyrite in aqueous acidic ferric sulphate solutions, *Trans. Am. Inst. Min. Eng.* **245**, 955–959 (1969).

231. R.C.H. Ferreira and A.R. Burkin, Acid leaching of chalcopyrite, in *Leaching and Reduction in Hydrometallurgy*, Ed. A.R. Burkin, Institution of Mining and Metallurgy, London (1975), 54–66.

232. E. André, German Patent 6048 1877.

233. D.J. Mackinnon, Fluidised-bed anodic dissolution of chalcocite, *Hydrometallurgy* **1**, 241–257 (1976).

234. D.J. Mackinnon, Fluidised-bed anodic dissolution of covellite, *Hydrometallurgy* **2**, 65–76 (1976).

235. T. Biegler and D.A. Swift, Dissolution kinetics of copper sulphide anodes, *Hydrometallurgy* **2**, 335–349 (1976/1977).

236. D.C. Price, Application of chronopotentiometric analysis to the anodic treatment of copper sulphides, *Metall. Trans.* **B12**, 231–239 (1981).

237. S.H. Cole, Thesis, Columbia University, 1972. Value reported in (236).

238. R.S. Parikh and K.C. Liddell, Anodic dissolution of chalcocite in chloride media, in *Electrochemistry in Mineral and Metal Processing II*, Eds. P.E. Richardson

and R. Woods, The Electrochemical Society, Pennington, New Jersey (1988), 325–339.

239. A. Etienne and E. Peters, Thermodynamic measurements in the Cu–S system in the temperature range 40–80°C, *Trans. Inst. Min. Metall.* (*Section C: Mineral Processing and Extractive Metallurgy.*), **81**, C76–181 (1972).

240. H.J. Mathieu and H. Rickert, Electrochemical-thermodynamic investigation of the system copper–sulfur at temperature from 15–90°C, *Zeitschr. Phys. Chemie, Neue Folge*, **79**, 315–330 (1972) (in German text).

241. R.J. Goble and D.G.W. Smith, Electron microprobe investigation of copper sulfides in the Precambrian Lewis Series of South-Western Alberta, Canada, *Can. Mineral.* **21**, 95–103 (1973).

242. J.D. Sullivan, Chemistry of leaching bornite, *Technical Paper, US Bur. Mines 486* (1931).

243. M.H. Stanczyk and C. Rampacek, Oxidation leaching of copper sulfides in acidic pulps at elevated temperatures and pressures, *Rep. Inv. US Bureau Mines 6193* (1963).

244. G.A. Kopylov and A.I. Orlov, Kinetics of dissolving of bornite, *Non-Ferrous Metall.* **6**, 86–95 (1963).

245. G.A. Kopylov and A.I. Orlov, Rates of bornite and chalcocite dissolution in ferric sulphate, *Tr. Irkutsk. Politeck Inst.* **46**, 127–132 (1969) (in Russian text).

246. J.E. Durtrizac, R.J.C. MacDonald and T.R. Ingraham, The kinetics of dissolution of bornite in acidified ferric sulfate solutions, *Metall. Trans.* **1**, 225–231 (1970).

247. J.E. Dutrizac, R.J.C. MacDonald and T.R. Ingraham, Effect of pyrite, chalcopyrite and digenite on rate of bornite dissolution in acidic ferric sulfate solutions, *Can. Metall. Quart.* **10**, 3–7 (1971).

248. F.J.G. Ugarte and A.R. Burkin, Mechanism of formation of idaite from bornite by leaching with ferric sulphate solution, in *Leaching and Reduction in Hydrometallurgy*, Ed. A.R. Burkin, Institution of Mining and Metallurgy, London (1975), 46–53.

249. R.H. Sillitoe and A.H. Clark, Copper and copper-iron sulfides as the initial products of supergene oxidation, Copiapo mining district, Northern Chile, *Am. Mineral.* **54**, 1684–1710 (1969).

250. H.E. Merwin and R.H. Lombard, The system Cu–Fe–S, *Econ. Geol.* **32**, 203–284 (1937).

251. *US Nat. Bur. Stand. Monogr.* 21 (1984). *JCPDS Powder Diffraction File*, File No: 35, 752 cards.

252. J.E. Dutrizac, T.T. Chen and J.I. Jambor, Mineralogical changes occurring during the ferric ion leaching of bornite, *Metall. Trans.* **B16**, 679–693 (1985).

253. N. Morimoto, Structures of two polymorphic forms of Cu_5FeS_4, *Acta crystallogr.* **17**, 351–360 (1964).

254. N. Morimoto, On the transitions of bornite, *Carnegie Institute in Washington Year Book 1961*, 139–141 (1962).

255. P.G. Manning, A study of the bonding properties of sulphur in bornite, *Can. Mineral.* **9**, 85–94 (1967).

256. D.C. Price and J.P. Chilton, The electrochemistry of bornite and chalcopyrite *Hydrometallurgy* **5**, 381–394 (1980).
257. D.C. Price and J.P. Chilton, The anodic reactions of bornite in sulphuric acid solution *Hydrometallurgy* **7**, 117–133 (1981).
258. S. Aguayo-Salinas and A.R. Burkin, A study of the leaching of bornite by electrochemical techniques, to be submitted.
259. K. Koto and N. Morimoto, Superstructure investigation of bornite, Cu_5FeS_4, by the modified partial Patterson function *Acta Cryst.* **B31**, 2268–2273 (1975).
260. S. Aguayo-Salinas, PhD Thesis, University of London (1985).
261. G. Frenzel, Idaite, a new mineral (preliminary note). *Neues Jahrb. Mineral. Monatsh.* **6**, 142 (1958).
262. G. Frenzel, Idaite, a new mineral, natural Cu_5FeS_6, *Neues Jahrb. Mineral. Abhandl.* **93**, 87–114 (1959).
263. C. Levy, Contribution a la minéralogie des sulfures de cuivre du type Cu_3XS_4, *Mém. Bur. Rech. géol. minièr es* **54**, 1967, 178 pages.
264. C.M. Rice, D. Atkin, J.F.W. Bowles and A.J. Criddle, Nukundamite, a new mineral, and idaite, *Mineral. Mag.* **43**(326), 193–200 (1979).
265. N. Wang, A contribution to the copper–iron sulfur system: The sulphidisation of bornite at low temperatures, *Neues Jahrb. Mineral. Monatsh.* (B), 346–352 (1984).
266. G. Frenzel and J. Ottemann, *Min. Deposita* **1**, 307–316 (1967). Paper not abstracted or seen.
267. A.J. Monhemius and F.J.G. Ugarte, personal communication, 1991.
268. F.J.G. Ugarte and A.J. Monhemius. Characterisation of high-temperature arsenic-containing residues from hydrometallurgical processes, *Hydrometallurgy* **30**, 69–86 (1992).
269. V.E. Klets and V.A. Liopo, Behaviour of chalcopyrite in salt leaching. *Tr. Irkutsk. Politelch. Inst.* No. 27, 123–130 (1966); *Chem. Abs.* **67**, 56248 (1967) (in Russian text).
270. F.P. Haver and M.M. Wong, Recovering elemental sulfur from nonferrous minerals. *Rep. Invest. US Bur. Mines*, 7474, 1971.
271. G. Björling and G.A. Kolta, Oxidizing leach of sulfide concentrates and other materials catalyzed by nitric acid, *Proc. 7th Int. Miner. Process. Congress.* New York 1964, Ed. N. Arbiter, Gordon and Breach, New York (1965), 127–138.
272. I.H. Warren, A. Vizsolyi and F.A. Forward, The pretreatment and leaching of chalcopyrite, *CIM Bull.* **61**, 637–640 (1968).
273. J.E. Hiller and K. Probsthain, Thermal and roentgenographic investigations of chalcopyrite, *Zeitschr. Kristallogr.* **108**, 108–129 (1956).
274. L.J. Cabri and S.R. Hall, Mooihoekite and haycockite, two new copper–iron sulfides, and their relationship to chalcopyrite and talnakhite, *Am. Mineral.* **57**, 689–708 (1972).
275. L. Cabri. New data on phase relations in the Cu–Fe–S system *Econ. Geol.* **68**, 443–454 (1973).
276. R.A. Yund and G. Kullerud, Thermal stability of assemblages in the Cu–Fe–S system, *J. Petrology* **7**, 454–488 (1966).

277. A. Vizsolyi, H. Veltman, I.H. Warren and V.N. Mackiw, Copper and elemental sulfur from chalcopyrite by pressure leaching, *J. Metals* **19**(11), 52–59 (1967).

278. R. Le Houillier and E. Ghali, Contribution à l'étude de la lixiviation de la chalcopyrite, *Hydrometallurgy* **9**, 169–194 (1982).

279. J.D. Prater, P.B. Queneau and T.J. Hudson, The sulfation of copper–iron sulfides with concentrated sulfuric acid, *J. Metals* **22**(12), 23–27 (1970).

280. K.J. Jackson and J.D.H. Strickland, The dissolution of sulfide ores in acid chlorine solutions: A study of the more common sulfide minerals, *Trans. Am. Inst. Min. Eng.* **212**, 273–279 (1958).

281. T. Uchida *et al.*, Leaching of copper from copper-bearing ores with dilute solution of sulphuric acid and ferric sulphate, *Hakko Kyokaishi* **25**(4), 168–172 (1967); *Chem. Abs.* **67**, 56269 (1967).

282. D.L. Jones and E. Peters, The leaching of chalcopyrite with ferric sulphate and ferric chloride, in *Extractive Metallurgy of Copper, International Symposium on Copper Extraction and Refining*, Las Vegas, 1976, AIME, New York (1976), 633–653.

283. P.B. Munoz, J.D. Miller and M.E. Wadsworth, Reaction mechanism for the acid ferric sulfate leaching of chalcopyrite, *Metall. Trans.* **B10**, 149–158 (1979).

284. F.A. Forward and I.H. Warren, Extraction of metals from sulphide ores by wet methods, *Metall. Rev.* **5**, 137–164 (1960).

285. K.N. Subramanian and P.H. Jennings, Review of the hydrometallurgy of chalcopyrite concentrates, *Can. Metall. Quart.* **11**, 387–400 (1972).

286. R.J. Roman and B.R. Brenner, The dissolution of copper concentrates, *Min. Sci. Eng.* **5**, 3–22 (1973).

287. J.E. Dutrizac and R.J.C. MacDonald, Ferric ion as a leaching medium, *Min. Sci. Eng.* **6**, 59–100 (1974).

288. J.E. Dutrizac, The dissolution of chalcopyrite in ferric sulfate and ferric chloride media, *Metall. Trans.* **B12**, 371–378 (1981).

289. J.E. Dutrizac, Ferric ion leaching of chalcopyrites from different localities, *Metall. Trans.* **B13**, 303–309 (1982).

290. L.E. Muir and J.B. Hiskey, Kinetic effects of particle size and dislocation density on the dichromate leaching of chalcopyrite, *Metall. Trans.* **B12**, 255–267 (1981).

291. B.R. Palmer, C.O. Nebo, M.F. Rau and M.C. Fuersteneau, Rate phenomena involved in the dissolution of chalcopyrite in chloride-bearing lixiviants, *Metall. Trans.* **B12**, 595–601 (1981).

292. H. Majima, Y. Awakura, T. Hirato and T. Tanaka, The leaching of chalcopyrite in ferric chloride and ferric sulfate solutions. *Can. Metall. Quart.* **24**, 283–291 (1985).

293. T. Hirato, H. Majima and Y. Awakura, The leaching of chalcopyrite with ferric sulfate *Metall. Trans.* **B18**, 489–496 (1987).

294. T. Hirato, M. Kinoshita, Y. Awakura and H. Majima, The leaching of chalcopyrite with ferric chloride. *Metall. Trans.* **B17**, 19–28 (1986).

295. F.K. Crundwell, Kinetics and mechanisms of the oxidative dissolution of a zinc sulphide concentrate in ferric sulphate solutions, *Hydrometallurgy*, **19**, 227–242 (1987).

296. C. Hoepfner, US Patent 507130 (1893).

297. C. Hoepfner and H. Orth, US Patent 704639 (1902).

298. J.P. Wilson and W.W. Fisher, Cupric chloride leaching of chalcopyrite, *J. Metals* **33**(2), 52–57 (1981).

299. J.J. Fritz, Representation of the solubility of CuCl in solutions of various aqueous chlorides, *J. Phys. Chem.* **85**, 890–894 (1981).

300. J.J. Fritz, Solubility of cuprous chloride in various soluble aqueous chlorides, *J. Chem. Eng. Data* **27**, 188–193 (1982).

301. J.J. Fritz, Heats of solution of cuprous chloride in aqueous HCl–$HClO_4$ mixtures, *J. Sol. Chem.* **13**, 369–382 (1984).

302. T. Sato and T. Kato, The stability constants of the chloro complexes of Cu(II) and Zn(II) determined by tri-*n*-octylamine extraction, *J. Inorg. Nucl. Chem.* 1977, **39**, 1207–1208.

303. Anon, New copper process from Cyprus is billed as "technological breakthrough". *Eng. Min. J.* **178**, 33 (October 1977).

304. Anon, Cyprus reveals details of new copper process. *Eng. Min. J.* **178**, 30–39 (November 1977).

305. M. Bonan, J.M. Demarthe, H. Renon and F. Baratin, Chalcopyrite leaching by $CuCl_2$ in strong NaCl solutions, *Metall. Trans.* **B12**, 269–274 (1981).

306. T. Hirato, H. Majima and Y. Awakura, The leaching of chalcopyrite with cupric chloride, *Metall. Trans.* **B18**, 31–39 (1987).

307. H. Majima, Electrochemistry of the oxidative leaching of chalcopyrite, in *Electrochemistry in Mineral and Metal Processing, II, Conf. Proc.* Eds. P.E. Richardson and R. Woods, The Electrochemical Society, Inc., Pennington, New Jersey (1988), 303–324.

308. A.R. Burkin, The stabilities of coordination complex compounds, *Quart. Rev.* (London) **5**, 1–21 (1951).

309. T. Biegler and D.A. Swift, Anodic electrochemistry of chalcopyrite, *J. Appl. Electrochem.* **9**, 545–554 (1979).

310. A.J. Parker, R.L. Paul and G.P. Power, Electrochemical aspects of leaching copper from chalcopyrite in ferric and cupric salt solutions, *Aust. J. Chem.* **34**, 13–34 (1981); *J. Electroanal. Chem.* **118**, 305–316 (1981).

311. R.S. McMillan, D.J. MacKinnon and J.E. Dutrizac, Anodic dissolution of n-type and p-type chalcopyrite, *J. Appl. Electrochem.* **12**, 743–757 (1982).

312. H.G. Linge, A study of chalcopyrite dissolution in acidic ferric nitrate by potentiometric titration, *Hydrometallurgy* **2**, 51–64 (1976).

313. G.W. Warren, M.E. Wadsworth and S.M. El-Raghy, Passive and transpassive anodic behaviour of chalcopyrite in acid solutions, *Metall. Trans.* **B13**, 571–579 (1982).

314. T. Biegler and M.D. Horne, The electrochemistry of surface oxidation of chalcopyrite, in *Electrochemistry in Mineral and Metal Processing, Conf. Proc.* Eds. P.E. Richardson, S. Srinivasan and R. Woods, The Electrochemical Society, Inc., New Jersey, Pennington (1984), 321–335.

315. M. Ammou-Chokroum, P.K. Sen and F. Fouques, Developments in mineral processing, *Proc. 13th Int. Min. Proces. Congress*, Warsaw, Poland, June 1979, Part A, Ed. J. Laskowski, Elsevier, New York (1981), 759–809.

316. J.H. Chen and W.W. Harvey, Cation self-diffusion in chalcopyrite and pyrite, *Metall. Trans.* **B6**, 331–339 (1975).

317. S.R. Hall and J.M. Stewart, The crystal structure refinement of chalcopyrite, $CuFeS_2$, *Acta Cryst.* **B29**, 579–585 (1973).

318. L. Pauling, *The Nature of the Chemical Bond*, 3rd edition, Cornell University Press, Ithaca, New Jersey (1960).

319. T. Teranishi, Magnetic and electric properties of chalcopyrite, *J. Phys. Soc. Jap.* **17**, Suppl. B-1, 263–267 (1962).

320. G. Donnay *et al.*, Symmetry of magnetic structure: Magnetic structure of chalcopyrite, *Phys. Rev.* **112**, 1917–1923 (1958).

321. A.K. Zhetbaev and D.K. Kaipov, Parameters of Mössbauer spectra of iron sulphides, *Izv. Akad. Nauk Kaz. SSR, Ser. Fiz-Mat.* **6**(6), 78–84 (1968); *Chem. Abs.* **70**, 110328 (1969).

322. D. Raj, K. Chandra and S.P. Puri, Mössbauer studies of chalcopyrite. *J. Phys. Soc. Jap.* **24**(1), 39–41 (1968).

323. R.L. Adams, P. Russe, R.S. Arnott and A. Wold, Preparation and properties of the systems $CuFeS_{2.00-x}$ and $Cu_{1.00-x}Fe_{1.00+x}S_{2.00-y}$ *Mater. Res. Bull.* **7**, 93–99 (1972); *US Nat. Bur. Stand.*, Special Publication No. 364, 713–719.

Index